An Introduction to

STATISTICS

for Canadian Social Scientists

An Introduction to STATISTICS for Canadian Social Scientists

Michael Haan
Jenny Godley

Third Edition

OXFORD
UNIVERSITY PRESS

Oxford University Press is a department of the University of Oxford. It furthers the University's objective of excellence in research, scholarship, and education by publishing worldwide. Oxford is a registered trade mark of Oxford University Press in the UK and in certain other countries.

Published in Canada by
Oxford University Press
8 Sampson Mews, Suite 204,
Don Mills, Ontario M3C 0H5 Canada

www.oupcanada.com

First Edition published in 2008
Second Edition published in 2013

Library and Archives Canada Cataloguing in Publication
Haan, Michael, 1974–, author
An introduction to statistics for Canadian social scientists / Michael Haan and Jenny Godley. — Third edition.

Includes bibliographical references and index.
ISBN 978-0-19-902059-1 (paperback)

1. Social sciences—Statistical methods—Textbooks. I. Godley, Jenny, author II. Title.

HA29.H22 2016 300.72'7 C2016-902883-6

Cover image: © iStock/sorbetto

Oxford University Press is committed to our environment. Wherever possible, our books are printed on paper which comes from responsible sources

Printed and bound in Canada

5 6 7 — 22 21 20

Contents

NUMBERED BOXES

EVERYDAY STATISTICS

Preface

When Michael Haan began writing the first edition of this text back in the fall of 2006, he had a very specific goal in mind: to produce a uniquely Canadian textbook to introduce undergraduates to the topic of social statistics. That undertaking was successful enough to invite Jenny Godley on board to create this third edition, which now includes new content and features that reflect the evolving needs of the country's young social scientists.

Most undergraduate social science students are required to take at least one statistics course as part of their training, usually in their second or third year. Traditionally, there has been a shortage of high-quality, non-technical Canadian textbooks in this area, and professors have almost invariably chosen an American text for their courses. The upside of this is that they have had a wide choice of excellent examples from which to choose; the downside is that the substantive content is largely foreign to students—examples and data sets are almost always from the United States. I believe that this is one of the reasons that students complain about the difficulty and, of greater concern, the irrelevance of statistics for their studies.

But it's not just students who suffer. Using a US text compounds the dismay of professors of subsequent courses about the overall lack of familiarity among students with the basic statistical characteristics of Canadian society. Things like age distributions, city populations, median income, and population changes over time could easily be used to exemplify statistical terms and concepts (such as means, medians, variances, and normal distributions), but with so many textbook offerings, students learn these characteristics about US society. Teaching students with Canadian content promotes greater familiarity with these topics while at the same time teaching the universal language of statistics. It has been quite satisfying to discover, then, that its ground-up development as a Canadian text has been one of the features instructors and students have liked best about *An Introduction to Statistics for Canadian Social Scientists*. The "Everyday Statistics" boxes and real-world Canadian examples throughout the third edition further the mission of presenting students with unique Canadian content.

Since the best way to learn statistics is "hands-on" (particularly with the ongoing proliferation of computers in social science departments), a "knowledge-through-discovery" approach is ideal. Instead of providing the platform for lectures with little or no application, effective pedagogy requires students to apply what they learn in each chapter through practice questions and laboratory exercises (included in the built-in manual for SPSS and STATA, two of the most popular statistical software packages). Additional chapter-end questions, revised lab manuals that use real data from the 2013 Alberta Survey, and multiple text-wide technical reviews for accuracy ensure the third edition thoroughly engages the knowledge-through-discovery approach.

Finally, in this third edition the authors also wanted to continue to offer a text that fits students' budgets. Several statistical textbooks retail for over $200 CDN, an exorbitant amount, made worse when instructors don't come close to covering all of the topics in a semester. To remedy this, we've kept the text to such a length that there is approximately enough material to fill one semester. Consequently, the third edition is shorter than most other introductory statistics texts, yet there is still plenty of material for a semester or year-long course.

This text is broad enough to be used in several social science departments. Although the authors are both trained as sociologists, the goal has been to keep examples diverse enough to be of interest to both sociologists and non-sociologists. We hope you'll enjoy (or at least not loathe) what the authors consider to be one of the most fascinating and useful topics in academia today.

Highlights of the Third Edition

- **"Everyday Statistics" boxed feature.** Developed to demonstrate how statistics figure into the day-to-day workings of Canadian society, "Everyday Statistics" looks at topics such as the use of bivariate statistics by the media during election campaigns. This feature also concludes with probing questions crafted to stimulate critical thinking.
- **Revised end-of-chapter practice questions.** Questions at the end of each chapter, with answers found in the end-of-text answer key, provide students with hands-on opportunities to put their learning to the test.
- **Updated lab manuals.** Both the SPSS and the STATA lab manuals found at the conclusion of the text have been revised, keyed to real data collected in the 2013 Alberta Survey. Working with real data affords students the opportunity to see actual social science in action.
- **Quick-reference guides.** Located on the inside front and back covers, guides to frequently used formulas and symbols serve as handy reference tools placed right at students' fingertips.
- **Technical checks.** Equations, solutions, and concepts throughout the text and ancillary material have undergone multiple independent technical checks and have been revised based on peer feedback, in order to ensure an edition that is as accurate as possible.
- **Ancillaries suite.** A fully automated test generator, an instructor's manual, and revised PowerPoint slides are available for instructors. Students will be aided by key concept cue cards and further statistical resources located on the text's companion site.

About the Authors

Michael Haan

Michael Haan is Associate Professor of Sociology and Canada Research Chair in Migration and Ethnic Relations at Western University. Professor Haan's recent research focuses on the economic and social implications of migration within and international migration to Canada. He teaches demography and public policy courses at Western University.

Jenny Godley

Jenny Godley is Associate Professor in the Department of Sociology at the University of Calgary where she also currently serves as the Director of Undergraduate Studies. Professor Godley's research interests include the social determinants of weight-related health, social networks, the life course, demography, and gender. She teaches Social Statistics to undergraduates as well as a graduate seminar on quantitative research methods.

Acknowledgements

For the recent edition, Amy Gordon, our development editor from Oxford University Press, has been incredibly patient, helpful, and encouraging. Although new to the text, she made the process of adding a co-author, updating and improving content, and providing us with flexible deadlines and encouraging feedback seamless, and for this we thank her.

We would also like to thank the reviewers of the text, whose detailed feedback helped guide our edits to this edition:

- Mark Flaherty, University of Victoria
- Lori Kirkpatrick, Brock University
- Edward Anthony Koning, University of Guelph
- Colin Reid, University of British Columbia

Michael Haan

Since writing the first edition, I moved from the University of Alberta to the University of New Brunswick and am now at Western University. At each university, I have had to enlist an entirely new "village" of bright young people to help me. That list has become too long to provide here, but I hope each person sees him- or herself somewhere in these pages. Jenny Godley has improved this book in countless ways, and I thank her for saying yes when I asked her to co-author the third edition nearly two years ago.

On a personal note, Dale Ballucci has remained a strong source of support for me with this edition. Our daughters, Evelyn and Abigail, have made life a little bit busier and a lot more interesting for us, all the while providing the details for some of the practice questions in this text.

Jenny Godley

I would like to thank Michael Haan for the opportunity to collaborate with him on this new edition and for all the work that went into writing the original text. A big thank you to my statistics professors, at Cambridge, Berkeley, and Chapel Hill, who answered my many questions and helped me make sense of it all when I was first learning. An even bigger thank you to my statistics students, at Haverford and Calgary, who asked their many questions and refined my understanding over the years. And finally, to Bill and Liam, for teaching me what is truly significant.

PART I

Introduction and Univariate Statistics

1 Why Should I Want to Learn Statistics?

LEARNING OBJECTIVES

This chapter will help you understand why you should want to take a statistics course. It will accomplish this by:

- describing some of the concerns that might be causing you to worry about this course;
- putting each of these concerns into context;
- discussing the merits of learning to think statistically;
- reassuring you that you have the tools to learn the material in this course; and
- linking exercises and practice questions to the material presented in this chapter.

Introduction

You may be reading this book involuntarily. You may have enrolled in introductory social statistics in one of the social science departments (history, political science, psychology, sociology, etc.) at your college or university, perhaps because the course might be required for your degree or diploma. Indeed, you might not be taking this course if it wasn't required. If you're like many other students, you've worried about taking statistics for some time now; you may be in the final year of your program, even though the course is listed as a second- or third-year course.

You are not alone. Very few people actually want to learn statistics. We didn't, and neither did most of our colleagues (many of whom now teach statistics courses). Like you, we were forced by our university bureaucracies to enroll in stats. We'd never met the person, or people, who made this decision for us; they didn't follow our progress through the course, and after we finished they didn't ask us if we agreed with their decision to require us to take the course.

That was probably a good thing because we likely would have told them that we dreaded statistics more *after* taking the course than we had before taking it. Sure, we'd learned a few things, but the information was so abstract and boring that it didn't seem relevant to our day-to-day lives. Up to that point, we'd survived without knowing what a standard deviation or a *z*-score was, and we were quite certain that we would have continued to survive without that knowledge. After completing our undergraduate statistics courses, we still hated statistics—and so did most of the people in our classes.

Why Do So Many People Dislike Statistics?

Looking back, we think we had at least four reasons for disliking statistics:

- First, we found it to be little more than useless math and equations, which made the material impenetrable and unintelligible for us. In our minds, the abstract equations discussed in class bore very little relevance to the practice of doing statistics.
- Second, we found the logic, and the assumptions, to be shaky at times. Why, for example, did we often have to assume that variables are normally distributed (we'll discuss what this means later), when this is so rarely true?
- Third, if statistics are so important and "objective," how can people on both sides of a debate use them to support their claims to knowledge?
- Finally, our professors (who will remain nameless), although capable statisticians, were not very good at, or perhaps not very interested in, making the course content palatable to 19-year-olds.

You may share some, or all, of these reasons for hating statistics. It is likely, however, that someone in your college or university has decided that you too must learn statistics, making the debate about whether you "should" learn statistics moot. This brings us to our discussion of whether you should want to learn statistics.

Let's see if this is going to be as bad as you fear. You're probably dreading the math and the equations, but they're not as hard as you may think. We'll review the necessary math in Chapter 2, but for now we can assure you that to use statistics well you don't need to know a lot of math or to understand equations that are any more complicated than those you learned in elementary or high school. That is not to say that the math underlying statistics isn't difficult, only that we are going to be focusing on the basics here. Similarly, while some of the equations look complex, the principles behind them are generally quite easy to understand. Whenever a new equation is introduced in the text, we try our best to explain, in everyday language, what that equation does. These explanations should be easy to understand because most social statistics concepts are "results driven." That is, once you understand what a certain procedure is designed to do, using it will be easier.

Another reason many people dislike statistics is that they don't trust them. Darrell Huff's *How to Lie with Statistics* (1954) is the bestselling statistics book of all time, so statistical doubters are plentiful. Sure, statistics can be used in misleading ways, but this can be said about any type of argument. (How many athletes have claimed to be "the best there ever was and there ever will be" over the years, even though at most only one of them can be correct?) It is your responsibility to learn how to assess the validity of any claim, and you can only do that if you understand the tools (be they statistics, rhetoric, logic, etc.) that are being used. Statistics is a set of tools for constructing a narrative, and one of the goals of a college or university education is to help you use these tools and to be able to assess whether others are using them effectively.

Let us illustrate with an example from everyday life. Say that your friend tells you that he read in the paper that Moose Jaw is an affluent city. Before believing his claim, you might want to analyze and evaluate it yourself. You might ask yourself the following questions:

1. Do I accept the definitions being used? (What exactly is "Moose Jaw"? Does it include the outlying suburbs? Is it correct to classify Moose Jaw as a "city"? What does "affluent" mean? How much of the population of Moose Jaw must be considered affluent before Moose Jaw can be called affluent?)
2. Do I agree with the underlying assumptions? (Is affluence a meaningful concept? Can I determine the affluence of one place without comparing it to the affluence of other places?)
3. Do I accept the methods that are being used to arrive at the claim? (Where did the information come from? Did someone do a study? Which residents of Moose Jaw answered questions about their affluence when the data were being collected?)

You might choose to reject your friend's claim about the affluence of Moose Jaw because you are not satisfied with the answers to one or more of those questions. When we use statistics in the social sciences we have a framework for assessing such claims in a systematic manner. People do lie with statistics, but that makes it more important for you to understand statistics so that you can identify how the claim was made and decide whether you agree with it. Understanding statistics helps you do what you already do countless times every day (evaluate information) with greater rigour.

Remember, statistics is just as hard to teach as it is to learn. Students are often in class not because they want to be there but because they have to be, and teaching students who would rather be just about anywhere else is difficult. Anecdotally, many statistics professors don't bother trying to make the subject matter more bearable because they don't think that they can get students to like the class no matter what they do.

So, to liven things up, we have used the most interesting and/or relevant examples we could find. Canadian introductory statistics courses are often hurt by the use of US texts, which contain examples to which Canadian students can't always relate. We have tried to rely exclusively on Canadian examples and content, with the hope that you will find the examples relevant.

Each chapter starts with a list of its objectives. Words that you may find unfamiliar will be **bolded** throughout the text. You can find the definition of each highlighted word in the glossary of statistical terms at the back of the book. If you still find the material difficult and require further help, there are several soothing reads on statistics, including *Statistics without Tears* (Rowntree, 2000) and *Statistics for the Terrified* (Kranzler and Moursund, 1999). However, as long as you read this text, listen to your professor's lectures, and do the practice problems at the end of each chapter, we hope you won't need those resources.

This book adopts a "knowledge-through-discovery" approach. Instead of studying abstract lessons on statistics with little or no application, you will apply what you learn in each chapter with laboratory exercises using the Alberta Survey. We hope that these exercises will help you see the links between what you learn in the classroom and what you do in the lab, or in your other social science courses. These exercises will close the gap between the theory of statistics and its practice.

When Did People Start to Think Statistically?

Using **statistics** as a toolkit to make claims about the world has become common practice only in the last 200 years or so. In 1975, Ian Hacking, a prominent Canadian philosopher of science, published an influential book titled *The Emergence of Probability* (Hacking, 1975). Hacking claims that one of the defining characteristics of the nineteenth century was that people began to see the world less in terms of indeterminism and chance, and more in terms of laws and **probabilities.** One of the consequences of this was what Hacking refers to as the onset of an "avalanche of printed numbers." In fewer than 20 years (from around 1820 to 1840), there was "an exponential increase in the number of numbers being published" (186). The newfound popularity of numbers caused a shift in the public's view, and understanding, of the world.

Hacking documented the change from not collecting statistics to collecting them, and, more importantly, the change in the nature of knowledge. He showed that the adoption of statistical methodology changed the dominant way of thinking, learning, and, ultimately, knowing about the world.

Learning statistics is thus as much about learning a new way of thinking as it is about mastering new subject matter.

Since that "avalanche" began roughly 200 years ago, the topics of statistical inquiry have ranged widely. Numbers and statistics form the basis for our understanding of many topics: aging, Americanization, apartheid, apple growth patterns, astronomy—you name it. Although the focus here is on social statistics, or those that are used to understand the behaviours and characteristics of people, the sheer diversity of topics that statistics are used to study shows that many aspects of the world can be understood from within the framework of stats.

Consider a study of cellular phone usage among teenage girls (Campbell, 2006). Teenagers in North America crave style, friendship, and individuality. Companies that sell cellphones are well aware of this, and present their product to teens as a way to enhance individuality, while at the same time promoting conformity to the norms of teens' peer groups. Most teenagers can choose whether or not to have a cellphone, and they see their choice as an act of individuality (Campbell, 2006). They and many of their friends have already chosen to have a cellphone—in 2003, the year the study data were collected, roughly half of all teens had a cellphone (that number is certainly higher now). Social scientists try to explain this sort of social behaviour. Statistics may or may not help.

If I Don't Plan to Use Statistics in My Career, Should I Still Learn about Them?

Of course you should! Within your discipline, there are conversations occurring between several communities. Some of them will assume that you have a certain level of statistical competency. To understand those discussions you need to be at least somewhat statistically savvy.

We also have a philosophical reason for believing that everyone should learn some statistics. Often there are internal divisions within Canadian social science departments—those who don't use statistics in their research can choose not to engage with those who do use statistics (and vice versa). This division hurts all of us within the academic community. It stymies the intellectual cross-pollination that occurs when different methodological allegiances are combined. To prevent these "methodological silos" from forming, and to get people with different methodological beliefs to read each other's work, it's important to understand how and what all members of your discipline think. It is equally important, and beneficial (although not the focus of this course), to learn how qualitative researchers think and to be familiar with the methods they use to do research.

Organization of This Book

We begin slowly, and in Part I of this book we cover the basics of what you need to know to survive a statistics course. Chapter 2 looks at some of the mathematical concepts you'll need to know in order to learn statistics. Many of these are basic, but you should still take the time to review them. They are the building blocks for the course, and they will help you decode and demystify subsequent chapters. The topics of Chapter 2 (including order of operations, fractions and decimals, exponents, and logarithms) are all commonly used in statistics, and you must become familiar with them.

We'll look at **univariate** (one variable) **statistics** in Chapter 3, probability in Chapter 4, and the Gaussian or normal curve in Chapter 5. The methods discussed in Chapter 3 (frequencies, percentiles, etc.) give vital information about a particular variable, including the **arithmetic average** (mean), the median, or the mode. This information can be an end in itself—for example, it is useful to tell people that the median total income of individuals in Canada in 2004 was $24,400 (Statistics Canada, 2006). Once we begin to think in terms of distributions, as we do in Chapters 5 through 7, we begin to think about further analysis of data.

Chapters 8 and 9 focus on sampling. When social scientists conduct a study, they are usually interested in a certain population. The **population** of interest can be any size. We may be interested in studying the whole population of Canada, or just the population of Manitoba. We may be interested in studying adolescents who live in Calgary (the population aged 13–19 in Calgary). We may be interested in studying all the students at one university. Whatever our population of interest, it is usually almost impossible to talk to everyone in the population. Therefore we often use data on a portion of the population, called a sample. There will always be a "mismatch" between the group being used in analysis (the sample) and the entire population. This is known as **sampling error**. However, if the sample is taken properly (randomly) we can carefully estimate how large that sampling error might be. These strategies will form the basis for Chapter 8. Chapter 9 will focus on techniques that can help to determine how closely a sample resembles the population it was drawn from. It will also cover the different ways to extract a sample from a population.

Once you understand how to describe single variables, the next logical step is to begin identifying **relationships** between variables. This is known as **hypothesizing relationships**, and in Part II of this book (Bivariate Statistics) we'll look at how to cast and test hypotheses with statistics. You probably already hypothesize and test for the existence of relationships all the time, but in Chapter 10 we'll begin to look at how to do this a little more systematically.

BOX 1.1

A Note on Causation

When we are looking at relationships between variables, it is important to distinguish between **correlation** and **causation**. If two variables are correlated, we know that they are related, but we cannot say which variable causes the other variable to change. In order to show causation, we need to demonstrate three things:

1. The **independent variable** (cause) must precede the **dependent variable** (effect) in time.
2. There must be an empirical relationship between the two variables.
3. There must be no other third variable that can explain the relationship.

Frequently, social scientists only have data that were collected at one point in time. (This is called cross-sectional data.) With cross-sectional data, it is often very difficult to prove causation. Social scientists try to be very careful not to use the word *cause* when they cannot determine causation. (Instead, we use phrases like "correlated with" or "related to.")

Chapter 10 also covers methods to test for relationships between two variables (such as gender and income, ethnicity and place of residence, rural/urban living and pickup truck ownership). These methods are useful, because they allow you to establish the existence of relationships. They also allow you to quantify the magnitude and direction of those relationships—we know, for example, that Canadian men, on average, earn more money than Canadian women do, but how much more?

Since we're almost always dealing with **samples**, we need the skills to determine whether the difference between two groups is statistically significant—that is, is the difference in the sample large enough that we can be sure it actually exists for the entire population? Assessing statistical significance allows you to determine whether the differences you observe in your samples could simply be due to **sampling error**, or if you could expect to find similar differences in the total population. This will be the focus of Chapters 10 through 15.

For an introductory statistics course, many instructors will choose to go through chapters 1 to 15. In Parts III and IV, we have added four additional chapters for more advanced courses or for those instructors who wish to expose their students to slightly more advanced statistical techniques in the social sciences.

In Part III, Multivariate Techniques, Chapter 16 will cover techniques for moving beyond studying two variables and focus on **multivariate statistics**. Multivariate statistics are often superior to **bivariate statistics** because they allow you to examine the relationship between two variables while controlling for the effect of one or more other variables. For example, you could use multivariate statistics to determine if men make a higher average wage than women because of differences in their educations. On average, people with a higher level of education earn more money, so maybe men earn more than women because they go to school longer or get more advanced degrees (they don't). Alternatively, perhaps people who live in the country are more likely to own pickup trucks because many of them are farmers. Since pickup trucks are useful on farms, perhaps the rural/urban difference in vehicle choice disappears when we **control** for occupation choice (it doesn't). Chapter 16 will look at multivariate techniques for continuous outcomes, allowing us to answer questions such as this: *Are gender pay differentials due to an education gap across genders?* Chapter 17 will focus on binary outcome variables, that is, variables with only two possible answers. An example of a binary variable would be whether a person owns a pickup truck or not.

In Part IV, Advanced Topics, Chapter 18 discusses techniques for diagnosing regression results, which compare many variables at once. Finally, Chapter 19 covers techniques for dealing with **missing data**, a situation that occurs when an individual is unable or unwilling to provide you with some, or all, of the information you request in a survey.

Once you have worked through these chapters, the statistics that you have learned will allow you to perform basic statistical analyses on most data sets and to understand much of what is published in the journals of your discipline. Since there is a lot to absorb, each chapter except this one has a series of practice questions and exercises, both at the end and scattered throughout the chapter. We suggest keeping up with the practice questions and exercises throughout your course. Doing the problems will help you identify where you are having problems and where you need further clarification from your professor. In addition to the main body of the text, there are a series of appendices. These are critical to understanding the information in several chapters.

This third edition includes several pedagogical features designed to help you as you move through your course. "Everyday Statistics" activities are taken from statistical interpretations in newspapers, magazines, and government reports. This edition also includes plenty of practice questions and exercises. Countless improvements have been made to this edition, made possible by the comments and helpful suggestions we've received since the first edition was published.

Conclusion

By the end of this course, we hope you will at least appreciate (if not enjoy) statistics. We also hope that we have allayed some of your fears about statistics and that we have given you some things to look forward to as you begin your statistical journey.

Glossary Terms

Arithmetic average (p. 6)
Bivariate statistics (p. 7)
Causation (p. 7)
Control (p. 7)
Correlation (p. 7)
Dependent variable (p. 7)
Hypothesizing relationships (p. 6)
Independent variable (p. 7)
Missing data (p. 8)
Multivariate statistics (p. 7)
Population (p. 6)
Probabilities (p. 5)
Relationships (p. 6)
Samples (p. 7)
Sampling error (p. 7)
Statistics (p. 5)
Univariate statistics (p. 6)

2 How Much Math Do I Need to Learn Statistics?

LEARNING OBJECTIVES

To learn statistics you will need some background knowledge in basic mathematical techniques. Luckily, most of you have already been exposed to all the information you will need in other courses, usually in high school. However, you may have forgotten some of this material, especially if you have not used it in a long time. In this chapter, we will review basic mathematical concepts. We will then introduce important concepts related to how we describe variables. We will cover:

- the order of operations;
- fractions, decimals, and logarithms;
- data, variables, and observations; and
- levels of measurement.

BEDMAS and the Order of Operations

Mathematics is the foundation of statistics and has its own internal logic that is not always intuitive; why, for example, must you multiply *before* you add? One of the most basic of all mathematical principles is the **order of operations.** We'll review it briefly here because it will provide a roadmap for solving problems and equations. The order of operations tells you how to go about solving a mathematical equation. **BEDMAS** stands for *brackets, exponents, division, multiplication, addition,* and *subtraction.* When faced with a mathematical equation, you must solve whatever is inside the brackets first, then calculate the exponents. For multiplication and division, the order in which you solve the problem doesn't matter. The same applies for addition and subtraction, so BEDMAS could just as easily be BEMDAS, BEMDSA, BEDMSA, etc. It is still important to remember that both multiplication and division must be done before addition and subtraction.

Here are some examples:

a. $2 + 3 * 4 = 14$
b. $(2 + 3) * 4 = 20$

For equation (a), above, we multiply the $3 * 4$ first (because multiplication, *M*, comes before addition, *A*, in BEDMAS). Then we add the 2.

$$2 + 3 * 4 = ?$$
$$2 + 12 = ?$$
$$? = 14$$

For equation (b), above, we add the 2 + 3 first (because brackets, *B*, come before multiplication, *M*, in BEDMAS. Then we multiply by 4.

$$(2 + 3) * 4 = ?$$
$$5 * 4 = ?$$
$$? = 20$$

Here are some more examples using brackets and exponents:

c. $(2 + 2)^2 = 16$
d. $2 + 2^2 = 6$

For equation (c), above, we add the 2s first since they are in brackets (*B* before *E*), and then square the results.

$$(2 + 2)^2 = ?$$
$$(4)^2 = ?$$
$$? = 16$$

For equation (d), above, we square the second 2 first (*E* before *A*) and then add the first two.

$$2 + 2^2 = ?$$
$$2 + 4 = ?$$
$$? = 6$$

For more practice using BEDMAS, see the problems at the end of the chapter.

EVERYDAY STATISTICS

An Order of Operations by Any Other Name . . .

In the United States, the order of operations is often abbreviated as PEMBDAS, which stands for Parentheses, Exponents, Multiplication, Division, Addition, and Subtraction (and is sometimes remembered with the phrase "Please Excuse My Dear Aunt Sally"). In the United Kingdom, BODMAS is often the acronym used for the order of operations, which stands for Brackets, Orders, Division, Multiplication, Addition, and Subtraction. Although these acronyms differ from BEDMAS, which we use in Canada, all of the variations of the order of operations mean the same thing.

Fractions and Decimals

Decimals and *fractions* are just different ways of stating the same thing. For example, the fraction $\frac{1}{4}$ is equal to 0.25; the fraction $\frac{1}{2}$ is equal to 0.50. You'll probably prefer to work with decimals because they're easier to use in these equations.

Exponents

To successfully use statistics, you also need to understand exponents, which take the following form: X^m, where m (the exponent) stands for how many times X is being multiplied by itself. The exponent always appears in the top right-hand corner. The larger character, X, is called the base. Using exponents is also called "raising to a power," where the exponent is the "power." It would not be uncommon to hear X^m referred to as "X raised to the power of m," or "X raised to the mth power." To illustrate, if $m = 3$, then $X * X * X$ would be represented as X^3.

If you're working with two or more exponents with identical bases and you need to multiply them, you can add the exponents together. Rather than relying extensively on multiplication, you can rely more heavily on addition. For example, $X^a * X^b$ is the same as X^{a+b}. This can only be done when the bases are the same. $X^a * Y^b$ is *not* the same as XY^{a+b}.

The **inverse function** of an exponent is the root: $2^2 = 4$, then $\sqrt{4} = 2$.

Logarithms

A **logarithm** is a special form of exponent, where the base is either 10 or 2.718 (the transcendental number, or *e*). Transcendental numbers are not algebraic numbers nor are they numbers that are expressed in any form of fraction; therefore, they can also be considered irrational numbers. When the base is 10, we're dealing with common logarithms, whereas a base of 2.718 (also called the base *e*) brings us to natural logarithms, which is what we will use in this text. As an example, let's look at the logarithm of 1000. Usually, in mathematics, if the base is not stated, it is assumed to be 10, so it could be expressed as $\log 1000 = 3$ or $\log_{10} 1000 = 3$. In this example, the logarithm is a way of stating that when 10 is the base it must be multiplied by itself three times (10^3) to obtain the product of 1000.

For now, you only have to remember that, just as the inverse function of an exponent is the root, the inverse of a log is the base. We'll return to this discussion in Chapter 17 when we discuss logistic regression.

Data, Variables, and Observations

Social scientists collect data using many different methods, including interviews, surveys, and observations. We can collect data on individuals, on couples, on households, on neighbourhoods, on schools, on cities, etc. The level at which we collect the data is called the **unit of analysis**. Each object (individual, couple, household, etc.) we collect data on becomes an observation, or a case, in our data set. The observations are usually listed along the rows in a data set, with each observation taking one row. **Variables** are the pieces of information we collect on our units of observation. They are usually listed in the columns. Each observation, or case, has a value for each variable. The example in Table 2.1 shows the raw data from a (very!) small data set that contains data collected from five individuals. The individual is the unit of analysis, there are five cases, and we show data for four variables: sex, age, major, and GPA.

TABLE 2.1 | Observations and Variables in a Data Set

Name	Sex	Age	Major	GPA
Jenny	Female	24	Sociology	3.0
Michael	Male	22	Sociology	3.6
Bill	Male	23	Engineering	2.8
Antoine	Male	22	Art	3.0
Sofia	Female	21	Math	3.1

In order to use this data in statistical analysis, all of the information that is currently provided in text form needs to be put into numbers. Table 2.2 is the same table reproduced with numbers instead of text data.

TABLE 2.2 | Observations and Variables in a Data Set in Numerical Form

CaseID	Sex	Age	Major	GPA
101	1	24	1	3.0
102	2	22	1	3.6
103	2	23	2	2.8
104	2	22	3	3.0
105	1	21	4	3.1

You can see now that each observation or case has an ID number instead of a name. (These ID numbers can be anything—they are just used to identify the case.) Sex is now coded as 1 = Female, 2 = Male. Major is coded as 1 = Sociology, 2 = Engineering, 3 = Art, and 4 = Math.

When you first examine a data set, it is important to understand how it is set up. You should always look at the data set (in your software package) and have a look at the number of cases, the number of variables, and the types of variables. The codebook should tell you how the variables are coded. (For example, in the example in tables 2.1 and 2.2, the codebook would show you that 1 = Female for the Sex variable, and 1 = Sociology for the Major variable.)

Levels of Measurement

When you first examine your variables in a data set, it is useful to classify each variable by the type of quantitative information it contains. Variables can be classified into four categories, or **levels of measurement**. These categories are **nominal**, **ordinal**, **interval**, and **ratio**. The primary distinction between the levels stems from the relationship that exists between the different possible values of the variable. For example, take the variable sex, which in most data sets has two possible values: male and female. If female is coded as 1 and male is coded as 2 (as in the example above), it does not mean that men are "twice as good" as women. You cannot use the numbers 1 and 2 for any purpose other than identification in this example. It is not possible to rank respondents or judge them based on whether they got a 1 or a 2 on the sex variable.

Variables that have values that have no relationship to one another mathematically are called *nominal*—that is, the numbers are really only names of things. With nominal data, it isn't possible to rank subjects based on value or to usefully measure the distances between response categories. For statistical purposes, nominal variables, such as religion, social insurance number, or hair colour, are used to distinguish among respondents.

The next level of measurement is *ordinal*: variables where the values can be placed into an order. A good example of this is the order that students finish in a foot race. If you knew how a person placed, you would be able to assess how well they did relative to other students, but you would *not* be able to assess the exact differences in their athletic skill. The first-place student could have finished a second ahead of the second-place student, and the third-place student could have finished five minutes behind the second-place student, but they would still be ranked "1, 2, 3." Ordinal variables can be ranked, but the exact distance or difference between the values of the variable is not the same across all values.

In another example of ordinal data, when you respond to a telephone survey on the degree to which you agree, for example, that *Game of Thrones* is the best television show ever, you might encounter a scale from 1 (strongly disagree) to 5 (strongly agree). This is an ordinal-level question (the intermediate scores would represent more moderate levels of agreement or disagreement). We know that there is a difference in the level of agreement between "strongly agree" and "strongly disagree," but we cannot accurately measure this distance, and we cannot assert that the difference between "strongly agree" and "agree" is exactly the same as the difference between "strongly disagree" and "disagree."

Interval data can be organized into an order and can be added or subtracted. However, they cannot be multiplied or divided because, although there are equal distances or intervals between a pair of values, you cannot say that a value of four is twice as much of what happens to be measured than a value of two. Temperatures in Fahrenheit are interval data because the zero point is statistically arbitrary. If the temperature is zero degrees, it does not make sense to say that there is no temperature. Interval levels of measurement are much less common than those at the ratio level.

With *ratio* data, you *can* rank individuals and accurately measure the distance between them. You can also add, subtract, multiply, or divide ratio data. In ratio data, zero means that there is none of what you are measuring. Ratio data can be in fractions or decimals—unlike some kinds of interval data, ratio data don't have set intervals. In the racing example mentioned earlier, a variable revealing the time it took each participant to complete the race in minutes would be a ratio variable. Before a race begins, the stop watch is set to zero because no time has yet passed in the race. The measurement of how long it took a competitor to win the race can be considered a ratio measure because the zero has statistical meaning. With ordinal data, it is not possible to measure how close the first- and second-place contestants were when they finished. Knowing the time of completion, however, allows us to not only identify the first- and second-place finishers but to determine how big the gap was between them. As with the other levels of measurement, examples of ratio data abound: age, income, airfare ticket prices, and so on. The list of ratio variables is almost endless.

Interval and ratio data can be classified further according to whether they are **continuous** or **discrete**. Discrete data have categories that cannot be subdivided, whereas continuous data *can* be subdivided.

See Table 2.3 for a chart showing the different levels of measurement.

TABLE 2.3 | Descriptions of Levels of Measurement and Examples

Level of measurement	Description	Can you rank response categories?	Can you measure distances between response categories?	Example	Example of codes
Nominal	Identifies the categories to which the observations belong, but does not allow them to be ranked or the difference between observations to be calculated.	No	No	Which country is each marathon runner from?	1 = Kenya 2 = Great Britain 3 = Canada, etc.
Ordinal	Provides information about how each observation is ranked.	Yes	No	Who came in first, second, and third place in the marathon?	1 = First place 2 = Second place 3 = Third place, etc.
Interval	Allows for the difference between observations to be measured but the value of zero has no qualitative difference when compared to the value of one.	Yes	Yes	What was each competitor's heart rate as he or she crossed the finish line?	160 = First runner's heart rate 155 = Second runner's heart rate 152 = Third runner's heart rate, etc.
Ratio	Has a meaningful zero value and allows for the exact difference between observations to be measured.	Yes	Yes	How much time did it take to cross the finish line?	31 minutes = First runner's time 32 minutes = Second runner's time 37 minutes = Third runner's time, etc.

When Four Levels of Measurement Become Three . . . or Even Two

Many believe that maintaining the four levels of distinction is excessive. Few people see any purpose in maintaining the distinction between interval and ratio levels, so these variables are often collapsed, leaving us with three categories for level of measurement: (1) nominal, (2) ordinal, and (3) interval/ratio.

Conclusion

This chapter reviewed the math skills you will need to be successful in this course and introduced the concept of level of measurement. You must learn the distinctions between the different levels of measurement. Whenever you examine a data set for the first time, you need to identify the level of measurement of each of the variables you will be using. For most statistical techniques, the type of analysis you are able to do depends on the level of measurement of the variables. Variables measured at a higher level of measurement can always be recoded to a simpler level of measurement (for example, the ratio variable "age" could be recoded into an ordinal variable, age categories), but variables measured at the lower level of measurement cannot be recoded to a higher level of measurement.

Glossary Terms

BEDMAS (p. 9)
Continuous variable (p. 14)
Discrete variable (p. 14)
Interval—level of measurement (p. 13)
Inverse function (p. 11)
Levels of measurement (p. 13)
Logarithm (p. 11)
Nominal—level of measurement (p. 13)
Order of operations (p. 9)
Ordinal—level of measurement (p. 13)
Ratio—level of measurement (p. 13)
Unit of analysis (p. 11)
Variables (p. 12)

Practice Questions

Please replace the question marks below with the appropriate answers:

1. 10 + 15 = ? 25
2. 10 + 15 - 5 = ? 20
3. 10 - (-2) = ? 12
4. (10 + 15) - 5 = ? 20

5. $(10 - 15) - 2 = ?$

6. $10 \times 15 = ?$

7. $10 \times 15 - 5 = ?$

8. $10 \times (15 - 5) = ?$

9. $10 \times 15 - \frac{15}{5} = ?$

10. $\frac{10}{5} \times 15 - 5 = ?$

11. $(X \times Y)^a + b = X^? \times Y^? + ?$

12. $(X^a)(X^b) = X^?$

13. $\sqrt{X} = X^?$

14. If $\ln 5 = 1.61$, then $e^{1.61} = ?$

Identify the levels of measurement (nominal, ordinal, interval, or ratio) for the following:

15. Percentage scores on a math exam

16. Letter grades on a math exam

17. Flavours of ice cream

18. Fitness training levels on an exercise machine classified as Easy, Difficult, or Impossible

19. Ethnic origins

20. Political parties

21. Commuting distances to school in kilometres

22. Years between important historical events

23. Age (in years)

24. Amount of money in your savings accounts

25. Temperature on the moon, measured in degrees Celsius

Answers to the practice questions for Chapter 2 can be found in Appendix H.

3 Univariate Statistics

LEARNING OBJECTIVES

In this chapter, we'll look at how to describe the characteristics of variables by using:

- frequencies;
- rates and ratios; and
- percentages and percentiles.

Frequencies

In this course, you will learn how to make large, unwieldy sets of numbers more understandable, or intuitive, allowing for comparisons of data to be made. Part of your role as a social statistician is to do exactly this, to simplify and translate numbers. It is your job to turn numbers that only a few people can understand into something that's easy to explain to anyone.

One of the most basic translation tools is the **frequency** table. Typically used for data measured at the nominal or ordinal level, frequency tables tell us the number of times an item, or a response category, comes up in a sample. If, for example, we wanted to know how many males and females there are in Canada, a good way to present this information would be in the form of a frequency table such as Table 3.1.

There are five columns of information in the table. The first, labelled "Sex," is the name of the variable: the sex of the respondent. Since "Sex" is a nominal variable, the order in which the response categories are presented is arbitrary (the information for males could appear before females). The next column provides the frequencies for each of the categories. Now we know that as of 16 May 2006 (the census reference date), there were 16,136,930 females and 15,475,970 males living in Canada. As the column "%" tells us, this translates into 51.05 per cent of the Canadian population being female and 48.95 per cent being male. "Cumulative frequency" is a running total of the frequency of observations in each category (thus, the 31,612,895 in the "Cumulative frequency" column beside "Male" equals the total number of males *and* females), and "Cumulative %" tells us the same about the percentage of observations in each category (since everyone is either male or female in the census, the total in the "Cumulative per cent" column beside "Male" is 100 per cent).

Notice that this table is quite different from the tables in Chapter 2, where we looked at the raw data. This table is just showing the summary data for one variable. The rows show the

TABLE 3.1 | Number of Males and Females in the Canadian Population, 2006 Census of Canada

Sex	Frequency	%	Cumulative frequency	Cumulative %
Female	16,136,930	51.05	16,136,930	51.05
Male	15,475,970	48.95	31,612,895	100.00

Source: Age and Sex Highlight Tables. 2006 Census. Statistics Canada Catalogue no. 97–551–XWE2006002. Ottawa. Released July 17 2007.
Note: Includes all Canadian citizens and landed immigrants who have a usual place of residence in Canada, or who are abroad either on a military base or attached to a diplomatic mission.

values that the variable can take, not individual observations. If we were to look at the raw data from the census, it would contain 31,612,895 rows!

There are several ways to present the same data. In Figures 3.1 and 3.2, **bar charts** are used instead of numbers. Many people prefer that format because it provides a visual that can be quickly and easily understood. Figure 3.1 shows the number of observations in each category.

Figure 3.2 shows the percentage of observations in each category. As you can see, the horizontal axis (commonly referred to as the ***x*-axis**), is the same for both graphs (sex of respondent), as is the height of the bars. The only real difference between the two charts is that Figure 3.1 has the number of observations in each category as the unit for the vertical axis, and Figure 3.2 has the per cent of total population as the vertical axis (commonly referred to as the ***y*-axis**).

In both of the charts it appears as though females outnumber males because the bar that corresponds with the frequency, or percentage, of women is almost three times higher than the bar for males. This is deceptive, because the y-axis begins at 48 per cent. A more representative chart would begin the *y*-axis at 0, or 0 per cent.

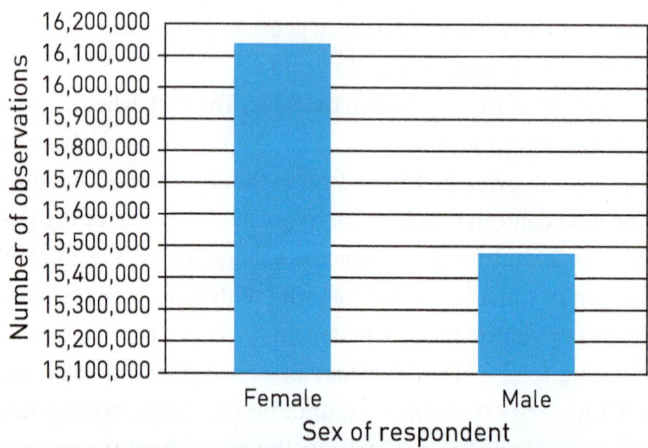

FIGURE 3.1 | Bar Chart of the Number of Males and Females in the Canadian Population, 2006 Census of Canada

Note: Includes all Canadian citizens and landed immigrants who have a usual place of residence in Canada, or who are abroad either on a military base or attached to a diplomatic mission.
Source: Age and Sex Highlight Tables. 2006 Census. Statistics Canada Catalogue no. 97–551–XWE2006002. Ottawa. Released July 17 2007.

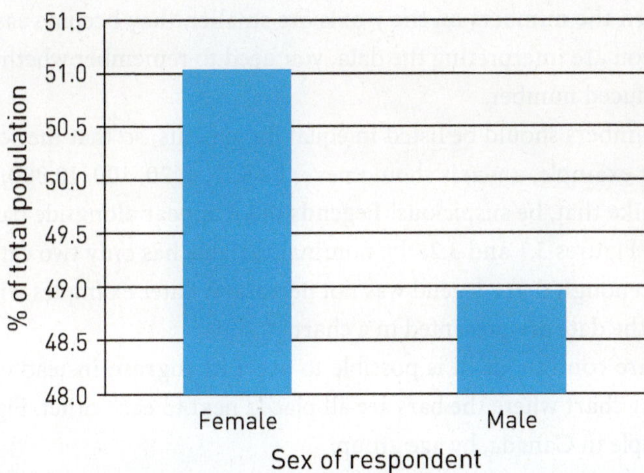

FIGURE 3.2 | Bar Chart of the Percentage of Males and Females in the Canadian Population, 2006 Census of Canada

Note: Includes all Canadian citizens and landed immigrants who have a usual place of residence in Canada, or who are abroad either on a military base or attached to a diplomatic mission.

Source: Age and Sex Highlight Tables. 2006 Census. Statistics Canada Catalogue no. 97–551–XWE2006002. Ottawa. Released July 17 2007.

Translating Frequencies

Frequencies can often be misleading and/or difficult to work with. The numbers are often large and difficult to simplify. Presenting data in percentages, as in Figures 3.2 (above) and 3.4 (in the following section), is one method of simplifying frequencies. Percentages are a good way to translate simple statistics. Knowing that 51.05 per cent of all Canadians are female is more informative than knowing that the number of women in Canada is 16,136,930. The number alone is not informative because we do not know the total population size.

Rules for Creating Bar Charts

You need to know a few things about using bar charts.

First, the response categories should always appear on the *x*-axis, and the frequencies (whether stated as a percentage or as the number of observations) should always be on the *y*-axis.

Second, the title should only describe the data presented; it should not impose an interpretation of the data. **Axis titles** should be brief and non-repetitive. For example, it would not be necessary to use the full title "sex of respondent in the Canadian census" as the *x*-axis for Figure 3.2, because that information is in the title of the chart. **Axis scales** should present data as efficiently as possible, without using too many numbers. In Figure 3.1, it would be acceptable to remove some of the zeros in the scale, and express the numbers in units of 10,000, 100,000, or even 1,000,000—so the first number would be 14,300, or 1430, or 14.3, respectively. When the numbers on the y-axis are smaller, they become easier to interpret. However, when you are interpreting the data, you need to remember whether you are using the full or the reduced number.

Third, the numbers should be listed in equal increments, so that the scale is consistent for each axis. For example, a y-axis should never be 0, 2, 4, 20, 100, 10,000, etc. If you see a chart that looks like that, be suspicious! **Legends** often appear alongside bar charts and are always useful. In Figures 3.1 and 3.2, the nominal variable has only two categories, and the data were simple enough that a legend was not necessary. Later examples will use legends to help clarify how the data are presented in a chart.

When data are continuous, it is possible to use a **histogram** instead of a bar chart. A histogram is a bar chart where the bars are all placed next to each other. Figure 3.3 shows a histogram of people in Canada, by age group.

EVERYDAY STATISTICS

History of the Bar Chart

William Playfair first developed the bar chart in his 1786 book *The Commercial and Political Atlas*. In this book, Playfair was interested in graphing the imports and exports from different countries over a period of several years. Accordingly, the first bar chart was a representation of Scotland's imports and exports to and from several different countries in 1781.

Q: Why do you think that a bar chart would be helpful to graph Playfair's observations?

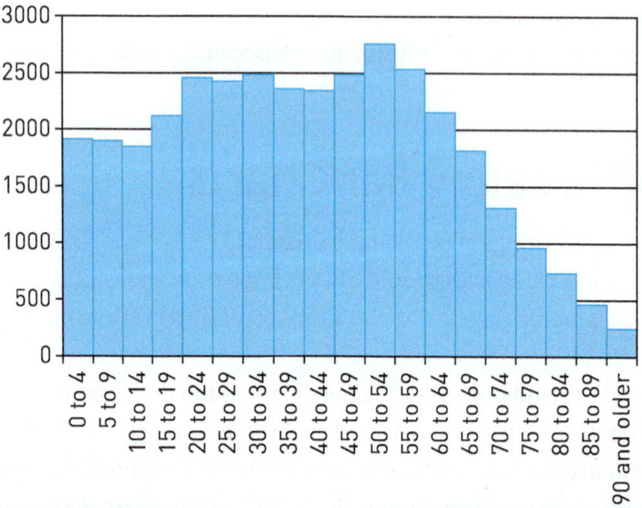

FIGURE 3.3 | Age Group by Persons (in Thousands) in Canada, 2014

Source: Adapted from Statistics Canada, CANSIM, Table 051-0001, accessed at www.statcan.gc.ca/tables-tableaux/sum-som/l01/cst01/demo10a-eng.htm

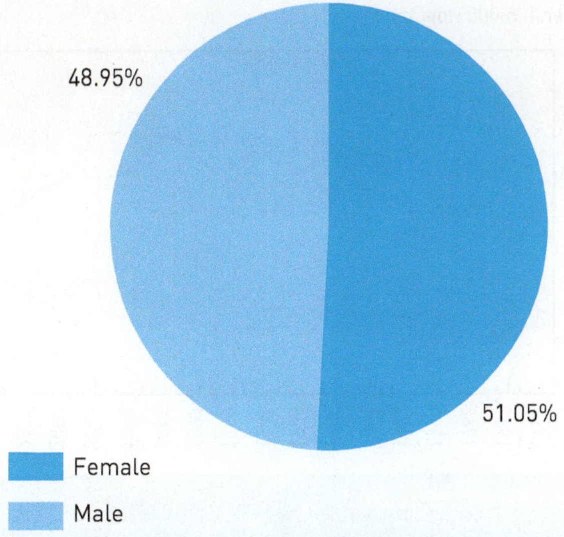

FIGURE 3.4 | Pie Chart of the Percentage of Males and Females in the Canadian Population, 2006 Census of Canada

Note: Includes all Canadian citizens and landed immigrants who have a usual place of residence in Canada, or who are abroad either on a military base or attached to a diplomatic mission.

Source: Age and Sex Highlight Tables. 2006 Census. Statistics Canada Catalogue no. 97-551-XWE2006002. Ottawa. Released July 17 2007.

Fourth, list the data source, usually in a smaller font, below the chart. Readers will need to know the source of the numbers if they want to replicate the results. If there are notes about your sample, include those, too. Notice the notes regarding the section of the population included in the census data at the bottom of Figures 3.1 and 3.2.

Data can also be presented in a **pie chart**, as in Figure 3.4. The rules for bar charts also apply to pie charts. Legends are helpful, titles should be brief and descriptive, and data sources should be listed beneath the figure.

Rates and Ratios

Two other numerical translation tools are **rates** and **ratios**. A ratio is the number of observations in one category compared to the number of observations in another category. For example, we could report the 2006 census data as a ratio and say that there are 16,136,930 women for every 15,475,970 men in Canada, but we would probably want to reduce these numbers to make them easier to digest. That can be done by expressing the numbers as fractions and cancelling out common factors in the numerator and denominator. By dividing the number of men and of women by 100,000, 155 can replace 15,475,970 and 161 can be substituted for 16,136,930 $\left(\frac{16{,}136{,}930}{100{,}000} \approx 161; \frac{15{,}475{,}970}{100{,}000} \approx 155\right)$. Thus, we would have a ratio of 161:155.

Rates are closely related to ratios, but the denominator is usually a round number (1000 people, 100,000 people, etc.) or an intuitive number (one hour for kilometres per hour, one minute for heartbeats per minute, etc.). For example, we could say that there are approximately 511 women for every 1000 Canadians. Ratios can be used to compare categorical or

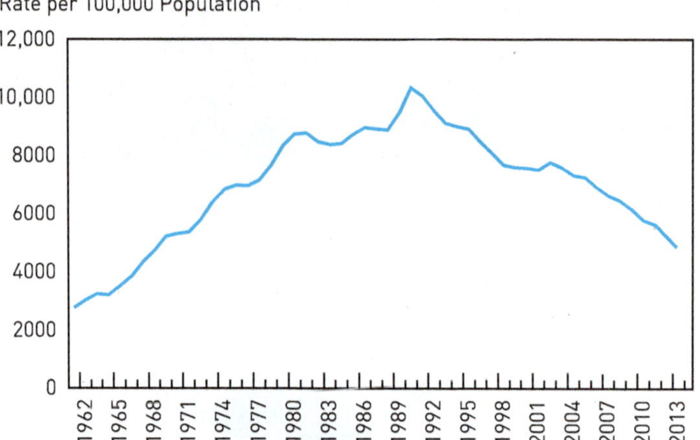

FIGURE 3.5 | Crime Rate, 1962 to 2013
Source: Statistics Canada, Canadian Centre for Justice Statistics, Uniform Crime Reporting Survey; accessed at www.statcan.gc.ca/pub/85-002-x/2014001/article/14040/c-g/desc/desc02-eng.htm

continuous data, while rates are usually only used to present continuous data. Rates are one of the most common methods in the social sciences for presenting univariate data (data that summarizes a single variable); examples include crime rates, death rates, birth rates, fertility rates, unemployment rates, and inflation rates. To further illustrate the use of rates, look at the Canadian crime rate for the period 1962–2013, as shown in Figure 3.5, which is a **line graph**, where the rates for each year are connected by a line.

In 1962, the crime rate was approximately 2800 for every 100,000 people. It continued to inch upward until about 1993, when it reached 10,000 per 100,000. After 1993, there was a dramatic reversal. In 2002, the rate stood just below 8000 per 100,000 people, and by 2013 the rate had declined to 5000 per 100,000 people.

Rates are a more elegant way of presenting data because they use the same denominator (100,000 people in this case). Presenting crime statistics (or birth statistics, death statistics, unemployment statistics, etc.) as a frequency would not tell us much because it would not take into account the size of the total population (which changes over time).

Rates, ratios, and percentages are attractive because they are **standardized**. That is, they use the same unit of measurement, and can be compared across countries, over time, or any other way you choose. When using rates, you must ensure that the denominator is the same across comparison groups (i.e., make sure that all ratios that you compare are based on the same "per *X* population"). With percentages that are already done for you, the denominator is always 100.

Percentages and Percentiles

The *Canadian Oxford Dictionary* defines *percentile* as "one of 99 values of a variable dividing a population into 100 equal groups as regards the value of that variable." You determine the percentiles by slicing your sample into 100 groups, making sure that each group has exactly the same number of people (the best you can, without slicing a person in half). If there were 30,000,000 people in Canada and we wanted to sort them into 100 age groups, each group would have 300,000 people. We would list each person by age, and the youngest 300,000 would form the first percentile, the second 300,000 would form the second percentile, etc. Each of those groups would represent one percentile of the data, and the age values used to delineate the 100 groups would form the cut-points for the percentiles, the points at which each percentile is differentiated.

Returning to the 2006 census, the age value for the first percentile is 0 because the youngest 1 per cent of the population are babies who haven't had a birthday yet. Showing all 100 percentiles in a table makes for a large, clumsy table, so you usually only show a few of them, as in Table 3.2. It is also common to show deciles (10, 20, . . .) or quartiles (25, 50, 75, and 100). Table 3.2 shows that age 19 is at the twenty-fifth percentile (or first quartile), meaning 25 per cent of the total population is below the age of 19. The fiftieth percentile, which is also called the median, is where the Canadian population is evenly divided in half. Half the people are older than this age and half are younger. The median age is 40.

The fiftieth percentile is the median, since half of the population is above 40, and half is below. The seventy-fifth percentile is those who are over the age of 54, and the one-hundredth percentile cut-off is age 85. This number is lower than you might expect because Statistics Canada recodes all values above age 85 so that those who are 85 and those who are older are all labelled as being 85, to ensure confidentiality.

Here's an example of percentiles that you are very familiar with: standardized test scores. If you scored 91 per cent on a standardized test, you would know that you were 9 percentage points from 100, and 91 points from zero. This would give you an idea of how you did on the test. Let's suppose that lots of people did well on the test, and that 91 per cent was actually the score for the fiftieth percentile (or the median). You could be proud that you beat half of your classmates, but the other half would also have beaten your score. Depending on the **measure of central tendency** (the average or typical score for those who wrote the test), your 91 per cent might only be average, not the exceptional grade that you thought it was. Percentiles rank you in relation to your peers, or everyone else who took that

TABLE 3.2 | The Age of the Canadian Population by Percentile Cut-Offs, 2006, Canada

Percentile	Value
1	0
25	19
50	40
75	54
100	85

Source: 2006 Census of Canada

standardized test. They are often used in standardized testing, such as the Scholastic Aptitude Test (SAT), the Law School Admission Test (LSAT), the Graduate Record Examination (GRE), the Medical College Admission Test (MCAT), or the American College Test (ACT).

Conclusion

This chapter introduced you to some ways to describe and display the distributions of variables in your data set. Even the most experienced statistical analyst must start with these procedures—examining the distribution of each variable before moving on to more complicated statistical techniques. In the next chapter, you will learn some further tools you can use to describe and summarize variables.

Glossary Terms

Axis scales (p. 19)
Axis titles (p. 19)
Bar charts (p. 18)
Frequency (p. 17)
Histogram (p. 20)
Legends (p. 20)
Line graph (p. 22)
Measure of central tendency (p. 23)
Pie chart (p. 21)
Rates (p. 21)
Ratios (p. 21)
Standardized (p. 22)
x-axis (p. 18)
y-axis (p. 18)

Practice Questions

1. Jorge placed one-hundred-and-thirteenth out of 1432 people in a national spelling bee by correctly spelling 37 of the 40 words he was given.
   a. What is the ratio of correct to incorrect responses?
   b. What is his score stated as a percentage?
   c. What is his rank in percentiles?

2. Ethel is conducting a telephone survey to determine the approximate number of people who are interested in lobbying the federal government about recent changes to the Canada Pension Plan. Over the course of a week, she calls 541 people. Of those people, she contacts 432 and identifies 112 individuals who are willing to participate in the lobbying effort.
   a. What is her contact rate per 1000 people?
   b. What is her contact/non-contact ratio in lowest terms?
   c. What is her lobbying participation rate as a percentage, using only contacted individuals?

3. Charles, a woodchuck with ego problems, thinks that he can chuck more wood than any other woodchuck. But he is insecure and decides that he'll ask 30 of his closest woodchuck friends if they agree. To his chagrin, he reaches only 22 of his friends, and only 5 agree with him. Although he didn't reach 8 of his woodchuck friends, he is certain that they'd

agree that he's the best wood-chucker out there. To celebrate his accomplishment, he decides to make a poster to hang on the wall of his dwelling, but he doesn't know how to calculate any univariate statistics. Help Charles out. Using his assumption, calculate the following:

a. The number of woodchucks who think he is the king of all woodchucks

b. The percentage of woodchucks who think he is the king of all woodchucks

c. The ratio of woodchucks who think he is the king of all woodchucks to those who disagree.

Answers to the practice questions for Chapter 3 can be found in Appendix H.

4 Introduction to Probability

LEARNING OBJECTIVES

Learning statistics relies on a basic understanding of mathematical probability. In Chapter 4, we'll cover:

- necessary terminology, including the law of large numbers; and
- types of probability.

Introduction

A standard deck of cards has four suits (spades, hearts, diamonds, and clubs) and 13 cards in each suit (ace through king). Let's assume there are no jokers in the deck. If each card is equally likely to be drawn, is it possible to calculate the likelihood of drawing a particular card?

The answer is yes. Any time you identify how likely an event is to occur, what you're really doing is calculating a probability. A **probability** is a number between 0 and 1 (or 0 to 100 per cent, when stated as a percentage), where 0 refers to an event that never occurs; and 1, to an event that definitely occurs. The event can be anything imaginable, from a coin landing heads-up, to a car accident on the way home from work. It doesn't really matter. What does matter is that a calculable chance, or probability, can be attached to the competing outcomes.

Why is probability important for learning statistics? The short answer to this question is that because we are typically working with a population subset or sample, we use probabilities to determine the likelihood that the patterns we see in our data set (or sample) would also be seen in our entire population of interest. For example, the National Household Survey of Canada is based, roughly, on a 3 per cent sample of the entire Canadian population. Without probabilities, we'd have no way of generalizing to all Canadian adults what we find in the survey. We use the sampling distribution and hypothesis testing (topics that are both based on theories of probability, and will be covered in chapters 9 and 10, respectively) to generalize from the sample to the population. In the next few chapters, we'll investigate how to use sample data to generalize to the population. First, we'll start with a discussion of what probabilities are and how to calculate them. Then, we'll discuss the normal curve before moving on to how these things relate to one another.

This chapter will cover the basic types of probabilities and will identify some of the laws that make probabilities work in statistics.

Some Necessary Terminology

To discuss probabilities, you need to know some vocabulary. These concepts will be used throughout the rest of the text, so take the time to learn them and understand them.

Sample Space

A **sample space** contains all of the theoretically possible outcomes of an event. Each probability is a fraction of the sample space. The sum of the probabilities of all possible outcomes equals 1. The probability of the occurrence of an event is 1 minus the probability that it won't occur.

For example, if the probability of picking a green apple from a barrel containing 44 green apples and 56 red apples is $\frac{44}{100}$, or $\frac{11}{25}$, then the probability *of not* picking a green apple is equal to $1 - \frac{11}{25}$, or $\frac{14}{25}$. When there are only two possible outcomes (picking a green apple or a red apple), $\frac{14}{25}$ is also the probability of picking a red apple. The probability of picking a red apple plus the probability of picking a green apple is $\frac{14}{25} + \frac{11}{25}$, or $\frac{25}{25}$. If you pick an apple, you can be certain that it will either be green or red. The sum of all probabilities of an outcome will always be 1.

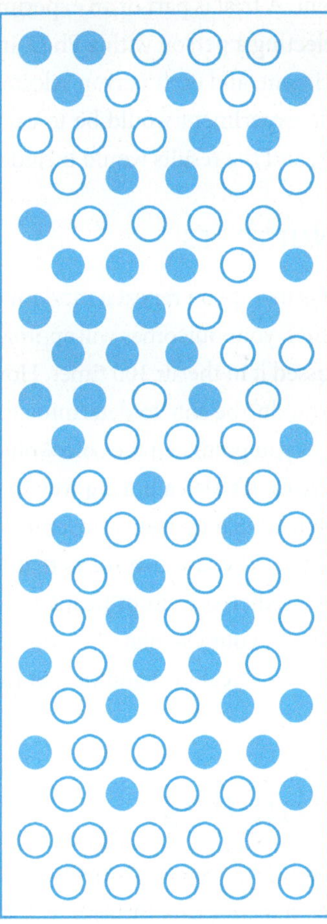

Random Variables

What makes a variable "random" is that the value is subject to variation from known or unknown sources. The value of a random variable is not predictable, strictly speaking, but the probability of a variable being a certain value can often be calculated. A random variable has a value that is the result of a process or experiment, such as tossing a coin or splitting an atom. Like other types of variables, random variables can take on different values, but the entire sample space is usually already known.

Imagine a coin-tossing experiment. Create a variable (let's call it X) for whether the coin will come up heads or tails. Since it (1) was created by a particular process or experiment (tossing a coin), (2) has a known sample space (heads and tails), and (3) is not possible to predict outcomes perfectly (that is, whether a toss lands heads or tails), we know that we are dealing with a random variable.

Trials and Experiments

Discussions about probability, or statistics, are usually conducted in terms of trials and experiments. A trial is an individual exercise that, when taken alongside other exercises, will collectively form the data for the experiment. A trial is part of an experiment. Suppose you were calculating the probability of randomly selecting a person with a Ph.D. in your student union building. The exercise would form the experiment, and each person selected would count as a trial. If you did this 100 times, the point of the experiment would be to calculate the probability of randomly selecting a person with a Ph.D., and the results would include the data from the 100 trials.

The Law of Large Numbers

The **law of large numbers** states that if you repeat a random experiment (such as tossing a coin or rolling a die) many, many times, your outcomes will approach a level of "stability." Pretend that you have a coin and that you tossed it in the air 100 times. How many heads would you expect to get? In a 100-toss event, you wouldn't be able to determine that perfectly, but you could make a pretty good "guess." Let's say that you predicted the coin would turn up heads half the time.

One hundred is a lot of tosses. It's also a boring way to spend your time. The law of large numbers states that you'll probably turn up heads closer to half of the time tossing it 100 times than if you tossed it only 10 times. The law also states that if you tossed it 100,000 times, you would get even closer to your predicted ratio of heads to tails (1 to 1). The more times you toss the coin, the closer you'll be to your calculated theoretical probability value of half heads. This is because empirical and theoretical probabilities, both discussed below, converge as the number of trials increases.

Types of Probabilities

There are numerous types of probabilities—so many that there are university courses dedicated just to that subject. We won't cover all of them. Instead, we'll focus on some of the more common types that you will need to know in order to understand the statistics presented in this book.

Empirical versus Theoretical Probabilities

There are two ways to calculate the probability of an event.

First, you can conduct an experiment and use the results to calculate the probability. If, for example, you wanted to know the probability of drawing a spade from a full deck of well-shuffled cards, you could draw single cards over and over again and use your results to calculate the probability of drawing a spade. If you conducted 100 trials, or draws, and pulled a spade 26 times, you could conclude that the probability of pulling a spade is equal to roughly $\frac{26}{100}$, or 26 per cent. Since you are using real data, that number is known as an **empirical probability**.

Second, you can calculate a **theoretical probability** using your powers of logical reasoning. Instead of conducting an experiment, you could determine the total number of spades in a deck (13) relative to the total number of cards (52). That would yield the theoretical probability of $\frac{13}{52}$, or 25 per cent. This means that you should expect to pull a spade about 25 per cent of the time.

You can see that there is a difference in the results for the empirical and theoretical probabilities. This can be due to either a random or a non-random error in the empirical probability. An example of non-random error could be if the deck of cards happened to be missing a few cards or contained more or less of a suit than you had initially believed. The probability would be affected because of the poor design of the experiment, which is not random. Perhaps the deck wasn't shuffled perfectly, or there weren't enough draws. If there were more trials (draws), the law of large numbers says that the empirical probability will converge with the theoretical probability.

For now, to make things simple, let's pretend that we always have a sample large enough to make the differences between empirical and theoretical probabilities negligible. We'll discuss what to do when sample size becomes an issue in Chapter 9 and beyond, but for now let's pretend that sample size is not an issue.

Discrete Probabilities

A discrete probability has clearly defined, non-overlapping outcomes. Variables with discrete probabilities often have an equal probability of occurrence for each value. For example, the roll of a single die has six discrete possible outcomes (turning up as 1, 2, 3, 4, 5, or 6). We can assign equal probabilities to each outcome. In this case, the probability of any one outcome would be $\frac{1}{6}$, the frequency of each outcome over the total possible number of outcomes.

Tossing a coin is another example of a variable with a discrete outcome. The coin will land either heads or tails, and if there is no reason to suspect that the coin will come up more often one way than the other, we can assign the probability of $\frac{1}{2}$ to each of the outcomes. The outcomes are discrete because a coin that comes up heads cannot also come up tails.

In both examples, each outcome has an equal probability, but it isn't hard to imagine examples of where this isn't so. For example, if there is a 15 per cent chance of rain, we assign a probability of 0.15 to the chance of rain and a probability of 0.85 to no rain. Once again, the occurrence of rain is discrete because it either rains or it doesn't.

Discrete probabilities are easy to work with because of their intuitiveness. If all of the possible outcomes are known, the sum of probabilities will equal 1 (that is, at least one event in the sample space will occur). For example, the die-tossing scenario has six outcomes, each with a probability of $\frac{1}{6}$. If we sum the values of our sample space, we get $\frac{1}{6} + \frac{1}{6} + \frac{1}{6} + \frac{1}{6} + \frac{1}{6} + \frac{1}{6}$, which equals $\frac{6}{6}$, or 1.

Even when probabilities are not equal, the sum is 1. The chance of rain in the earlier example is 0.15 and the chance of no rain is 0.85, so the probability of either rain or no rain is 1 ($0.15 + 0.85 = 1$). State the probability as a percentage and it's 100 per cent.

EVERYDAY STATISTICS

Probability and Car Insurance

Often, car insurance companies set premiums based on the probability that a customer will make a claim against his or her insurance. The perceived amount of risk that an individual poses is influenced by age, gender, the length of time the person has had a licence, etc. Based on the probability obtained by the insurance company, a profitable rate is calculated to charge the customer. Therefore, young drivers are charged more for car insurance than middle-age drivers because the probability that young drivers will get in an accident (and therefore make a claim against their insurance) is higher.

Q: What types of probabilities do you think insurance companies would use? Why?

Discrete probabilities are also called simple probabilities because they involve only one set of outcomes. (See Box 4.1.) In scenarios where more than one set of outcomes is likely, calculating probabilities is more difficult.

The Probability of Unrelated Events

If one event does not affect the probability of another event, the events are independent. Suppose you're calculating the probabilities of two unrelated activities. An example would be rolling two 6s in a row with a six-sided die. The probability of rolling a single 6 would be equal to $\frac{1}{6}$. But what about rolling two 6s in a row?

To calculate the probability of two independent events occurring, first, calculate the two discrete probabilities, p(A) and p(B). Then use the **multiplication rule of probabilities**, which states that observing two independent outcomes in succession is equal to the product of the probability of the two individual outcomes. In our example, we have two independent probabilities, p(A) and p(B), with respective probabilities of $\frac{1}{6}$ and $\frac{1}{6}$. Each of these refers to the independent probability of rolling a 6, whereas multiplying the two together provides the probability of the two events occurring in succession. To identify the probability of rolling two successive 6s, we multiply these numbers together:

BOX 4.1

The Steps: Calculating a Discrete, Theoretical Probability

1. Identify the experiment of interest (coin tossing, card drawing, etc.), and be sure that the outcomes are mutually exclusive.
2. Determine the total number of possible outcomes. This will be the denominator of your probability calculations.
3. Determine the frequency of occurrence of your outcome of interest (a coin landing on heads, drawing a spade, etc.). This will be the numerator of your calculation.
4. Divide the numerator by the denominator to determine the discrete probability.
5. Convert the probability to a percentage by multiplying the number by 100 if you wish, or leave it as a decimal.

$$\begin{aligned} \mathbf{p(\mathit{A} \text{ and } \mathit{B}) = p(\mathit{A}) * p(\mathit{B})} \\ = \frac{1}{6} * \frac{1}{6} \\ = \frac{1}{36} \\ = 0.028 \end{aligned}$$

The equation yields 0.028, so there's about a 3 per cent ($0.028 * 100 \cong 3\%$) chance of rolling two 6s in a row.

The Probability of Related Events

In the example above, the events are independent because rolling a 6 on the first toss has no impact on the outcome of the second toss. If, for example, you are at a slot machine, hoping to hit a jackpot, you might continue feeding the machine because you believe that your ongoing losses must eventually affect the odds of getting a winner. (You may even tell yourself that you are "due" for some good luck!) This isn't true, though, and is perhaps one of the greater misconceptions of gambling. The fact is, gambling outcomes are independent of one another. You are just as likely to hit the jackpot the first time you put money in the machine as you are the one-hundredth time you put money in the machine.

If one event affects the probability of another event, then the events are dependent. Solving problems involving related events is more complicated because an intermediate calculation is required. Let's use a bag containing 40 marbles to illustrate. There are 10 green marbles, 10 red, 10 yellow, and 10 blue. Suppose that we wanted to know the probability of pulling out a green marble, then pulling out a red marble. How would we figure it out?

Pulling out the red marble is only of interest when the first marble that's pulled out is green. Additionally, the number of marbles changes across experiments, from 40 to 39, thereby altering our probability calculations. Event A is pulling out a green marble first. Since 10 of the 40 marbles are green, $p(A) = \frac{10}{40} = \frac{1}{4}$. If the first marble is green, what is the probability that the second marble will be red? Of the 39 remaining marbles, 10 are red, so $p(\frac{B}{A})$, or the probability of *B*, given *A*, is $\frac{10}{39}$. The probability of *A* then *B* is therefore equal to

$$\begin{aligned} p(A \text{ then } B) &= p(A) * p(B \mid A) \text{ (this is read as the probability of } B \text{ given } A) \\ &= \frac{1}{4} * \frac{10}{39} \\ &= \frac{10}{159} \\ &= 0.064 \end{aligned}$$

There's a 6.4 per cent chance of pulling a green and then a red marble in succession.

Dependent probabilities like this can be thought of as successive because interest in the second outcome is contingent on the outcome of the first. If your first marble isn't green, you don't really care what happens next, because your first condition wasn't met.

Mutually Exclusive Probabilities

Mutually exclusive events are events that cannot occur together. To determine the probability of either of two mutually exclusive events occurring, you need to add the independent probabilities. If you wanted to know the probability of a die roll yielding *either* a 1 or a 6, you would need to sum the two independent probability calculations:

$$p(1) = \frac{1}{6}$$

$$p(6) = \frac{1}{6}$$

$$p(1 \text{ or } 6) = \frac{1}{6} + \frac{1}{6} = \frac{1}{3}$$

This is the **addition rule of probabilities** and is useful for calculating outcomes when you don't care what the outcome is. Suppose, for example, that you have a bag with 2 candy bars and 3 bags of chips inside it. Suppose you wanted to calculate the probability of getting the candy bars, and that you didn't care which candy bar you received. (You just don't want a bag of chips.) The probability would be calculated as

$$p(A) = \frac{1}{5}$$

$$p(B) = \frac{1}{5}$$

$$p(A \text{ or } B) = \frac{1}{5} + \frac{1}{5} = \frac{2}{5}$$

Naturally, underpinning this calculation is the assumption that you are equally likely to grab a candy bar or a bag of chips, and which you get is completely random.

Non–Mutually Exclusive Probabilities

When two events can occur simultaneously, they are considered non–mutually exclusive, or interchangeable, probabilities. With non–mutually exclusive categories, there is a danger of double-counting. When that occurs, values that have been double-counted need to be subtracted. Of the probabilities that we'll cover, this is the most difficult type to grasp. (See Box 4.2 for further help on non–mutually exclusive probabilities.)

Consider the following scenario: In Dodge City in 1810—a town with a population of 200—there are 40 people who smoke but don't drink and 60 people who drink but don't smoke. There are also 98 people who both smoke and drink. What is the probability of randomly picking a smoker p(A), a drinker p(B), a smoker and a drinker p(A and B), or a smoker or a drinker p(A or B)?

- For a smoker, the outcomes are independent of one another (a person either smokes or doesn't smoke). So, p(A) is $\frac{138}{200}$ (40 smokers, and 98 smokers and drinkers).
- For a drinker, the outcomes are once again independent, so p(B) is equal to $\frac{158}{200}$ (60 drinkers, and $\frac{98}{200}$ smokers and drinkers).
- For a smoker and a drinker, we are already given the information in the description. It is p(A and B), which is equal to $\frac{98}{200}$.
- For a smoker or a drinker, it becomes more complicated because of the risk of double-counting. In that case, we'd need to add p(A) to p(B), then subtract the duplicates p(A and B), yielding:

$$\begin{aligned} \mathbf{p}(A \,\mathbf{or}\, B) &= \mathbf{p}(A) + \mathbf{p}(B) - \mathbf{p}(A \,\mathbf{and}\, B) \\ &= \frac{138}{200} + \frac{158}{200} - \frac{98}{200} \\ &= \frac{(138 + 158 - 98)}{200} \\ &= \frac{198}{200} \\ &= 0.99 \end{aligned}$$

These are also called cumulative probabilities only because the outcomes overlap to some extent.

BOX 4.2

How to Calculate Probabilities for Non–Mutually Exclusive Events

1. Calculate the probability of event A.
1. Calculate the probability of event B.
2. Subtract the number of duplications. The formula takes the form:

$$\mathbf{p}(A \text{ or } B) = \mathbf{p}(A) + \mathbf{p}(B) - \mathbf{p}(A \text{ and } B)$$

3. If you prefer to see probabilities expressed as percentages, multiply the result by 100.

Continuous Probabilities

There is another class of probabilities for variables that don't have exact values, such as time or height, called continuous probabilities. There are no discrete, exact measures of these variables, because there are an infinite number of possible values. Consequently, the probabilities are also continuous. Although it is possible to calculate continuous probabilities, these are beyond the focus of this text.

Conclusion

To check the plausibility of your calculations, keep in mind the following:

1. The probability of an event that cannot occur is 0.
2. The probability of an event that must occur is 1.
3. Every probability is a number between 0 and 1, inclusive. As a percentage, the probability will range between 0 per cent and 100 per cent.
4. The sum of the probabilities of all possible outcomes of an experiment is 1.
5. When thinking about probabilities, remember what Aristotle said: "The probable is what usually happens."

It is useful to think of statistics as an elaborate way of calculating probabilities.

Glossary Terms

Addition rule of probabilities (p. 32)
Empirical probability (p. 29)
Law of large numbers (p. 28)
Mutually exclusive (p. 32)
Multiplication rule of probabilities (p. 30)
Probability (p. 26)
Sample space (p. 27)
Theoretical probability (p. 29)

Practice Questions

1. Which of the following is the sample space when two coins are tossed?
   a. H, T, H, T
   b. H, T
   c. HH, HT, TH, TT
   d. H, H, T, T
2. At the University of Regina, three out of five students graduate with a Bachelor of Arts degree. The remainder receive a different degree, such as a Bachelor of Science. What is the probability that a randomly chosen graduating student will *not* be getting a Bachelor of Arts degree?

3. A pair of dice is rolled. What is the probability of getting a sum of two?

4. There are two soccer teams in Winnipeg with 30 players in total. One team (Team A) has 16 players and the other (Team B) has 14. Of these players, 5 are left forwards, 3 of whom are on Team A. If a player is chosen at random, what is the probability of choosing someone on Team A that is a left forward?

5. In Canada, roughly 52 per cent of people wear a seat belt while driving. If 2 people are chosen at random, what is the probability that both of them are wearing a seat belt?

6. Three cards are chosen at random from a deck *without* being replaced. What is the probability of getting a 3, a 9, and a jack, in that order?

7. In poker, one of the better hands is a flush. A flush is 5 cards of the same suit (all hearts, all clubs, etc.). Calculate the likelihood of being dealt 5 consecutive cards of the same suit (hint: there are 4 suits in total, 13 cards to a suit, and 52 cards in a standard deck). State the likelihood as a decimal, rather than a fraction.

8. Ryan only had time to study six of the eight essay questions that could be on his sociology exam. His professor will be choosing two of the eight questions for the exam. What are the chances that both of the questions that Ryan didn't study will appear on the exam?

9. Every year, Julie usually dresses up as either a male pirate or a female pirate for Halloween. So rarely does she stray from her costume that you believe it's possible to predict her costume for the coming year before even seeing her. Since you know that of the last 10 Halloweens, she was a male pirate 5 times, a female pirate 3 times, a dog once, and a llama the other time, what is the probability that Julie will appear this year dressed up as something other than a pirate?

10. Suppose that you have few friends but that you've decided to use your newly acquired statistical knowledge to make some. Since you like to gamble, you decide to head to the casino to try to meet people at the craps table. What would you say to people who say the following?
    a. "I haven't had any luck all night! I'm due for some good fortune!"
    b. "Nobody around me is winning!"
    c. "The probability of both of these two dice turning up fives is nil!"

Answers to the practice questions for Chapter 4 can be found in Appendix H.

5 The Normal Curve

LEARNING OBJECTIVES

Now that you've learned about probabilities and how to describe the distribution of a single variable, Chapter 5 will expand on those topics. You will learn:

- how data are distributed;
- the principles of the normal distribution;
- how probability is related to the normal curve; and
- some useful terms for describing distribution.

The History of the Normal (Gaussian) Distribution

Imagine that you're a gambler and want to know how frequently an outcome, such as the number of times a coin will turn up heads in 100 tosses, will occur. When there are only two possible outcomes (heads or tails), a safe guess would be to predict that the coin would turn up heads 50 per cent of the time. Usually that's pretty accurate, but gamblers make money by betting on something *other* than the most obvious outcome. Since coin tossing is a **random process**, the outcome won't always be the most obvious or intuitive one, although it is the most likely one. There could be 53 heads and 47 tails in one set of 100 tosses, 65 heads and 35 tails in another, and 75 heads and 25 tails in yet another. It is theoretically conceivable that you could get 95 heads and 5 tails in one round of tosses. In fact, anything between 0 heads and 100 tails, and 100 heads and 0 tails is possible.

Imagine that you've been hired to predict the likelihood of outcomes for an avid gambler, someone not schooled in how probabilities work. Your gambler boss wants to know the probability of the outcome that you think would be most likely (and therefore which outcome he/she should bet on). Depending on how risk averse your boss is, you probably wouldn't suggest choosing 95 heads and 5 tails, given the over-representation of heads and the low likelihood of that outcome. Instead, you might suggest placing a bet on a more evenly divided outcome because it makes more sense to you, and everyone else, which is also why it wouldn't pay as well as a more extreme bet. By making this more conservative suggestion,

BOX 5.1

The Normal Curve: History of a Term

The normal curve is an example of a histogram, as discussed in Chapter 3. The reason the normal curve is sometimes called the bell curve is because of its bell-like shape. It is often (mistakenly) called the Gaussian curve, to pay tribute to Carl Friedrich Gauss's work on the distribution of errors in an ordinary least squares regression equation. This equation fits a line on top of data that best represents the relationship between variables. (This will be covered in greater detail in Chapter 16.) Gauss argued that the distribution of errors in the equation is random and that their shape, therefore, assumes a normal distribution. Since it was Abraham de Moivre who used the curve first, it should bear his name. The fact is that the Gaussian curve follows the "law of eponymy," coined by statistician Stephen Stigler, which states that "no scientific discovery is named after its original discoverer."

you're applying the **central limit theorem** and the **normal curve**. See Box 5.1 for a description of the normal curve.

Applying the principles of probability is what de Moivre, one of the Western world's earliest statisticians, did for a living. He was born in 1667 in Vitry-le-François, France, and moved to London around 1685. There, in addition to being a tutor and mathematician, he served as a gambling consultant at a local coffeehouse. It is likely that de Moivre frequently faced problems like the coin toss example, above, because he derived an equation that allowed him to estimate the probability of any of the 100 possible outcomes. He could do that because he noticed that when the number of events (coin flips) increased, the distribution of outcomes approached a smooth and predictable bell-like curve. This observation later became known as the central limit theorem, and the curve he saw is known today as the **Bell**, **Gaussian**, or **normal curve**.

Illustrating the Normal Curve

Next, let's look at the central limit theorem and discuss its relation to the normal curve. A simplified version of the central limit theorem states that if any variable (such as one that contains the number of times a coin toss shows heads) has a known range, then the distribution of outcomes will increasingly approximate the normal curve as the number of samples increases.

To illustrate the normal curve, let's continue using the coin-toss example. You would start out with a small exercise, such as performing 100 tosses once. Once you've flipped the coin 100 times (say it yielded 55 heads and 45 tails), you have completed one **experiment**. This probably took some time to do, but it is not even *close* to the number of tosses needed to see the normal curve. In fact, in Figure 5.1 the 100-toss experiment was repeated 100 times (that's 10,000 tosses!).

The type of graph in Figure 5.1 is a histogram. Histograms are often used to assess **distributions**. The x-axis represents the frequency with which heads was observed in

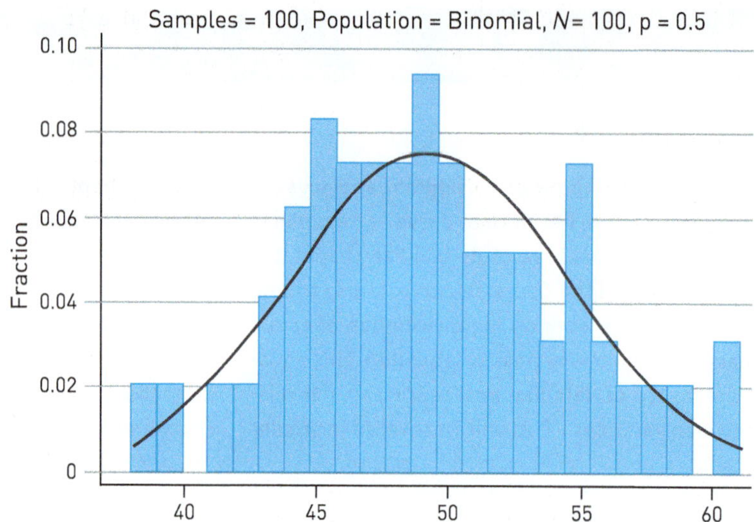

FIGURE 5.1 | Results from 100 Coin Toss Experiments, 100 Tosses per Experiment

each experiment; the values on the *y*-axis are the proportion of the experiments where each frequency of heads was observed. The *y*-value of the tallest bar (which is the mode, and has an *x*-value of 49 heads) is 0.09, meaning that in 100 experiments, 49 heads were observed 9 per cent of the time, that is, in 9 experiments. On either side of this value are shorter bars with values of 0.07, meaning that 48 and 50 heads were observed in 7 experiments.

The thin black curvy line in Figure 5.1 represents the normal curve. The normal curve will be bell-shaped, as the black line shows. As de Moivre noted in the central limit theorem, the fewer the number of tosses in each experiment, the farther away or worse the approximation of the normal curve will be, creating more gaps between the bars of the histogram and the normal curve line. Figure 5.1 represents 100 experiments, or 100 sets of tosses. Each set contains 100 tosses. You can see that the tails are thicker than the normal curve would suggest and that there are several instances where the bars don't align with the curve. The values of 40 and 59 don't even occur. In Figure 5.2, the number of samples or experiments is increased to 1000 (each experiment still contains 100 tosses). Notice how much closer the distribution of data is in the normal curve.

There are far fewer gaps between the normal curve and the data in Figure 5.2, but the graph still doesn't follow the curve exactly.

When the number of samples or experiments is increased to 10,000 (Figure 5.3), the fit shows even more improvement. There is very little difference between the results and the overlaid normal curve. This figure illustrates the results of 10,000 * 100 coin tosses, or 1,000,000 coin tosses!

As the exercise demonstrates, the distribution of data, in this case the proportion of heads in each set of 100 tosses, resembles the normal curve more closely as sample size (the number of sets of 100 tosses, or the number of experiments) increases. This is the crux of the

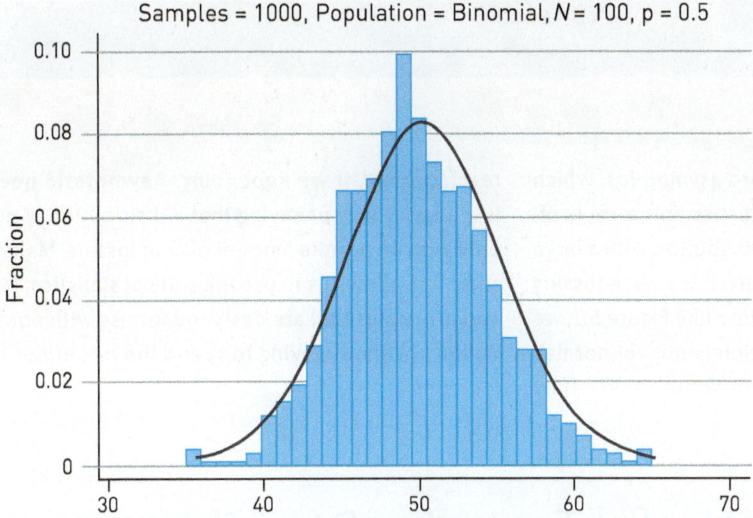

FIGURE 5.2 | Results from 1000 Coin Toss Experiments, 100 Tosses per Experiment

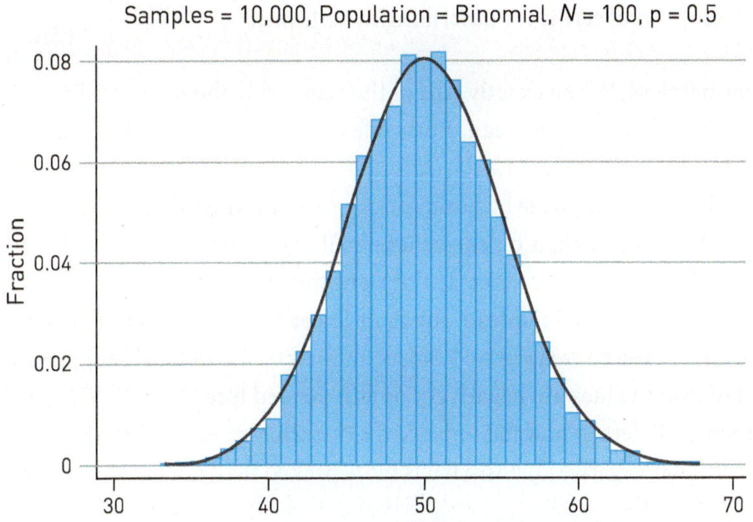

FIGURE 5.3 | Results from 10,000 Coin Toss Experiments, 100 Tosses per Experiment

central limit theorem, which states that the sampling distribution approaches the normal curve as the number of observations increases. Knowing this made it possible for de Moivre to predict how often a set of tosses would return any combination of heads and tails. The curve can also be used to predict other kinds of outcomes. If there are enough trials, it's possible for the distribution of a sample statistic to approach the normal distribution.

As Figure 5.3 suggests, the normal curve can also be used as an approximation (see Box 5.2) of outcome for a finite set of coin-toss trials.

BOX 5.2

What Is an Asymptote?

Asymptote comes from the Greek word *asymptotos*, which means "never falling together." In statistics, this term is often used to describe the shape of a distribution with a large sample and an infinite number of trials. If we were tossing coins, for example, and got a distribution like Figure 5.1, we would say that although the distribution is not yet normal, it would become increasingly so, but would never fully reach normal, if we kept going. **Asymptotic normality** is just a fancy way of saying that a distribution is normal but only with an infinite amount of coin tossing. Making this assumption allows us to use the suite of statistical techniques and principles that are designed for use with normal distributions (without having to spend the rest of our lives tossing coins!).

Some Useful Terms for Describing Distributions

Below are several useful terms for describing histograms:

- **Symmetrical**: When exactly half of the scores fall above the mean, and exactly half of them fall below the mean. Both sides of the mean have the same pattern of distribution. (See Figure 5.4.)
- **Skewness**: The opposite of symmetrical. Occurs when there are more scores on one side of the mean than on the other, resulting in one of the tails of the histogram being longer than the other. If the right tail is longer than the left (meaning that high values are more spread out than lower values), we say that the histogram is positively skewed (see Figure 5.5a), or skewed to the right. Histograms with a longer tail of lower values are negatively, or left, skewed (see Figure 5.5b).
- **Kurtosis**: Refers to how flat or peaked a distribution is. If a distribution is flatter than usual, it has negative kurtosis; if it is more peaked than normal, it has positive kurtosis.

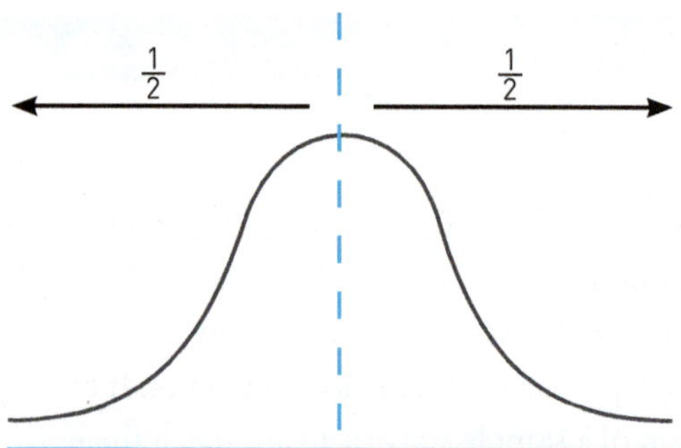

FIGURE 5.4 | Symmetrical Curve

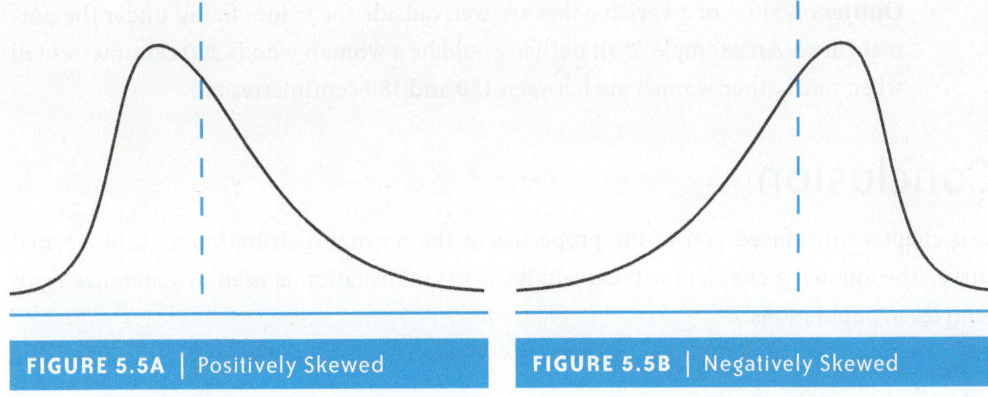

FIGURE 5.5A | Positively Skewed

FIGURE 5.5B | Negatively Skewed

- **Unimodal**: A distribution that has only one mode (the most frequently occurring value in your data set, on a variable of interest). Histograms of unimodal distributions will have only one major "hump" in them (see Figure 5.6).
- **Bimodal**: A distribution with two modes. A bimodal distribution will have two major "humps" (see Figure 5.7).
- **Multimodal**: Any distribution that has more than two modes (see Figure 5.8).
- **Bell curve**: The normal curve, which is shaped like a bell (as in Figure 5.9); as a result, it is often called a bell curve.

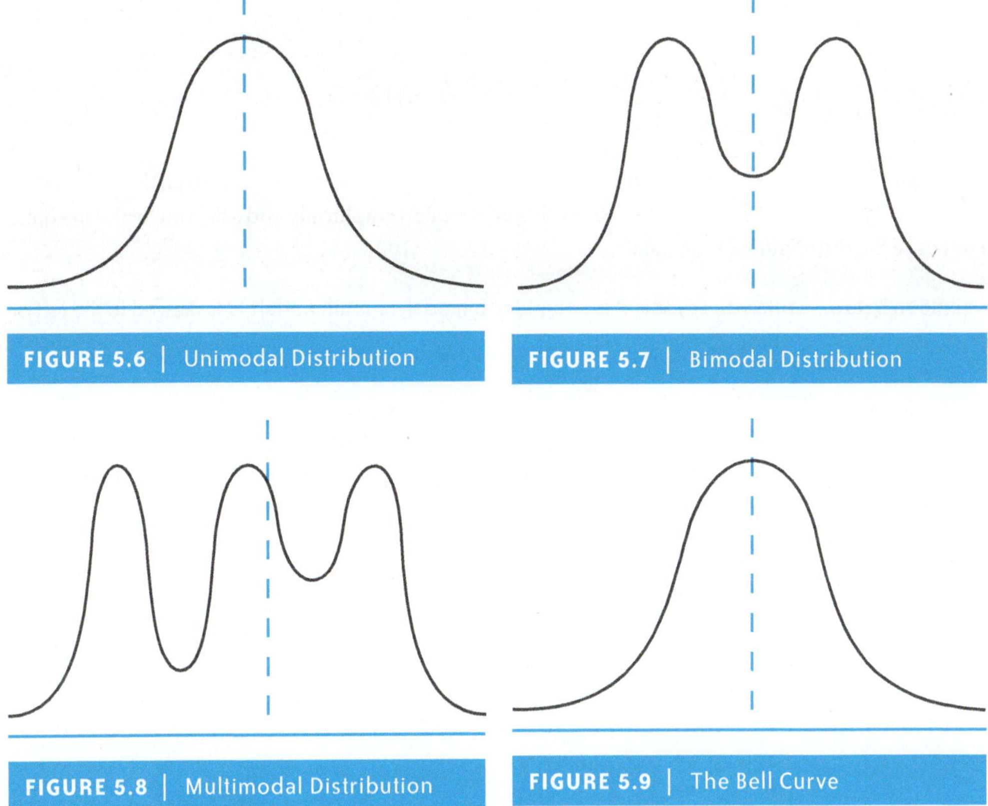

FIGURE 5.6 | Unimodal Distribution

FIGURE 5.7 | Bimodal Distribution

FIGURE 5.8 | Multimodal Distribution

FIGURE 5.9 | The Bell Curve

- **Outliers**: Values of a variable that are well outside the values found under the normal curve. An example of an outlier would be a woman who is 200 centimetres tall when most other women are between 150 and 180 centimetres tall.

Conclusion

This chapter introduced you to the properties of the normal distribution and the normal curve. The following chapters will explain how this information is used to generalize from samples to populations.

Glossary Terms

Asymptotic normality (p. 40)
Bell curve (p. 37, 41)
Bimodal (p. 41)
Central limit theorem (p. 37)
Distributions (p. 37)
Experiment (p. 37)
Gaussian curve (p. 37)
Kurtosis (p. 40)
Multimodal (p. 41)
Normal curve (p. 37)
Outliers (p. 42)
Random process (p. 36)
Skewness (p. 40)
Symmetrical (p. 40)
Unimodal (p. 41)

Practice Questions

1. Dr Knifewell performs pancreatic surgeries and has found that although most patients are released quickly, sometimes infections occur, necessitating a longer stay. She wants to see this information graphically and asks you to plot the data. How would you describe the distribution of the data?

2. Describe the distribution below. Would you say that it is unimodal, bimodal, or multimodal? Is it skewed to the left or the right?

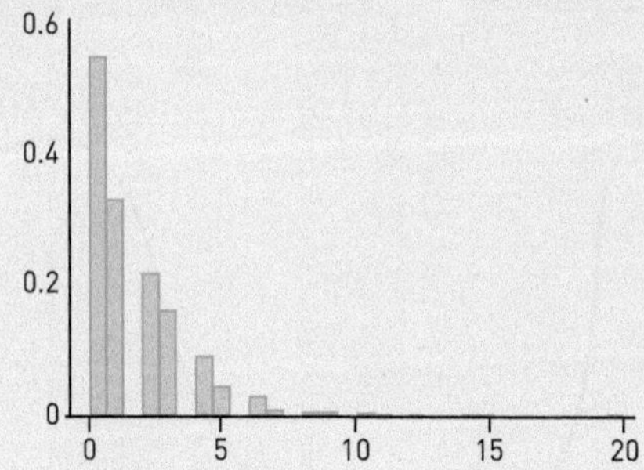

3. Look at the distribution of the variable below, with a super-imposed normal curve. What is the value for the largest outlier? What effect do you think it's having on the kurtosis value of the normal curve?

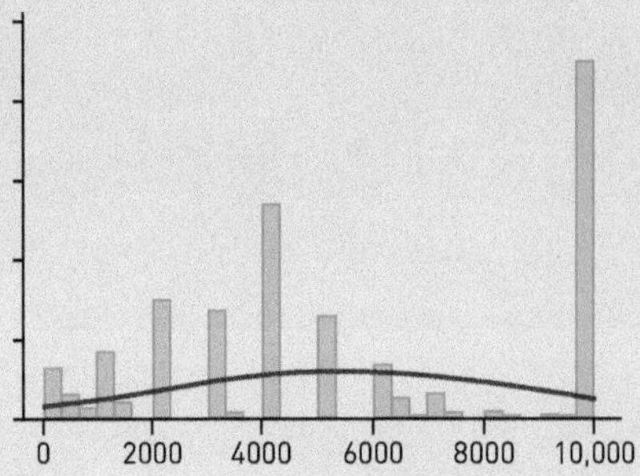

4. In recent years, the gap between rich and poor has been growing in Canada, suggesting that the middle class has been shrinking. Describe what impact this will have on a histogram of income in Canada.

5. Because of climate change, weather in Canada has become more volatile. What impact will this have on the tails of a histogram that plots daily temperatures? Will they get thicker or thinner?

6. A major cellular phone provider has been having problems with clients calling to complain that they're continually exceeding their monthly allotment of 500 air-time minutes, even though the average usage is well below the allotment amount. They hire you to determine why they receive so many calls. As an adept statistician who sees merit in collecting data before making recommendations, you decide to run a small survey of usage among clients. These are the reported minutes used in a sample of 20 respondents:

50	124	130	23	77	1012	102	1750	900	499
998	890	12	42	35	223	399	239	1200	0

What would you tell the cellular provider? How would you describe the distribution of data to the provider?

7. If you wrote a statistics exam and found out that the distribution was skewed to the left, would you expect that there are outliers in the higher grade range or the lower?

8. Try to think of examples of data in your life that could be plotted as a histogram. Would you expect these data to be skewed, bimodal, etc.? Why?

Answers to the practice questions for Chapter 5 can be found in Appendix H.

6 Measures of Central Tendency and Dispersion

LEARNING OBJECTIVES

This chapter will examine the statistical measures used to understand the distribution of a single variable. We'll be studying:

- measures of central tendency; and
- measures of dispersion (or measures of variability).

Introduction

In order to fully describe the distribution of a variable, you need to be able to talk about the most common, or most central, value of the variable, but also about how much other values vary around that common or central value. To do this we use measures of central tendency and measures of dispersion (also called measures of variability).

This chapter will cover measures of central tendency and dispersion in order of level of measurement, first presenting measures that are appropriate for nominal variables, then measures that are appropriate for ordinal variables, and then measures that are appropriate for interval/ratio variables. Remember that measures can always be used for a variable at a higher level of measurement, but not vice versa. (Therefore, a measure that can be used for nominal variables can also be used for ordinal and interval/ratio variables, but a measure used for interval/ratio variables cannot be used for ordinal or nominal variables.)

Measures of Central Tendency

A common word for a measure of central tendency is an *average*. People often equate the average with the arithmetic mean. However, there are other forms of averages besides the arithmetic mean. **Measures of central tendency** describe "typical" observations in your data set. In addition to the arithmetic mean, these measures may include the value that falls in the middle of an ordered set (the median), or the most frequently occurring value (the mode). Which measure of central tendency you choose to report will depend on the level of measurement of your variable and, if your variable is an interval/ratio variable, the distribution of your data.

Mode

One measure of central tendency is the **mode**, which is the most frequently occurring value of a variable of interest in your data set. Suppose we are looking at ethnic origin in the 2006 Canadian census. The mode, or most frequently occurring answer among all ethnic-origin categories, would be Canadian (32.2 per cent of all respondents checked this box). More census respondents identified as Canadian than as any other ethnic origin. Sometimes there is more than one category that appears as the most frequent value. In this case, data are said to be "bi- or multi-modal," and you would list all the categories that occur most frequently as the modes. Modes are most commonly used for nominal or ordinal data, although they can be used with any level of measurement. The mode is the only measure of central tendency that can be reported for nominal data. It can also be reported for ordinal and interval/ratio data.

Median

The second measure of central tendency is the **median**. The median divides the sample in half. In any distribution, half the values lie above the median and half lie below it. Let's suppose that we have five yearly income values: $5000, $10,000, $15,000, $20,000, and $25,000. If we sort the incomes by their value and find the value that falls exactly in the middle, that's the median. In this example we have five observations, so the third individual is the one in the middle, and that person's income ($15,000) is the median of this sample. Two people have incomes that are higher than $15,000 and two people have incomes that are lower than $15,000. If there were an even number of observations (say, six instead of five), the calculation would be more complex. We would need to find the middle pair of numbers (the third and fourth observations), and then find the value that's halfway between them by adding the values together and dividing by two.

Mean

The most commonly used measure of central tendency is the **mean**.

Suppose that you are in a statistics class with nine other students. You and your classmates just received the grades from your first Intro to Stats exam. The grades were 25 per cent, 35 per cent, 45 per cent, 47 per cent, 53 per cent, 64 per cent, 67 per cent, 75 per cent, 85 per cent, and 95 per cent (your grade is the 67 per cent). You want to know how you compare to your classmates. One way to do this would be to compare your grade to the arithmetic average, or mean, for the class. To do this, you need to add up all of the individual scores ($25 + 35 + 45 + 47 + 53 + 64 + 67 + 75 + 85 + 95 = 591$), and divide the sum by 10, the number of observations. This would yield 59.1 per cent ($591 \div 10 = 59.1$ per cent). So your score was almost 8 percentage points above the mean.

The mean can be expressed as the sum of all values of a particular variable, where the level of measurement is interval/ratio, divided by the total number of observations used to calculate the sum. As you know from your past exam scores, the mean tells you the value of a "typical" person.

Let's put that information more formally. The equation for deriving the mean of the population (remember in this example the whole population of the class is only 10 people) is as follows:

$$\bar{x} = \frac{X_1 + X_2 + X_3 + X_4 + \cdots + X_N}{N}$$

Or, more simply:

$$\bar{x} = \frac{\sum X_i}{N}$$

In these equations, $\bar{x}$ is equal to the mean of a variable (e.g., age). X_1, X_2, X_3, etc., are equal to the individual values of X (e.g., in the example above, person 1 is 25, person 2 is 35, person 3 is 45). X_i indicates the value for individual i (e.g., X_{10} would be the value for the tenth person—in the example above, 95). N is equal to the total number of observations (e.g., $N = 10$ in this example). The second equation merely simplifies the first. Thus, $\sum X =$ the sum ($\sum$) of the values for the sample (replacing the $X_1, X_2, X_3, \ldots, X_N$, etc.). The denominator does not change.

Let's return to the income example we used for the median, above. Let's suppose that there are five people in our population and we have five yearly income values: \$5000, \$10,000, \$15,000, \$20,000, and \$25,000. The mean is obtained by summing the five values (\$5000 + \$10,000 + \$15,000 + \$20,000 + \$25,000 = \$75,000) and dividing the sum by five, or the number of observations. This gives us \$15,000, which is the income of the average person in this group of five.

In this example, the median and the mean are the same. However, in many cases (especially with variables like income) there will be extreme values (very high or very low ones), which have a much stronger impact on the mean than on the median. The total median income earned by everyone age 15 and over in the 2006 Canadian census was \$25,615.00, but the individual values ranged from negative \$50,000 to \$200,000.[1] Why do we sometimes present the median instead of the mean? Because there is a substantial number of Canadians earning a large income, the mean will be affected by the extreme values. With variables that have extreme values (whose distribution is skewed), like income, the median is the preferred measure of central tendency because it provides a truer picture of the average person. How typical is a person earning \$200,000 or more? The answer is "not very typical," but these high earners have a huge impact on the calculation of the mean. When you have extremely high outlying values, the median will be lower than the mean; when there are extremely low outlying values, the median will exceed the mean. In the income example, since we have a cluster of people earning \$200,000 or more, the median will be significantly lower than the mean. It is crucial to emphasize that the median can only be used with variables with an ordinal level of measurement or higher.

The average person is an abstraction. (See Box 6.1 for a discussion of the origin of this abstraction.) There does not need to be someone in your sample with the mean value. A test average could be 59 per cent, even if no one received that score. The average value is useful because you can examine how any individual score ranks compared to the average. Remember that the mean can only be calculated for variables that are at the interval/ratio level of measurement.

EVERYDAY STATISTICS

Middle Matters

Every year, Statistics Canada releases an analytical report that summarizes the income dispersion of Canadians. In the analysis and the subsequent final report, the income information of Canadian families is reported on the basis of medians. The median is the point at which half of the families have a higher income and the other half have a lower income.

Q: Why do you think Statistics Canada releases income information using the median instead of the mean or the mode?

BOX 6.1

Means and Medians: History of Two Terms

Whose idea was it to compute summary statistics, and for what purpose? *Mean* is a very old term, sharing its roots with *median* from the Latin *mediānus*, for "middle." Although the concept of an average is quite common, the use of the mean also has a long history. The mean is largely a descriptive and utilitarian measure (Stigler, 1986). In the eighteenth century, astronomers would average measurements, but only a small number and only if they were all taken under similar conditions. These measurements of celestial bodies were taken for the purposes of navigation. Pierre-Simon Laplace (1749–1827) worked on the mean in that context by comparing three observations, and he published a paper on the subject in 1776.

Adolphe Quetelet (1796–1874) advanced the work of Laplace. Quetelet was an astronomer, a statistician, and a sociologist. He was interested in the mean as more than a descriptive measure. The stability of statistical aggregates, and hence of mean values, was the foundation of the science of social physics that Quetelet announced in 1831. Its key concept was *l'homme moyen*, the average man (Gigerenzer et al., 1991: 41). Quetelet measured physical characteristics to find their distribution and mean. From that he equated the normal with "the good"; and deviation, with deviance. Quetelet believed that if God produced man in His image, but in an imperfect way, some men would come closer to the perfect, or divine, image of man than others. The more a person deviated from the average, the less perfect Quetelet considered that person to be.

Measures of Variability

When we examine the **distribution** of a variable, we are not only interested in measures of central tendency. We also want to know how much variation there is in our sample around the measure of central tendency. In order to describe a distribution, we must also understand measures of variability. In its simplest terms, variability measures capture where individuals are positioned relative to one or more of the measures of central tendency. Some measures of variability (like the range) are very simple to calculate, whereas others (like standard deviation) require more effort on your part.

Range

Of all of the measures of variability or dispersion, the **range** is the easiest to understand. The range is simply the lowest value subtracted from the highest value, where H is the highest value of your variable and L is the lowest:

$$\textbf{Range} = H - L$$

For the example of the five yearly income values, the highest minus the lowest would be \$25,000 − \$5000 = \$20,000. Note that you cannot calculate a range for nominal variables since the values have no inherent order to them.

The interquartile range (IQR) shows where the middle 50 per cent of the observations lie. It is obtained by subtracting the value of the twenty-fifth quartile (Q1) from the seventy-fifth quartile (Q3).

$$\textbf{IQR} = \textbf{Q}_3 - \textbf{Q}_1$$

The range and the IQR can be used for variables at the ordinal level of measurement or higher. If you need a reminder on how to calculate percentiles, please refer to Chapter 3.

Mean Deviation

The range and interquartile range are useful for revealing how "wide," or spread out, the values are. However, they cannot tell you how far the "average" person is from the mean value. For example, it is possible that all values but two (one extremely high and one extremely low) are tightly clustered around the mean, but you would not know that by looking at the range.

The simplest method of determining how an observation ranks in the sample is the mean deviation. Defined as the average "distance" that each variable is from the class mean, the mean deviation is superior to the range because it reveals more than the difference between the highest and lowest scores (i.e., the most extreme values); it shows how far the average observation is from the mean. This shows how similar the individuals in your sample are, rather than just the highest or lowest-scoring person. The equation for the mean deviation is:

$$\textbf{Mean Deviation} = \frac{\left|\sum X - \bar{x}\right|}{N}$$

In this equation, $\bar{x}$ is equal to the mean of a variable (e.g., age). X is each value for the variable. The two bars around the $X - \bar{x}$ indicate that you take the absolute value of each difference between X and the mean. The Σ indicates that you take the sum of these values. You then divide by N. N is equal to the total number of observations.

The mean deviation is the sum of the **absolute values** of the distances from the mean, divided by the total number of observations. Continuing with our income example, we'd subtract the mean from every value (5000 − 15,000 = −10,000; 10,000 − 15,000 = −5000; 15,000 − 15,000 = 0; 20,000 − 15,000 = 5000; 25,000 − 15,000 = 10,000), sum the absolute values (10,000 + 5000 + 5000 + 10,000 = 30,000), and divide by the total number of observations to get \$6000 ($\frac{\$30,000}{5} = \$6000$).

Although you've probably encountered absolute values already, it's good to be reminded that absolute value refers to the positive score of every value. The absolute values of −1, 2, −5, 11, and −100 would be 1, 2, 5, 11, and 100, respectively.

Remember that mean deviation can only be calculated for variables measured at the interval or ratio level of measurement, since it requires a mean.

BOX 6.2

The Steps: Mean Deviation

The mean deviation for any sample can be easily obtained:

1. Subtract the mean from each value.
2. Sum the absolute values.
3. Divide the sum by the number of observations. This is the mean deviation.

Variance and the Standard Deviation

The mean deviation is a useful measure of the **dispersion** of values for a variable, but since the absolute value has no straightforward mathematical relationship with the location of reported values, it is not usually used. Luckily, there are more desirable measures, and mean deviation is a useful foundation for understanding these measures.

Building on the mean deviation, we will now examine the standard deviation and variance. Both are closely related to the mean deviation, but neither uses absolute values. Instead, both use squared values.

The **variance** is the average *squared* distance (as opposed to absolute value) from the mean value. Squaring values eliminates negative values. Each observation contributes to the overall calculation of deviation. The equation is:

$$\sigma^2 = \frac{\sum (X - \mu)^2}{N}$$

In the equation, the numerator is also known as the **sum of squares**. It represents the sum of the squared deviations from the mean. (Note: we will use this frequently in the remainder of the text, so it's important to understand both this measure and standard deviation, below.).

You calculate the sum of squares by subtracting the mean of X from each X value, squaring the result, and summing them all. You then divide by N to get the variance. See the explanation of the mean deviation above if you've forgotten what X, $\bar{x}$, or N stand for.

The **standard deviation** is the square root of the variance. Here's the equation:

$$\sigma = \sqrt{\frac{\sum (X - \mu)^2}{N}}$$

As you can see, you can simply take the square root of the variance to get the standard deviation.

The standard deviation and variance build on the mean deviation. Researchers most often use the standard deviation for dispersion because it can easily be related to the normal curve, and it is reported in the same units as the original variable.

Like the mean deviation, the variance and standard deviation are measures of the average distance each observation, or person, in your data set is from the mean. (See Box 6.3 for a history of the terms.) In other words, it tells you how different, on average, all the observations in the data set are from *l'homme moyen*, although it takes a few steps to get there. These steps are outlined in Box 6.4.

BOX 6.3

Variance and Standard Deviation: History of Two Terms

The term "standard deviation" and the symbol σ (the Greek lower case sigma) were first used in 1893 (Walker, 1929; Pearson, 1894). Variance was first employed by R.A. Fisher in 1918 (Walker, 1929).

Pearson developed standard deviation in the context of evolution, using the error curve, or *normal curve*. When discussing biological matters, specimens are not uniform and universal. To understand and describe genetic and population variations, not only is it important to recognize the measures of central tendency at work, but it's also important to have a formalized way of discussing the ends of the curve and the population described by the curve's extremities.

BOX 6.4

The Steps: Variance and the Standard Deviation

The variance for any population can be obtained by following these steps:

1. Subtract the mean from each value.
2. Square the differences calculated in step 1.
3. Add the squared deviations together.
4. Divide the sum of the squared deviations by the number of observations.

To get the sample standard deviation, take the square root of the variance.

Remember that variance and standard deviation can only be calculated for variables at the interval or ratio level of measurement since they require a mean.

To understand what variance and standard deviation mean in everyday language, perhaps an example would be useful. Imagine that you are deciding between living in Toronto versus Calgary. Both locations have similar mean high and low temperatures (although Calgary is a tad lower throughout the year), but does that mean that the two cities have the same climates? Not at all. Aside from differences in humidity, Calgary has a much higher probability of experiencing summer temperatures that dip below 5 degrees Celsius and winter temperatures that top 10 degrees Celsius. Toronto, by comparison, has a climate that's heavily affected by its proximity to the Great Lakes. This means that these water masses bring an element of stability. If we were to compare the two cities statistically, we could say that the means are similar but that the standard deviation and variance are not. The variance and standard deviation of temperature in Calgary are much greater than in Toronto, thereby resulting in climates that are quite different. In Toronto, you can safely put away your winter coat in one part of the year, and shorts in another, whereas in Calgary you'd be less likely to be able to do so.

Since standard deviation and variance are so closely related (the variance is just the standard deviation squared), it is somewhat redundant to talk about them both. Since the standard deviation is expressed in the same units as the original variable (rather than squared units), researchers will often talk about just the standard deviation. The other reason for focusing on standard deviation is that it connects directly to the normal curve. That is, the standard deviation describes how data are distributed about the mean in a normal curve. This forms one of the more important topics of the next chapter.

Conclusion

In this chapter, you learned how to describe variables both in terms of what is the most central or most common value and also in terms of how widely or narrowly distributed the values of the variable are in your dataset. Univariate descriptive statistics form the basis of all statistical analysis and should always be your first step in data analysis.

Glossary Terms

Absolute values (p. 49)
Dispersion (p. 49)
Distribution (p. 48)
Mean (p. 45)
Measures of central tendency (p. 44)
Median (p. 45)
Mode (p. 45)
Range (p. 48)
Standard deviation (p. 50)
Sum of squares (p. 50)
Variance (p. 50)

Practice Questions

1. Last year, a small statistical consulting company paid each of its five clerks \$22,000, two statistical analysts \$50,000 each, and the senior statistician/owner \$270,000.
   a. How many employees earn less than the mean salary?
   b. What is the salary range?

2. The following 10 numbers represent the number of times a random sample of celebrities has signed autographs in the past month:

   46 57 68 2 4 14 0 0 2 101

   What is the mean, and mean deviation, for these numbers?

3. A sample of underweight babies was fed a special diet and the following weight gains (in lbs) were observed at the end of three months:

   6.7 2.7 2.5 3.6 3.4 4.1 4.8 5.9 8.3

   What are the mean, variance, and standard deviation of the weight gains?

4. If most of the measurements in a large data set are of approximately the same magnitude, except for a few measurements that are quite a bit larger, how would the mean and median of the data set compare, and what shape would a histogram of the data set be?

5. A sample of 99 distances has a mean of 24 metres and a median of 24.5 metres. Unfortunately, it has just been discovered that a value erroneously recorded as "30" actually has a value of "35." If we make this correction to the data, what would happen to the value of the mean? What about the median?

6. Whenever means and medians are compared for income in Canada, the mean is higher. Why do you think this is?

7. Over the past 100 to 200 years, several things have happened to the age distribution of Canadians. First, people have been having fewer children. Second, Canadians are living longer. Third, rates of premature death have declined. Discuss what you believe has happened to the mean, median, range, and standard deviation of the age distribution of Canada across this time period.

8. Given the discussion in this chapter on the relationship between standard deviation and the normal curve, do you think the tails would be longer or shorter for a distribution with a bigger standard deviation?

9. Jessica sends 55 text messages a day and has done so consistently since she got her phone 365 days ago. What is the mean deviation, standard deviation, and variance score for the number of text messages she sends every day?

10. When Helen becomes interested in something, she dedicates nearly all of her time to that activity. In the past, she's studied gardening, photography, sewing, scrapbooking, and crocheting. What usually happens, however, is that she eventually gets bored with the activity and drastically reduces the amount of time she dedicates to it. Imagine that the number of hours per day that she spends on each activity follows a normal curve, except that she typically loses interest in an activity faster than she acquires it. Would you predict that the median number of hours she spends on each activity is higher or lower than the mean? Why?

Answers to the practice questions for Chapter 6 can be found in Appendix H.

Note

1. Yes, some people in Canada have negative income! Can you think when this might occur? Hint: In some years, some businesses and investments lose money. Also, Statistics Canada recodes everyone earning more than $200,000 to $200,000 in the publically available data sets to prevent breaching confidentiality.

7 Standard Deviations, Standard Scores, and the Normal Distribution

LEARNING OBJECTIVES

Chapter 7 will delve more deeply into the normal distribution. You will learn:

- more about standard deviations; and
- how to calculate, and use, standard scores.

Introduction

The crime rate in Prince Edward Island was 5254.88 per 100,000 persons in 2014. That might seem high (or low), but how can you tell? How does PEI's crime rate compare to that of other provinces? How does it compare to the crime rate in US states? If the rate is high, how high is it relative to other places? If there is a difference in the crime rate between two places, is the difference meaningful? What if the difference is due to chance or the way that you chose your sample? These questions form the focus of this chapter. How do we compare the values of variables for different groups within a sample?

To answer those questions, beyond comparing the statistical means, we'll need to apply and expand on what was covered in the previous chapters. By the end of this chapter, you'll have enough statistical knowledge to place observations on a normal distribution and to rank values using a common metric known as the **standard score** (or ***z*-score**). Before we examine crime rates, you'll need to know how the mean and standard deviation relate to the normal curve.

How Does the Standard Deviation Relate to the Normal Curve?

First, let's discuss σ, the population standard deviation. The standard deviation is a unit of measurement that follows an already-known continuum. This continuum is the **normal curve**. The standard deviation and standard score are used to determine the rank of an observation. Understanding these concepts will allow you to discuss your data with other people, even if they don't know the particulars of your study area.

EVERYDAY STATISTICS

Standard Deviation and Your Local Weather

Standard deviation is often used in climatology to measure differences in temperature between two locations. For example, climatologists will compare two cities that have the same mean temperature and calculate the standard deviation for each city. This allows them to measure the difference in temperature variation between the two cities.

Q: Considering the above application of the standard deviation, do you think that the standard deviation would be higher or lower for a coastal city compared to an inland city with the same mean temperature?

It is always possible (with some background information) to compare two observations using standard deviation scores.

Using the mean and standard deviation, the normal curve provides information about the characteristics of a variable. As in the example about selling boats, it is not necessary to know anything about a variable or a study to see if its results show important differences between groups. You just need to know how to interpret the mean and standard deviation.

More on the Normal Distribution

If a variable is distributed approximately normally, most values of that variable will lie beneath the normal curve and are therefore subject to the rules of normal distributions. Because of that, there are standardized cut-points (standard deviations) that give a metric for determining the proportion of all values that lie at any predetermined distance from the mean (usually denoted by $\overline{x}$). There are six of these points that are commonly used, usually three above the mean and three below. To further illustrate this, consider Figure 7.1.

Figure 7.1 is a plot of the means of 10,000 samples of 200 cases, or the mean of 200 values on an interval/ratio variable (number of strikes at a plant, number of children, a heart rate, yield per hectare, or anything else), repeated 10,000 times. In the figure, each mean is treated like an observation. That might seem confusing, but don't worry about it too much at this point. We'll talk about it more when we get to **sample distribution of means** in Chapter 9. At this point, think of each mean as an observation.

As we saw in the previous chapter, the mean is useful because it is a measure of central tendency. If a variable is normally distributed, 50 per cent of all observations have values that exceed the mean, and 50 per cent have values that do not. By knowing the mean and the score on a particular variable, we can determine whether a person is in the top or bottom half of the sample. The mean provides a useful "cut-point" for assessing how an individual ranks. You likely already know this from exam scores in your previous courses because the mean allows you to determine if you scored above or below the average.

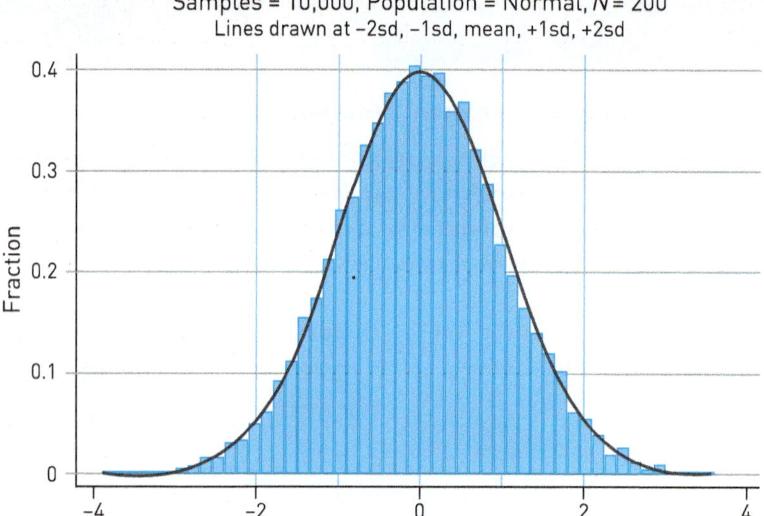

FIGURE 7.1 | The Normal Distribution

Suppose you want to find cut-points other than the mean because you require more information than the mean provides. You could use the standard deviation, which you learned about in the last chapter, or the standard score (also known as the *z*-score or **normal score**). If a variable is normally distributed, the standard score allows you to say definitively what proportion of observations lie between the mean and any specific standard score from the mean. The standard deviation is a coarser version of the standard score, so we'll talk about that first. The standard deviation shows what proportion of the sample, or population, lies on either side of certain key values. These values from a simulated normal distribution are denoted by vertical lines in Figure 7.1. The line in the centre of the histogram is μ (mu), the mean, and has a value of zero. The mean value of the raw scores is not necessarily zero (the mean age of the Canadian population, for example, is around 40), but when the values of a variable are standardized, zero is used for convenience. The vertical lines that move away from the mean are standard deviation markers. In the case of Figure 7.1, they are at $\pm 1\sigma$, and $\pm 2\sigma$. (Note: $\pm$ is shorthand for plus or minus, so $\pm 1\sigma$ is "plus or minus one standard deviation from the mean.")

If a variable is normally distributed, we know that approximately 68.26 per cent of all observations lie between ± 1 standard deviation from the mean, that about 95 per cent of all observations can be found between ± 2 standard deviations from the mean (the more accurate figure is 95.44 per cent), and that almost all observations (about 99.74 per cent) lie between ± 3 standard deviations from the mean. These values, often referred to in shorthand as 68-95-99, tell you what percentage of all observations lies between the positive and negative values of the standard deviation score and the mean. Remember that these values are hypothetical and that to see these exact percentages in your data, the sample size would have to be huge—larger than what's usually available. The data would also have to be normally distributed.

One variable which we know is normally distributed in the population is BMI (body mass index). We will use this variable to illustrate standard deviation and standard scores in the following sections of this chapter.

We will use data from the 2012 Canadian Community Health Survey microdata public use files. The Canadian Community Health Survey is a survey conducted by Statistics Canada every year. For the purposes of this example, we are going to examine two different samples from this data set. First, we will look at women aged 18–64 who were not currently pregnant at the time of the survey. This sample contains 20,211 women. Then we will look at men aged 18–64. The male sample contains 17,607 respondents.

Figures 7.2a and 7.2b show the distribution of the BMI variable for each sample.

You can see from Figure 7.2a that the mean BMI for the female sample is 26.08. The standard deviation is 5.88. The range of BMIs reported by women in this sample went from 13.57 to 54.96.

Figure 7.2b shows the data for the male sample. The mean BMI for the male sample is 27.13. The standard deviation is 4.88. The male BMIs ranged from 13.53 to 54.42.

As you can see from Figures 7.2a and 7.2b, both these distributions approximate the normal distribution. (This is not surprising since we know that BMI is normally distributed in the population, and we have two very large sample sizes.)

We know a lot about normal distributions. We know, for example, that about 99 per cent of all observations are between ±3 standard deviations from the mean. So approximately 1 per cent of all cases are outside of this range, because 100 per cent − 99 per cent = 1 per cent. We also know that roughly 68 per cent of all observations are within ±1 standard deviation from the mean, and that 32 per cent of all observations are likely to differ from the mean by

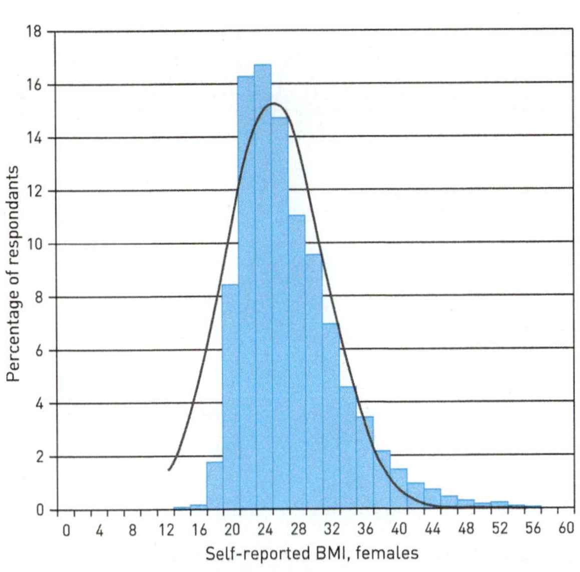

FIGURE 7.2A | BMI for the Female Sample

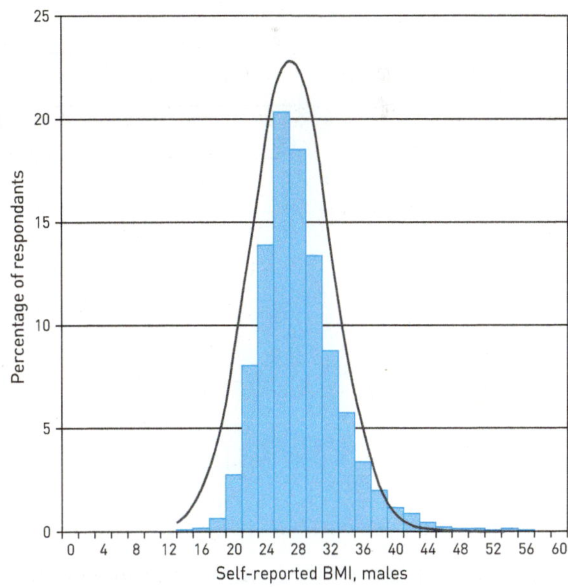

FIGURE 7.2B | BMI for the Male Sample

at least 1 standard deviation (100 per cent − 68 per cent = 32 per cent). Finally, the same procedure applies for 2 standard deviations (100 per cent − 95 per cent = 5 per cent).

With the standard deviation, it is easy to determine the proportion of people above or below the standard deviation cut-points. Figures 7.3a and 7.3b show the sample distributions of BMI, with the horizontal axis now representing standard deviations rather than points on the BMI scale. Following convention, the mean for Figures 7.3a and 7.3b is now given a value of 0. Values higher than the mean are positioned to the right of 0, and negative values are to the left. This is done so that anyone can understand the data without knowing anything about the subject matter. The numbers across the horizontal axis don't refer to actual BMI anymore but instead to BMI expressed in standard deviation units.

Using what we know about the normal distribution, we can use these histograms to tell us about the distribution of BMI in these two samples. We know that approximately 68 per cent of the female sample has a BMI of between 20.20 and 31.96 (mean of 26.08, ±1 standard deviation of 5.88) and approximately 68 per cent of the male sample has a BMI of between 22.25 and 32.01 (mean of 27.13, ±1 standard deviation of 4.88). We also know that approximately 95 per cent of the female sample has a BMI of between 14.32 and 37.84 (mean of 26.08, ±2 standard deviations of 5.88) and approximately 95 per cent of the male sample has a BMI of between 17.37 and 36.89 (mean of 27.13, ±2 standard deviations of 4.88). In addition, approximately 99 per cent of the female sample has a BMI of between 8.44 and 43.72 (mean of 26.08, ±3 standard deviations of 5.88) and approximately 99 per cent of the male sample has a BMI of between 12.49 and 41.77 (mean of 27.13, ±3 standard deviations of 4.88).

These calculations capture the observations that are in the centre portion of the normal distribution, but what if we're only interested in the left or right tail? That would be a one-tailed assessment and can be calculated by dividing the values by two. For example, if

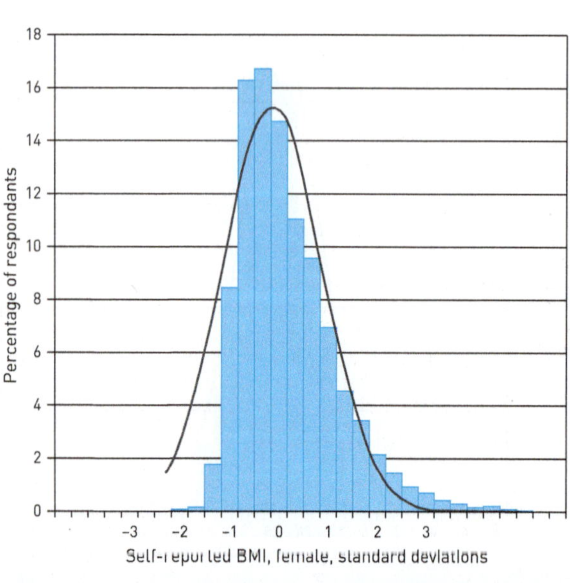

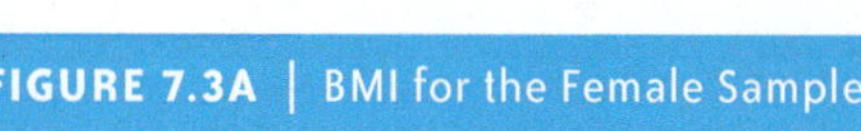
FIGURE 7.3A | BMI for the Female Sample

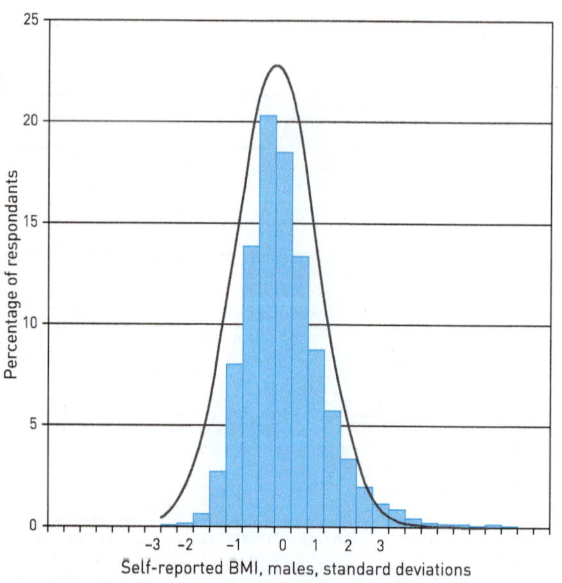

FIGURE 7.3B | BMI for the Male Sample

we want to know the number of cases above 3 standard deviations, we divide 1 per cent by 2. We know that roughly 1 per cent of all cases lie above or below 3 standard deviations from the mean, which means that half of 1 per cent, or 0.5 per cent, must lie above 3 standard deviations from the mean. Returning to our example, only 0.5 per cent of the female sample has a BMI above 43.72, and only 0.5 per cent of the male sample has a BMI above 41.77. We find this by using the "68-95-99 rule" mentioned above, which refers to the proportion of observations between 1, 2, and 3 standard deviations from the mean.

An Extension of the Standard Deviation: The Standard Score

The standard deviation is a great way to determine how many observations are on either side of $\pm 1\sigma$, $\pm 2\sigma$, and $\pm 3\sigma$, but what if you want to set your own cut-point, such as the value that separates the bottom 10 per cent of your observations from the top 90 per cent? Suppose you wanted to study the characteristics of the lowest achievers in elementary school and decided to look at the lowest 10 per cent. Or suppose you wanted to study the world's most volatile nation-states, and you chose to identify them by how many years of peace they've had. Whether you're interested in low blood pressure, high earnings, large families, small insects, or high mortality rates, the standard score allows you to compare a single score with those of the population of interest.

The equation for the standard score when you have data on a whole population is:

$$z = \frac{X - \mu}{\sigma}$$

Where:

z = the z or standard score (expressed in standard deviations)
X = an individual's raw score
μ = the population mean
σ = the population standard deviation

Remember it is rare to have data on a whole population. It is far more likely that you will have data on a sample. Extrapolating from a sample to a population has its own sets of issues, which will be discussed in Chapter 9. For now, it's important that you understand the difference between samples and populations (see Chapter 4) and that you keep in mind that samples require different equations. For example, when calculating the standard score using the sample standard deviation, you must also use the sample mean (rather than the population mean) as you can see below.

The equation for the standard score when you have data for a sample is:

$$z = \frac{X - \bar{x}}{s}$$

Where:

z = the z or standard score (expressed in standard deviations)
X = an individual's raw score
$\bar{x}$ = the sample mean
s = the sample standard deviation

See Table 7.1 for a reminder of the various symbols used for sample and population introduced in the text so far.

The standard score, or z-score, is interpreted as a standard deviation, where the sum and mean are 0. The formula for the z-score will convert the score of any individual observation into a z-value. This value is directly related to that of the standard deviation. A z-score of +1.0 is equivalent to 1 standard deviation above the mean. A z-score of −2.0 is the same as 2 standard deviations below the mean, etc. You can get some practice with standard deviations by taking a look at Box 7.1.

Converting raw scores (test averages, heartbeats per minute, etc.) to z-scores makes it possible to determine the rank of *any* score. This rank is expressed in **percentiles.** Like standard deviations, the standard score lets you place an observation on the normal curve so that you can express the proportion of the sample, or population, above or below a particular value. Unlike the standard deviation, there are more scores to remember than the 68-95-99 rule. Appendix A has a table for converting z-scores to percentile ranks.

TABLE 7.1 | Summary of Sample and Population Symbols

	Number of observations	Mean	Standard deviation	Variance	Proportion	Standard score
Sample	n	$\bar{x}$	s	s^2	p	z
Population	N	μ	σ	σ^2	P	

BOX 7.1

It's Your Turn: Determining the Proportion of Observations at Various Standard Deviation Cut-Points

Now that you know the "68-95-99 rule," can you determine approximately what percentage of all observations lie at the following cut-points?

1. Below −2 standard deviations?
2. Below −3 standard deviations?
3. Above the mean?
4. Above +1 standard deviation?
5. Below +2 standard deviation?

The solutions for Box 7.1 can be found in Appendix H.

TABLE 7.2 | The *z*-Table (Area under the Normal Curve)

A *z*-score	B Area between *z* and mean	C Area beyond *z*
0.0	0.000	0.500
0.1	0.040	0.460
0.2	0.079	0.421
0.3	0.118	0.382
0.4	0.155	0.345
0.5	0.192	0.309
0.6	0.226	0.274
0.7	0.258	0.242
0.8	0.288	0.292
0.9	0.316	0.184
1.0	0.342	0.159
1.1	0.364	0.136
1.2	0.385	0.115
1.3	0.403	0.097
1.4	0.419	0.081
1.5	0.433	0.067
1.6	0.445	0.055
1.7	0.455	0.045
1.8	0.464	0.036
1.9	0.471	0.029
1.96	0.475	0.025
2.0	0.477	0.023
2.1	0.482	0.018
2.2	0.486	0.014
2.3	0.489	0.011
2.4	0.492	0.008
2.5	0.494	0.006

Table 7.2 is a primer, containing a few *z*-scores. The first column in Table 7.2 lists the *z*-score, followed by the proportion of all observations that lie between the mean and the *z*-score. The third column lists the proportion of observations that are beyond the calculated *z*-score (the percentage of all observations in a tail). Since the *z*-score is a standardized measure, these values remain true for any normally distributed variable.

There will be times when you will need to determine the area between two *z*-scores. Continuing with the BMI example, if you wanted to know what proportion of women in the sample have a BMI between 20 and 30, you would need to calculate two z-scores, one to establish a lower bound, and one for the upper bound. The mean BMI for women is 26.08 and the standard deviation is 5.88. Calculate the lower bound first:

$$\begin{aligned} z &= \frac{X - \bar{x}}{s} \\ &= \frac{20 - 26.08}{5.88} \\ &= \frac{-6.08}{5.88} \\ &= -1.03 \end{aligned}$$

(1.03 standard deviations below the mean)

Next, the upper bound:

$$\begin{aligned} z &= \frac{X - \bar{x}}{s} \\ &= \frac{30 - 26.08}{5.88} \\ &= \frac{3.92}{5.88} \\ &= 0.67 \end{aligned}$$

(0.67 standard deviations above the mean)

Next, you need to find these scores on the z-table. Usually, only positive values can be found on a z-table, but because the normal curve is symmetrical you can look for the nearest absolute value; −1.03 becomes 1.03. Now you simply need to add the values together from column B in Table 7.2.

For 1.03, the closest z-value shown in this table is 1.1, and for 0.67 the closest value shown in this table is 0.7. Since we want to know the proportion of women *between* the two scores, we need to use information from column B. The value for 1.1 is 0.364, and for 0.7 it is 0.258. The first value tells us that 36.4 per cent of women will lie between 1.1 standard deviations above the mean and the mean. The second value tells us that 25.8 per cent of women will lie between 0.7 standard deviations below the mean and the mean. Adding these two values together yields 0.622, which means that roughly 62 per cent of women will have a z-score between −1.1 and 0.7, or will have a BMI between 20 and 30.

A z-score can also be translated back into its actual value (such as BMI). Suppose that we wanted to determine, with 95 per cent confidence, what the BMI of a working-age man in Canada would be. (Recall that Figure 7.2b showed the mean BMI for our male sample as 27.13.) To determine this, work backwards from the formula. Instead of calculating z, which is unknown in the example above, we would calculate the upper and lower values of X, the cut-points that 95 per cent of observations are found in.

First, look at the z-table. Use the value where 47.5 per cent of all observations lie between z and the mean (column B). This table represents the absolute values of z. If you want to know the area both above and below the mean, you'll have to place a negative sign in front of the lower value. Since we want the z-value for the point where 47.5 per cent of all cases fall between z and the mean, we find it in column B: 1.96. We find the value for 47.5 per cent because it's half of 95 per cent, the number we want to capture. By ensuring that 47.5 per cent of our cases are above the mean and that 47.5 per cent are below, we get a total of 95 per cent of all cases. We can insert the known values into our equation for the z-statistic, focusing first on the lower bound (where we assign a negative value to the z-statistic):

$$z = \frac{X - \bar{x}}{s}$$

$$-1.96 = \frac{X - 27.13}{4.88}$$

$$-1.96 * 4.88 = X - 27.13$$

$$17.57 = X$$

For the upper bound:

$$z = \frac{X - \bar{x}}{s}$$

$$1.96 = \frac{X - 27.13}{4.88}$$

$$1.96 * 4.88 = X - 27.13$$

$$9.56 + 27.13 = X$$

$$36.70 = X$$

We can be 95 per cent confident that a man selected from this sample will have a BMI between 17.57 and 36.70.

It can be difficult at times to know what numbers you are interested in. You might be wondering why you should subtract a value from column B from 1 at some times but not others. That is a good question, and there is no easy answer to it beyond simply understanding what each column represents (Figures 7.4a–7.4e can be helpful in this regard). One strategy that we employed when first learning this stuff was to draw out normal distribution and figure out which part of the curve we were interested in. Ask anyone who has taken our statistics classes, and they'll confirm that this is the advice we're likely to give to those in doubt.

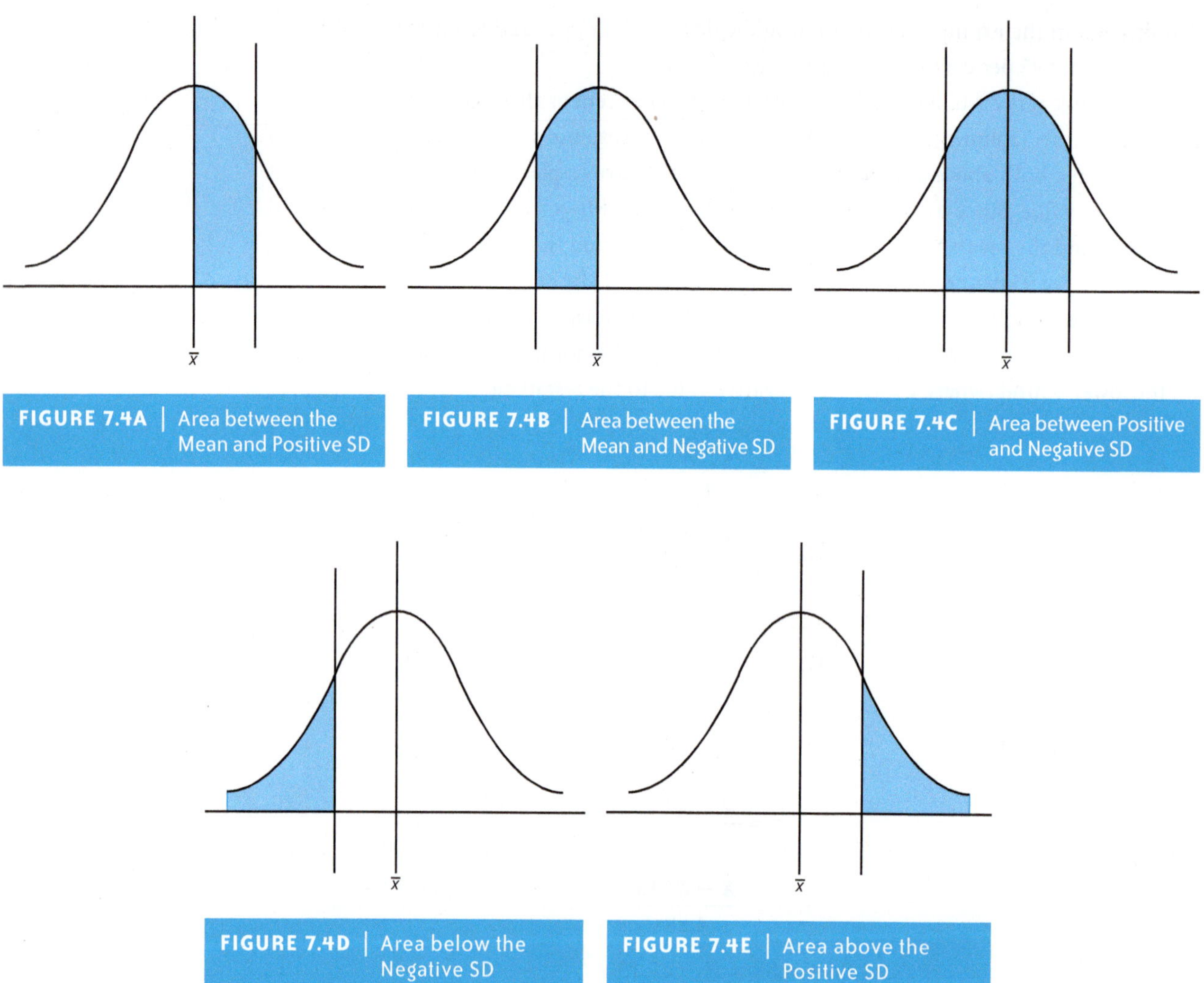

FIGURE 7.4A | Area between the Mean and Positive SD

FIGURE 7.4B | Area between the Mean and Negative SD

FIGURE 7.4C | Area between Positive and Negative SD

FIGURE 7.4D | Area below the Negative SD

FIGURE 7.4E | Area above the Positive SD

One-Tailed Assessments

So far, we have been finding ranges where we know the upper and lower points (e.g., "How many observations lie beyond points 1 and 2?"). We've been looking at the proportion of observations *between* two known values, as in Figure 7.5. This may not always be the type of information we seek.

It is also useful to know the percentage of observations above or below a specified point, with no upper or lower limit imposed. To use our BMI example, we might want to know the percentage of women who have a BMI below 15. Or maybe we want to know the percentage of women who have a BMI over 40. In either case, we are interested only in observations above or below a specified point rather than between two values. The difference is shown in Figure 7.6.

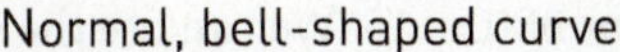

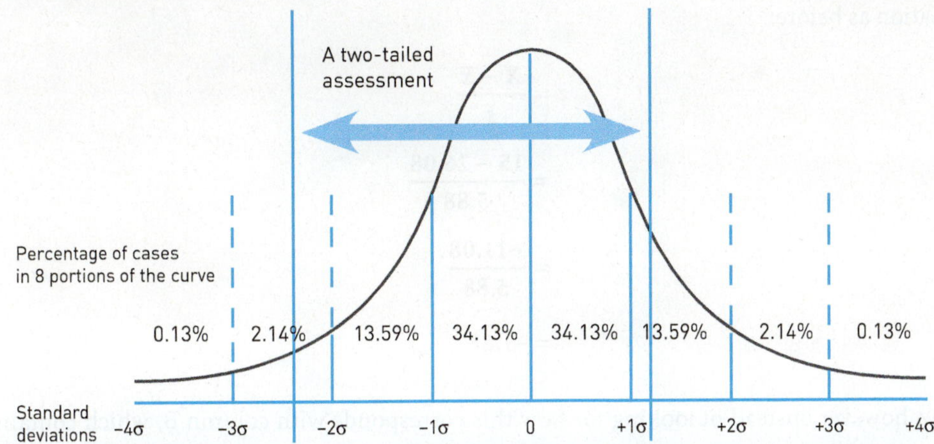

FIGURE 7.5 | A Two-Tailed Assessment

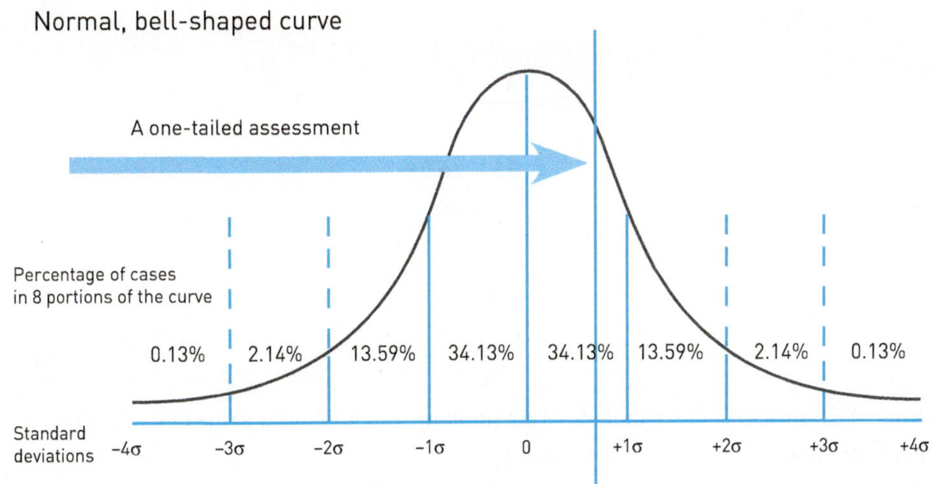

FIGURE 7.6 | A One-Tailed Assessment

When performing a **one-tailed assessment**, we must look at a different column on the *z*-table than for a **two-tailed assessment**. Instead of column B, we must now focus on column C. The change is very subtle, particularly since the number in column B plus the number in column C always captures exactly half of the normal distribution and is always equal to 0.5, representing half of all observations.

To illustrate a one-tailed assessment, let's continue with our BMI example. Suppose that we want to know what percentage of women have a BMI below 15.

We'd first need to calculate the z-statistic for women for BMI = 15. We can use the same equation as before:

$$z = \frac{X - \bar{x}}{s}$$

$$= \frac{15 - 26.08}{5.88}$$

$$= \frac{-11.08}{5.88}$$

$$= -1.88$$

Now, however, instead of looking for how this corresponds with column B, which contains the area *between* the mean and z, we need to look at the area *beyond* z, found in column C (as always, treat the z-statistic as an absolute value—look up 1.9). We see the number 0.029, which is very low. This suggests that only 2.9 per cent of women are likely to have a BMI of less than 15.

What if we wanted to find out what percent of women have a BMI over 35? The calculation would be the same:

$$z = \frac{X - \bar{x}}{s}$$

$$= \frac{35 - 26.08}{5.88}$$

$$= \frac{8.92}{5.88}$$

$$= 1.52$$

Looking at column C of the z-table for 1.52 (look up 1.5) gives us a value of 0.067. Only 6 per cent of Canadian women have a BMI over 35.

So far, we have assumed that we are always looking for low values when the cut-off is below the mean, or high values when the cut-off is above the mean. The final scenario is one in which we want to know the percentage of all observations above a value when that value is below the mean, or the percentage of all observations below a value when that value is above the mean. To do that, we need to modify the values in the z-table slightly. Let's suppose that we want to know what percentage of women have a BMI less than 35. The z-value would remain the same, at 1.52, and we would still need to look at the value in column C, but we'd need to subtract that value from 100, yielding 94 per cent of all observations. The difference between this calculation and the previous one is that 6 per cent refers to all values above a z-value of 1.52, or BMI = 35. There, we were interested in the information to the right of the cut-off point. Here, we are interested in the information to the left of the cut-off point, which is everything except 6 per cent. Since we begin with 100 per cent of all observations, we need only to subtract the portion we're not interested in.

Instead of subtracting the z-value from 100, you could get the same answer by taking the number in column B (0.43) and adding 0.50. This is possible because 50 per cent of all observations lie on each side of the mean, so column B refers to the proportion of all observations between a specified value and the mean. If you are conducting a one-tailed assessment and will be dealing with more than half of all observations, this is the procedure that you'll have to use. If you are using less than half of all observations, the simple technique will work.

To help keep all of this straight, it is useful to draw a histogram like the one in Figure 7.6. (Also see Box 7.2.)

BOX 7.2

How to Convert the Standard Score to a Ranking: An Example

Between 1980 and 2004, Canada admitted over 4.5 million immigrants. Suppose we have data for this entire population (all the immigrants to Canada from 1980 to 2004). The average years of schooling for this group is 11.42, with a standard deviation of 4.8 years. Suppose that a person has 10 years of education, and we want to know the proportion of people with more than 10 years of education (assume that years of schooling is normally distributed).

Since the distance between our observed value and the mean is less than one standard deviation, we need to convert our score with the following equation:

$$z = \frac{X - \mu}{\sigma}, \text{ or } z = \frac{10 - 11.42}{4.8}, \text{ or roughly } -0.296.$$

Plunking this number into our table gives us the value 0.382 in column C, meaning that an impressive 61.8 per cent (100 − 38.2) of all immigrants to Canada have more than 10 years of education.

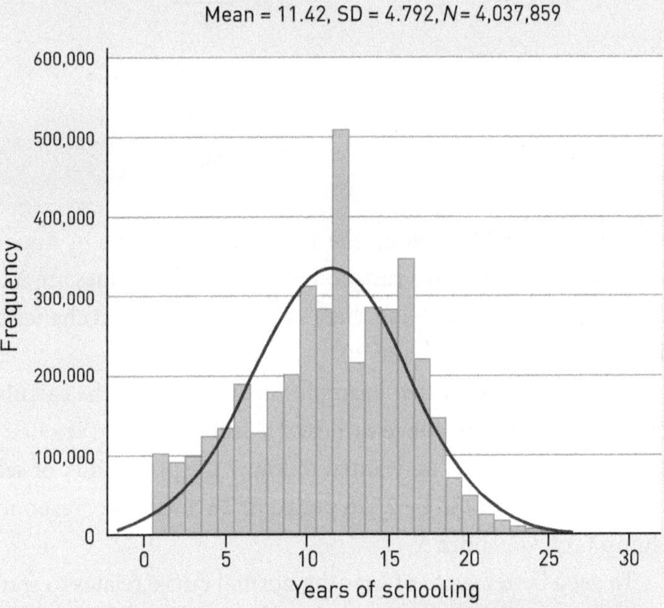

FIGURE 7.7 | Average Years of Schooling of Canadian Immigrants, 1980–2004

Source: Statistics Canada. 2003. Longitudinal Survey of Immigrants to Canada: Process, progress and prospects. Catalogue no. 89-611-XIE. Ottawa.

Probabilities and the Normal Distribution

In the sections above, we used standard scores and the normal curve to help us identify numbers of occurrences, which are measured outcomes. The normal curve has many other uses, such as calculating probabilities. Think abstractly about the normal curve for a moment; **outliers** aside, every individual in a sample or population should be located somewhere under the normal curve. This means that there's a 100 per cent chance that we'll be able to locate someone, and once we know a little bit more about the person, we can predict where they'll be under the normal curve. This links standard deviations, z-scores, and the normal curve to probabilities, and everything we learned up to now applies, except that we're dealing with probabilities instead of sample means, scattered across the normal distribution.

Suppose we want to know the probability of randomly selecting a man whose BMI is between 25 and 30. The first thing we need to do is calculate two z-statistics. The first is for 25:

$$z = \frac{X - \bar{x}}{s}$$

$$= \frac{25 - 27.13}{4.88}$$

$$= -0.436$$

And the second is for 30:

$$z = \frac{X - \bar{x}}{s}$$

$$= \frac{30 - 27.13}{4.88}$$

$$= 0.588$$

Next, we need to look up the values on the z-table in Appendix A (using column B because we're interested in a range). We find that the values are 0.1700 and 0.2224. Adding the scores together yields 0.39, so there's about a 39 per cent chance of randomly selecting a man whose BMI is between 25 and 30.

As with all of the other examples in this chapter, the calculated z-values can also tell us how many people are above or below a certain value. (See Box 7.3 for some practice questions.) If, for example, we wanted to know the probability of selecting someone whose BMI is above 30, that would be 27.76 per cent. This value corresponds with a z-value of 0.588 in column C of Appendix A.

To get a better sense of how the normal curve relates to standard deviation, cumulative percentages, and percentiles, consider Figure 7.8.

BOX 7.3

It's Your Turn: Converting Standard Scores to Percentile Ranks

Emily loved to eat out and go to movies with her friends, but her parents thought these activities were a waste of time and money. She was sure that everyone her age went out at least three times a week. Her parents did not agree—they were convinced that the majority of people only went out for special occasions, maybe once a month. Using the public-use micro data files collected as part of the 2004 General Social Survey, Emily looked for the number of evenings per month that Canadians between the ages of 18 and 29 reported they went to restaurants, movies, or theatres. She recoded people who went out "less than once a month" as going out once a month, because she did not want to exclude these people just because their response was not a whole number. She found the following:

- The mean number of nights per month people went out was 5.7.
- The standard deviation was 5.1. (Weighted $n = 5{,}183{,}000$ [rounded to thousands], excluding residents of Yukon, the Northwest Territories, and Nunavut, and full-time residents of institutions.)

Using the chart in Appendix A and the equation for the standard score, determine the following:

1. The range of values ± 1 standard deviation from the mean.
2. The value that the lowest 10 per cent of all observations fall below.
3. The value that the highest 40 per cent of all observations are above.
4. The percentage of cases that fall between the values of 4 and 9.
5. The value that 75 per cent of all observations fall below.

The solutions for Box 7.3 can be found in Appendix H.

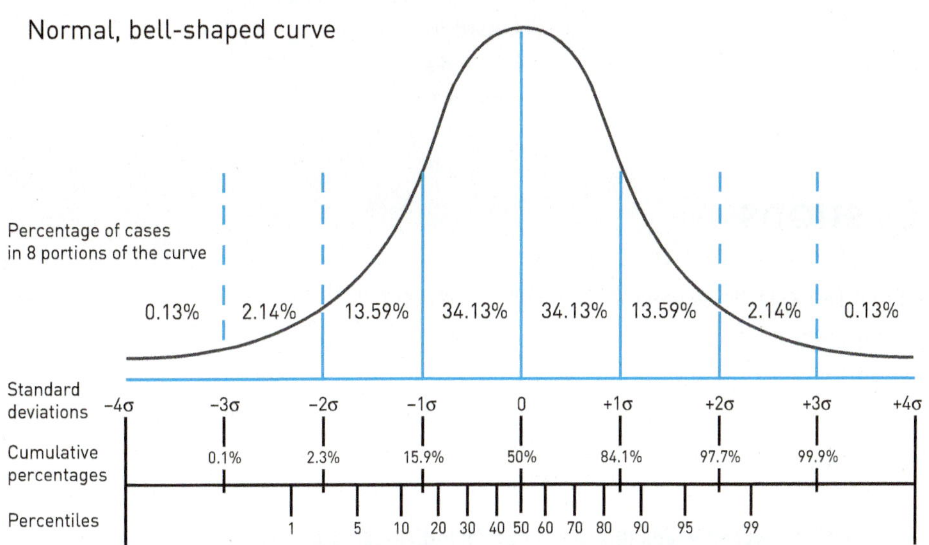

FIGURE 7.8 | Standard Deviations, Cumulative Percentages, Percentiles, and the Normal Curve

BOX 7.4

N versus *n*: A Note on Notation

Increasingly throughout this text you will notice that sometimes we use upper-case letters in our equations, sometimes we use lower-case letters, and at other times we use Greek symbols. As you will discover in Chapter 9, our primary reason for doing this is to distinguish between samples and populations. We try to use lower-case letters to denote samples, and Greek and upper-case letters to denote populations. As you'll see in the chapters ahead, there are often different equations for samples and populations, and it is important to use distinct symbols for each to avoid confusion.

Conclusion

This chapter further explored the normal curve and introduced you to standardized scores and sampling distributions. These topics are crucial to understanding how social scientists generalize from samples to populations.

Glossary Terms

Normal curve (p. 54)
Normal score (p. 56)
One-tailed assessment (p. 65)
Outliers (p. 68)
Percentiles (p. 60)
Sample distribution of means (p. 55)
Standard score (p. 54)
Two-tailed assessment (p. 65)
z-score (p. 54)

Practice Questions

1. Using Appendix A, for the numbers below, find the area between the mean and the z:
   a. $z = -1.18$
   b. $z = 0.84$
   c. $z = -2.06$
   d. $z = 1.36$
2. For the numbers below, find the percentile rank (the percentage of individuals scoring below z).
   a. $z = 2.25$
   b. $z = -1.67$
   c. $z = 1.43$
   d. $z = -0.44$

3. For the numbers below, find the percentage of cases falling above z.

   a. $z = 0.25$

   b. $z = -1.21$

   c. $z = 1.21$

   d. $z = -2.01$

4. For the numbers below, find the percentages of cases falling between the two z-scores.

   a. $z = -0.38$ and $z = 1.63$

   b. $z = 0.88$ and $z = 1.55$

   c. $z = -1.93$ and $z = 1.09$

   d. $z = -2.22$ and $z = -1.34$

5. Sigmund wrote a statistics exam and scored 45 (mean = 52, standard deviation of 5). What is his percentile rank?

6. Lesley wrote the same test and scored 54. What percentage of individuals received a higher score?

7. You believe that your child is a genius and decide to have him write a standardized achievement test. To your delight, he scores a 148 (mean = 125, standard deviation = 15). What is your child's percentile rank?

8. Feng and Lucy both took a spatial abilities test (mean = 80, standard deviation = 8). Feng scored a 76 and Lucy scored a 94. What percentage of individuals would score between Feng and Lucy?

9. Evelyn typically brushes her teeth for two minutes (120 seconds). Her younger sister Abigail finds it annoying that she doesn't vary much in this regard. Evelyn decides to prove Abigail wrong, and determines that she spends either less than 88 seconds or more than 152 seconds roughly 5 per cent of the time. Assuming a normal distribution, what is Evelyn's standard deviation?

10. Life in Ratroy is fairly boring, so much so that, out of boredom, Kody decides to spend every afternoon for a week counting how many crickets he hears chirping every hour in the field beside his house. Because he is statistically savvy, he knows that the week he chooses may not be representative, so he decides to compare his sample data to that of a normal population. Here are the sample data for the five afternoons.

	Monday	Tuesday	Wednesday	Thursday	Friday
1:00	53	34	65	54	47
2:00	44	23	55	33	54
3:00	32	27	47	45	53
4:00	66	19	59	39	45

   a. Calculate the mean and standard deviation score for the numbers above.

   b. How many hours fell beyond (or above and below) 1, 2, and 3 standard deviations from the mean?

Answers to the practice questions for Chapter 7 can be found in Appendix H.

8 Sampling

LEARNING OBJECTIVES

It is rare that we collect data on an entire population. In Canada, the only data we collect on the whole population is the short-form census, which includes only eight questions, regarding age, sex, marital status, and mother tongue. In reality, you are likely to be working with data from samples most of the time. Working with sample data requires a few additional considerations. In this chapter you will:

- learn how to identify probability samples and determine when to use them; and
- examine the non-probability/non-random sampling strategies and learn how to use them.

Introduction

Remember from Chapter 1 that the population is defined as the entire group you are interested in. If you are studying everyone who lives in Canada, the population would be all residents of Canada. If you are studying adolescents in Calgary, the population would be Calgarians aged 13 to 19. If you are studying students at your university, the population would be all students enrolled at your university. Usually, collecting data from the entire **population** is costly and cumbersome. Instead, social scientists usually gather data on a **sample**, or a fraction, of the entire population. Then they use this sample (along with some rules about probability and the central limit theorem) to generalize to the population of interest. If you wanted to know how much the average university student in Canada earns, for example, you would not need to ask every single Canadian university student. You could ask a sample of students and use the data you gather from that sample to generalize to all Canadian university students. A sample that is accurately and carefully selected allows a precise analysis without including the full population.

Sampling has a long history in Canada and was first used as part of data collection by Statistics Canada for part of the Canadian census in 1941. While every household received the short form of the census questionnaire, a longer 27-question questionnaire was given to a sample of Canadians. The long form was completed for every tenth dwelling by an enumerator. Data from the sample was used to obtain estimates of things like earnings and average levels of education (since these questions were asked only on the long form, not on the short form).

Our discussion in the last chapter of standard deviations, standard scores, and normal distributions also applies here, although the topic is sufficiently complex to warrant its own chapter (Chapter 9). For now, as you read this chapter, think about how treating the mean of a sample as a value allows you to plot a distribution of sample means in the same way that you might plot a distribution of individual values for a variable. Furthermore, each of the sampling techniques described in this chapter will have a different degree of error, that is, the accuracy with which a sample resembles a population. We'll discuss sampling error more at the end of this chapter.

We're going to look at some of the challenges and potential pitfalls of using samples in quantitative research by examining the different methods for deriving samples and identifying the strengths and weaknesses of each method. Then we'll cover the sources and consequences of bias in the sample.

Probability Samples

There are two broad types of sampling techniques: **probability** and **non-probability**. In a probability sample, each unit has a known chance of being selected. For example, if we take a random sample of 10 units out of a population of 100, each unit has a 10 per cent chance of being included. If we want to sample 100 people on a university campus, 50 students and 50 professors, and there are 100 professors and 5000 students, the probability of inclusion would be different for the two subpopulations. For professors, the selection probability would be $\frac{50}{100}$, or 50 per cent, and for students the chance of being selected would only be $\frac{50}{5000}$, or 1 per cent.

Probability samples have several desirable qualities. They are representative, allowing for generalization from sample to population. This means the following:

1. Sample means can be used to estimate population means.
2. If the population is normally distributed, the sample will usually be normally distributed, too.
3. It is possible to estimate the discrepancy between sample mean and population mean. This measurement is called the sampling error.
4. It is possible to test how well our results resemble what we could expect to see in the population by using **inferential statistical tests**.

In the next section, the four most popular types of probability samples are described. Keep in mind that while this chapter presents the sample types individually, any sample used by a statistical agency, such as Statistics Canada, is likely a combination of these types.

Simple Random Sample

Simple random samples are probably the most basic probability samples. To select one, list all possible units, number them consecutively, and then use a random number chart (like the one in Appendix F) to select a certain number, or percentage, of units. In a simple random sample, every unit has an equal probability of selection. The probability of selection is $\frac{n}{N}$, where n = sample size and N = population size. Once a sample is selected, a variety of methods may be used to contact respondents, including mail, telephone, and email.

Suppose that we have a population of 100 and we want to derive a sample of 10 observations. Each unit in a random sample would have a $\frac{10}{100}$, or 1 in 10, chance of being selected in the sample. This means that the probability of selection is 10 per cent.

Simple random samples are easy to grasp, and they are the only sampling technique that gives a *truly* random sample (all others include some other source of bias). The downside of simple random samples is that they can be cumbersome to generate. If you are collecting a sample of individuals, you are required to have a list of every individual in your population, then assign a random number to every individual in the population, and then contact those specific individuals who are randomly chosen to be in your sample. If an individual who is chosen to be in your sample does not respond to your letter, phone call, or email asking him or her to take your survey, you cannot simply substitute the next person on your population list for that person. Such a substitution would mean that your sample was no longer truly random. Therefore, you have to leave that individual out of your sample and consider that person part of your group of non-respondents.

Remember from Chapter 1 that the unit of analysis in your data does not always have to be individuals: it could be cities or countries. In that case, you would need to compile a list of all the cities or countries that you wish to study, assign a random number to each city or country in the list, and then analyze the data for only those cities or countries that are randomly chosen to be in your sample.

Systematic Random Sample

Instead of choosing units from the population list by using a random number chart, for a **systematic random sample**, units are chosen at regular intervals. The person deriving the sample can use a chart to select a random starting point or choose their own. Then they consistently choose every nth unit (such as every 10th observation) from the list of the whole population. As long as there is no inherent ordering in the data set, a systematic random sample will represent the population fairly accurately, though not as well as a simple random sample, since it is not quite as random. Imagine that you were selecting every tenth person from class lists where all classes had 49 people and the final person on the list was the instructor. You would end up selecting a disproportionate number of instructors and your sample would not be truly random. However, systematic random samples are a quick and accurate approximation of a simple random sample.

The defining characteristic of a systematic random sample is that it relies on a system of selection from the population list that has a pre-determined number of non-chosen observations between the chosen observations. For example, a one-in-ten sample would have nine non-chosen observations for every one observation. One of the benefits of this is that it is possible for others to replicate your sample as long as they have the same **sampling frame** (the population list) you used. The downside is that breaching **confidentiality** can become a concern. If an individual is able to identify your system of selection and has access to your population list, then that person can identify all individuals in the sample.

Stratified/Hierarchical Random Sample

A **stratified/hierarchical random sample** can best be described as a series of two or more simple random samples operating within the same population. If you wanted to compile a representative sample of the elementary school students in Calgary and you wanted to ensure that all schools were represented, you would need to select a random group of students from each school. With a stratified sample you would take a simple random sample of people *within* each school, instead of taking one simple random sample and hoping that each school was equally represented.

The key benefit of stratified sampling is that each population subgroup of interest is adequately represented in a sample. The disadvantage is that, like simple random sampling, generating a sample can be a lot of work, and you need to have your population list organized by the variable you are using for stratification.

As mentioned earlier, until 2011, Statistics Canada administered a long-form and a short-form questionnaire. The short-form questionnaire, administered to 80 per cent of Canadians, contained only eight questions, collecting data on age, sex, marital status, and mother tongue. The other 20 per cent of the population received a *much* longer questionnaire: the same 8 questions on the short form plus another 53 unique questions. In 2011 the long-form census was replaced by the National Household Survey, only to return for the 2016 headcount.

In each instance, the agency used stratified sampling to decide whom to include in its long-form questionnaire sample. In all self-enumeration areas (defined as those where the majority of people were able to complete the questionnaire themselves), a one-in-five random sample of occupied private dwellings was selected to receive a long questionnaire. The non-sampled occupied private dwellings received just the short questionnaire.

Recognizing that certain portions of the population would have to be "over-sampled" to enable meaningful analysis with samples, Statistics Canada decided to stratify or divide its sample into specific sub-populations. Over-sampling occurs when you include a disproportionate number of people from a certain group in your sample to ensure they are represented in your data. For example, most persons in collective dwellings received a long-form questionnaire in 2006, where a collective dwelling

> [r]efers to a dwelling of a commercial, institutional or communal nature. It may be identified by a sign on the premises or by a census representative speaking with the person in charge, a resident, a neighbour, etc. Included are lodging or rooming houses, hotels, motels, tourist homes, nursing homes, hospitals, staff residences, communal quarters (military bases), work camps, jails, missions, group homes, and so on. Collective dwellings may be occupied by usual residents or solely by foreign and/or temporary residents.

Source: http://www12.statcan.ca/census-recensement/2006/ref/dict/dwelling-logements002-eng.cfm

Cluster Sample

Cluster sampling is another way of gathering a large sample. Researchers use it when they cannot get a complete list of the population they wish to study but *can* get a complete list of groups, or "clusters," of the population. Rather than include individuals at random, this technique randomly includes clusters. Usually, everyone in a cluster is included in the sample.

Suppose you wanted to investigate the use of lawn pesticides by the residents of Kelowna, BC, but didn't have the resources to randomly sample the entire city. A cluster sample could be taken by identifying every street as a cluster. A random sampling of streets could be taken, and all residents of each selected street could be included in the sample. The main advantage of cluster sampling is that it is cost-effective. As long as the samples are chosen at random, researchers don't have to travel all over to get a representative sample. The disadvantage is that there is a higher, and more difficult to quantify, risk of sampling error.

Non-Probability/Non-Random Sampling Strategies

There is also a set of non-random techniques used to gather samples. They are less common in quantitative research, but you are still likely to encounter them, often when dealing with research done by polling firms or market research agencies.

Non-probability samples differ from probability samples in a few ways. They are not supposed to be representative of the population and are likely to be somewhat biased. Non-probability samples are often used when researchers aren't concerned with representing an entire population or when they lack the resources to select a probability sample. Following are some of the more common types of non-probability samples.

Convenience Sample

A **convenience sample** targets only individuals who possess characteristics that make them more accessible to the researcher. For example, if you live in Edmonton, it is easiest for you to derive a sample that contains only people living in Edmonton. Depending on the nature of your study, that might be okay, but you couldn't make generalizations about residents in Calgary or Toronto by only researching in Edmonton. Convenience samples are also useful for **pilot testing** a research instrument, like a questionnaire, when generalizability isn't a concern but finding unclear questions and non-mutually exclusive response categories is.

Snowball Sample

Typically, **snowball sampling** techniques are used for populations that are not easily identified, resistant to being studied, or otherwise hard to reach (for example, sex workers).

The term "snowball" is used because the sample increases in size as it rolls away from its source—just like a snowball. The sample will grow in size until it reaches the researcher's ideal. Researchers will make contact with a small group, asking members of that group to identify others who might be interested in participating in the study. Initial group members become informants, leading to others in their network.

Quota Sample

Quota samples are the non-probability counterpart to stratified samples. Often used in market research and opinion polls, they are a relatively cheap and quick way to get an adequate sample. As with a stratified sample, researchers decide on strata (such as levels of income), and then try to ensure that the sample is proportionately representative of the population in the categories of interest. Unlike a stratified sample, there is non-random sampling of each stratum's units. Since this is essentially a convenience sample, under-representation of less accessible groups is still a problem.

Sampling Error

Samples rarely match the population they are chosen to represent perfectly. No matter how carefully a probability or non-probability sample is selected, there will be some degree of "mismatch" between the sample and the population. This is called **sampling error**, and it is made up of some systematic error and some random error. To differentiate between samples and population, different symbols are used to represent mean and standard deviation for a sample, versus mean and standard deviation for the population. For the standard deviation of the *population*, the text will continue to use μ to describe the mean and σ to describe the standard deviation. For the standard deviation of a *sample*, the text will use $\overline{X}$ for the mean and s for the standard deviation.

EVERYDAY STATISTICS

The Right Sample for the Right Task

Each month, Statistics Canada conducts the Labour Force Survey, which estimates unemployment and employment in the Canadian population. To collect its data, Statistics Canada divides each province into layers and selects small geographical areas from within each one. Every dwelling unit in each of the selected small geographical areas remains in the sample for six months, and $\frac{1}{6}$ dwelling units are randomly sampled each month.

Q: What type of sampling method do you think Statistics Canada uses to conduct the Labour Force Survey?

Tips for Reducing Sampling Error

Every good social statistician who collects his or her own samples aims to reduce sampling error. Reducing sampling error will usually reduce standard error, too, so every good social statistician who collects his or her own samples aims to reduce sampling error, thereby reducing standard error.

Some error is to be expected in any sample that is smaller than the population. The error may not be problematic for your research, as long as it is random in terms of the variables you select for your research. Non-random sources of error are more serious, but there are some ways to avoid them when you are trying to derive a representative sample. Remember that probability samples are more representative than non-probability samples. When samples are not randomly selected, certain types of people are selected more often—the most agreeable, available, or even attractive—which also affects the accuracy of your results.

An inadequate sampling frame or population list is another potential problem. If you exclude or forget certain members of your population when you calculate the sampling frame, bias will be introduced in your research. Statistics Canada encounters difficulties when it is trying to enumerate Aboriginal and homeless populations. Not including these people will result in a misrepresentation of the Canadian population, thereby biasing any results that come from these data. Statistics Canada tries to make its sampling frames as complete as possible in an effort to represent the entire population.

Finally, non-response is also a problem. Some people refuse to respond to the survey at all. Other times, potential respondents will sometimes refuse to answer certain questions, either because they don't want to or because they fear the consequences. If the non-response is random (e.g., if respondents were all equally unlikely to answer a question) this would not be much of a problem. Unfortunately, most studies find regularities in non-response, suggesting that certain people (men, those who are single, those with low education, etc.) are more likely to not respond than are others. Researchers should always try to correct for non-response bias when presenting their results, either by acknowledging that their data do not represent the non-responding groups or by weighting the data so that they are more representative of the whole population.

Conclusion

This chapter introduced you to the various sampling techniques that social scientists employ. Whenever you read a scientific article or a newspaper article that reports the results from a scientific study, you should ask about the population the sample is supposed to represent and how the sample was taken. Without that information, you cannot evaluate the quality of the results being reported.

Glossary Terms

Confidentiality (p. 74)
Convenience sample (p. 76)
Inferential statistical tests (p. 73)
Non-probability samples (p. 73)
Pilot testing (p. 76)
Population (p. 72)
Probability samples (p. 73)
Quota samples (p. 77)
Sample (p. 72)
Sampling error (p. 77)
Sampling frame (p. 74)
Simple random samples (p. 73)
Snowball sampling (p. 76)
Stratified/hierarchical random sample (p. 75)
Systematic random samples (p. 74)

Practice Questions

1. You need to derive a sample of writing samples from a collection of grade 6 grammar classes. You have a list of all the children in the population, and your primary concern is with representativeness. Which sampling strategy would you use? Why?

2. The University of Alberta has 21 faculties, schools, and colleges. If you wanted to draw a representative sample of the entire student body but wanted to be sure that you had a sufficient number of observations from each faculty, school, and college, which sampling strategy would you use? Why?

3. You're working with a vulnerable, hard-to-reach population and would like to administer a questionnaire. Your primary concern is obtaining a sufficient sample size, even if representativeness is limited. Which sampling technique would you use? Why?

4. Frieda is having trouble choosing people to join her new orchestra, not because of a lack of interest but because she is getting too much interest. She decides that she'll sample her population to find suitable people. She feels that everyone has equal talent, but she needs to ensure that each instrument is adequately represented. She approaches you for advice on how to randomly select the right mix of musicians. Which sampling technique would you tell her to use? Why?

5. Draw a 10 per cent simple random sample (use the random numbers in Appendix F) and a 10 per cent stratified random sample from the numbers below. Think about the strengths and weaknesses of each approach.

37	93	43	35	77	99
55	52	12	86	32	12
43	16	57	3	95	52
39	99	24	17	36	36
34	65	28	72	48	33
66	82	60	50	57	2
66	78	8	35	53	44
61	62	93	35	6	83
63	36	28	42	3	26
81	3	92	99	17	64

6. The 2011 census did not have a long-form version of the questionnaire that was distributed to 20 per cent of the Canadian population. Instead, the government of the day decided to employ a national household survey (NHS) in place of the long-form questionnaire, and to administer it to 33 per cent of the Canadian population, hoping that increasing the sample size would offset the non-response rates of a voluntary survey (previously, both the short-form and the long-form of the census were mandatory). The NHS questionnaire was largely identical to the 2006 census long-form questionnaire, except that individuals could choose whether they wanted to complete it. Although it ended up being closer to 65 per cent, the expected response rate for the NHS was 50 per cent, which means that only half of all people issued an NHS questionnaire would complete and return it. For the long-form census in 2006 (which was mandatory), the comparable response rate was 94 per cent.
   a. If the long-form census questionnaire had still existed in 2011, and response rates were the same as in 2006, what percentage of the population would have filled out the long-form questionnaire in 2011? With a population of 34.5 million in 2011, how many Canadians would have completed a long-form census in 2011?
   b. Approximately 65 per cent of those selected for the 2011 NHS actually completed it. How many people is this?

Answers to the practice questions for Chapter 8 can be found in Appendix H.

9 Generalizing from Samples to Populations

LEARNING OBJECTIVES

In this chapter, we will discuss how we use samples to generalize to populations, by studying:

- the sample distribution of means;
- the central limit theorem;
- confidence intervals and how to calculate them;
- how *t*-distributions can be used for small samples;
- the sample distribution of proportions.

Introduction

Using the insights that can be gained from the normal curve, it is possible to estimate how closely a sample approximates the population. To do this, we need to review one concept, the central limit theorem, and introduce two new concepts: the sample distribution of means and the sample distribution of proportions. In case that wasn't enough, we'll also cover confidence intervals, how to modify standardized scores for small samples, and the sampling distribution of proportions. Hold on tight!

The Sample Distribution of Means and the Central Limit Theorem

First, let's start with the **sample distribution of means**, which is easiest to explain with an example. Suppose that you wanted to identify the mean body mass index (BMI) of all Canadian adolescents. You could measure all of the roughly 4.2 million people between the ages of 10 and 19 and get an accurate measure of the mean BMI for this group, but it would be prohibitively expensive and time-consuming to do so. Instead, you can use what you know about probability sampling and the normal distribution and get similar results in a more cost-efficient way. Rather than taking measurements from 4.2 million people (the entire population of Canadian adolescents), you could choose a random subset of that group, using one of the sampling techniques outlined in the previous chapter.

When choosing the subset, you will need to consider a couple of things:

- How many people do you need to measure to get an accurate depiction of the population?
- How many people can you afford to measure?

A bigger sample size will give a more accurate approximation of the true population value, but a sample that is too big will give you too many people to measure, which might be nearly as impractical as measuring the entire population, thereby defeating the purpose of sampling.

Hypothetically, you could take a series of samples and plot the mean of each sample to reveal the sampling distribution of means. Repeatedly resampling from the adolescent population and calculating the average BMI from each sample will provide you with a list of means, like the coin-tossing example in Chapter 4. (In the coin-tossing example you are producing a sample distribution of proportions, which we'll discuss after we discuss means.) On their own, each of these sample means may or may not be close to the population mean for BMI, but if you were to present the means as a histogram, you would get—you guessed it—an approximation of the normal curve, centred around the true population mean for BMI. The most noteworthy difference in this instance is that cases represent sample means instead of individual people.

It is the **central limit theorem** that allows us to assume that any sample statistic, such as the mean, that we generate from a known population will lie along a normal distribution. To continue with our example above, if we were to repeatedly draw a 5 per cent random sample from our 4.2 million 10- to 19-year-olds, and if we then plotted the mean BMI that we generate from each of our samples, the central limit theorem states that we'd have a distribution that is "normal enough" that we can consider it to be normal. With this information, we can use many of the equations that we've covered in the text so far.

The sampling distribution of means has two other notable qualities: First, the mean of the distribution (the mean of the sample means) will be equal to the population mean μ. So a mean BMI calculated using a series of population subset means (for example, 5 per cent random samples taken 1000 times) will be close to equal to the mean BMI for the whole population. The mean of means will be very close to the population mean.

Second, the distribution of means will be quite tightly clustered around the true population mean (and will follow a normal distribution). As long as the sample size is large enough, the variance and standard deviation of the sampling distribution of means is small so we can be fairly confident that the mean of any random sample will be very close to the true population mean. As a result, instead of taking numerous samples to approximate the population, this characteristic of the sampling distribution of means ensures that any one random sample mean should be close enough to the population mean. That said, the symbols we use to denote means, variance, and standard deviations differ between samples and population (Box 9.1).

In real life, we only ever take one sample. We do not take multiple samples, and we do not plot the means of the samples. We have only one sample, but we use what we know about the *hypothetical* distribution of sample means to generalize from that sample to the population.

BOX 9.1

Relevant Symbols for Describing Sample Characteristics

Most of the equations we have examined so far in this book relate to populations. Luckily, moving from populations to samples usually doesn't result in substantial changes to any equations. But some of the symbols, listed at right, are different:

	Sample	Population
Mean	$\bar{x}$	μ
Standard deviation	s	σ
Variance	s^2	σ^2

Since it's unusual to take more than one sample, it's unlikely that we will know much about the sampling distribution of means for our samples. (We don't know the mean of means, or the standard deviation of means.) If we assume that our population is normally distributed, we can estimate how closely the mean of our sample approximates the population mean by using the following equation:

$$\boldsymbol{\sigma}_{\bar{x}} = \frac{\boldsymbol{\sigma}_X}{\sqrt{\boldsymbol{n}}}$$

Where:

σ_X is the population standard deviation
***n* is the total number of cases (the sample size)**

This equation gives us the **standard error of the mean** ($\sigma_{\bar{x}}$) when the population standard deviation (σ) is known. The standard error is described as the standard deviation of the population divided by the square root of the sample size. As you can see by the denominator $\sqrt{n}$, the larger the sample size is, the smaller the standard error. This makes sense because we expect the mean to become more accurate as the sample size approaches the population. As sample size increases, the standard error should approach—but, until the sample and population are one and the same, never reach—zero. The standard error of the sample mean is also the standard deviation of the sampling distribution of means.

In reality, we will rarely have a sample that is close in size to our population (the 2006 census public-use sample, for example, is only 2.7 per cent of the population), so we need to attach a measure of how confident we are in our measurement of the mean. This is known as the **confidence interval**.

Confidence Intervals

Because we know that the mean we draw from a sample will almost never be exactly the same as the mean of the population we're interested in, we need to acknowledge that there is a chance of error, the standard error of the mean, in the estimate. Think of the standard error as

EVERYDAY STATISTICS

Confidence Intervals and Opinion Polls

In Canada, opinion polls are often conducted to gain an understanding of the opinion of the public on a specific issue, using a sample of a population. Since the polls are based on a sample of a population, the results are subject to a sampling error. The margin of error in opinion polls is usually represented as a confidence interval.

Q: Why do you think confidence intervals are used to represent the sampling error in opinion polls?

a type of standard deviation, except that instead of measuring how far an observation is from a mean, it refers to the distance that a sample mean is from a population mean (and remember from above that it represents the standard deviation of the sampling distribution of means).

To indicate level of confidence, we need to employ the standard error. Because the standard error is derived from the standard deviation, it can be used in almost the same way as the standard deviation. Instead of determining what proportion of observations lie between one, two, and three deviations from the mean, we are determining how much confidence we have in the accuracy of the mean taken from the sample.

Think of the sample mean as one of many in the sampling distribution of means. Each mean can be treated as an observation. Just as we can be certain that 95 per cent of all observations lie between ± 1.96 standard deviations from the mean in a normal distribution, we can also be certain that roughly 95 per cent of sample means lie between ± 1.96 standard errors from the population mean. Extending this, we can construct a "confidence interval" by providing the upper and lower ranges (or **confidence limits**) of the standard error calculations. We can also construct a 99 per cent confidence interval by postulating that our population mean is within ± 2.58 standard error units of the sample mean. This translates to $z_{critical}$ values of 1.96 and 2.58, respectively, from the sample mean. To understand where we got 1.96 and 2.58 from, we need to take another look at the *z*-distribution (see Box 9.2).

Looking at the equation for the standard error, notice that the standard error will vary, as it is based on both sample size and the size of the standard deviation. If the population standard deviation is large, the standard error is also likely to be large. Calculating confidence intervals requires multiplying the standard error by the appropriate *z*-value (1.96 for a 95 per cent confidence interval, and 2.58 for a 99 per cent confidence interval).

The *t*-Distribution

William Gosset was a mathematician and chemist who graduated from Oxford in 1899. Arthur Guinness Son & Co. Ltd. was looking for new ways to make a beer of consistent quality and decided that Gosset was the person to ask for help. There was considerable variability

BOX 9.2

The Steps: Estimating a Population Mean

1. Calculate the sample mean.
2. Assuming that the population standard deviation (σ_X) is known, calculate the standard error of the sample mean by using the following equation:

$$\sigma_{\bar{x}} = \frac{\sigma_X}{\sqrt{n}}$$

Where:

σ_X is the population standard deviation
n is the sample size

If you do not know the population standard deviation, you'll need to rely on the sample standard deviation and use this equation:

$$s_{\bar{x}} = \frac{s_X}{\sqrt{n}}$$

Where:

s_X is the sample standard deviation
n is the sample size

3. Find the relevant value of $z_{critical}$ in the *z*-table in column A of Appendix A that corresponds with a 95 per cent confidence interval (1.96 for 95 per cent, 2.58 for 99 per cent in a two-tailed test).
4. Insert the relevant values into one of the following equations.

If you know the population standard deviation:

$$\textbf{confidence interval} = \bar{x} \pm (z_{critical} * \sigma_{\bar{x}})$$

Where:

$\bar{x}$ is the sample mean
$z_{critical}$ is found in the z-table (Appendix A)
$\sigma_{\bar{x}}$ is the population standard error

Or, if you do *not* know the population standard deviation:

$$\textbf{confidence interval} = \bar{x} \pm (z_{critical} * s_{\bar{x}})$$

Where:

$\bar{x}$ is the sample mean
$z_{critical}$ is found in the *z*-table (Appendix A)
$s_{\bar{x}}$ is the sample standard error

Note that it will be necessary to do the calculation for both the upper and lower bounds, which means that you will need to solve the equation for a positive and negative value of $z_{critical}$.

in brewing quality across batches, making it difficult for Guinness to establish regularity in taste and standards. Guinness wanted to improve its consistency but didn't know how and had neither the budget nor the inclination to botch large batches of brew to attain consistency. Gosset was thus limited to working with a few small batches. At that time, most statistical work focused on very large samples, so Gosset had to forgo traditional methods and develop techniques for assessing small samples.

The histograms in Chapter 5 showed that small samples tend to produce distributions that deviate from the normal distribution. Most important, the tails in small samples are larger, and the kurtosis value is often lower, even though the variable might have a normal distribution. Gosset's problem—and, often, our problem—was that the sample size was too small to accommodate the normal curve, causing him to underestimate the standard error. This led Gosset to the **Student's *t*-distribution.** (See Box 9.3 for an explanation of why this name exists.)

BOX 9.3

Why Is It Called the Student's *t*-Distribution?

Before Gosset's arrival at the Guinness Brewery, a Guinness employee had published a paper revealing some of the company's brewing secrets. As a result, Guinness heavy-handedly forbade all of its employees from publishing articles of any kind!

When Gosset developed the *t*-distribution as an employee of Guinness, then, he couldn't share his discovery with the world (at least, not unless he wanted to lose his job), and selected the pseudonym "Student" to protect his anonymity. Since that time, the distribution that he discovered has been known as Student's *t*-distribution.

Before Gosset, statisticians knew that their standard error estimates were slightly too small, but they surmised (correctly) that the difference would be very slight in samples that were greater than 120. Gosset determined the exact relationship between small samples and the normal curve with his discovery of the *t*-distribution.

The *t*-distribution is actually an infinite number of curves, one for every sample size greater than or equal to two. As sample size increases, the *t*-distribution increasingly resembles the standard normal distribution. By the time sample size reaches 120, the differences between the two are difficult to detect.

Values for the *t*-distribution can be derived by using the following equation (when mean = 0 and the variance is greater than one):

$$t = \frac{\bar{x} - \mu}{s_{\bar{x}}}$$

or

$$t = \frac{\bar{x} - \mu}{\frac{s_X}{\sqrt{n}}}$$

or

$$t = \frac{\bar{x} - \mu}{\sigma_{\bar{x}}}$$

or

$$t = \frac{\bar{x} - \mu}{\frac{\sigma_X}{\sqrt{n}}}$$

Where, in all of the above,

- $\bar{x}$ **is the sample mean;**
- s_X **is the sample standard deviation;**
- $s_{\bar{x}}$ **is the estimated standard error of the sample mean;**
- μ **is the population mean;**
- σ_x **is the population standard deviation;**
- $\sigma_{\bar{x}}$ **is the standard error of a population mean**
- N **= total population size; and**
- n **= total sample size.**

The last two equations, though technically correct, are unlikely to be very helpful because they are used to calculate *t* when population characteristics are known, even though *t* is used for small samples.

In the numerator for each equation, we subtract the population mean from the sample mean and divide that number by the denominator, which is the standard error estimate of the sample mean. The second equation is the same, except that the denominator is expressed as the standard deviation of the sample, divided by the square root of the sample size minus one, instead of the standard error. If you recall the earlier equation for standard error, you will recognize that the four equations produce essentially the same result, and which we use depends on the information at our disposal.

What Is a Degree of Freedom?

When we are using the *t*-distribution, we need an understanding of **degrees of freedom**. Imagine that you live with five roommates and split the bills according to the size of your bedrooms. This month, the total is \$500, and your roommates claim \$435 of the bill. You don't need to know what each of the others owes to know that your share is \$65 because what you owe is constrained by the amount that your roommates owe. Any combination of values could be assigned to what your roommates owe. As long as the sum equals \$435, the amount that you owe does not change. You have the freedom to assign values (within reason) to everyone in the sample *except* the last observation. This is the main principle behind degrees of freedom. In this instance, there are five minus one, or four, degrees of freedom.

Typically, when working with the mean, the degrees of freedom (*df*) are equal to the number of observations in a sample, minus one ($n - 1$).

$$df = n - 1$$

Using the *t*-statistic, instead of *z*, is suitable when sample sizes are small (under 120) and you want to construct confidence intervals. When constructing intervals, knowing *t* and the degrees of freedom allows you to account for the slight differences between the distribution of a variable with small *n*, and a distribution taken from a larger sample. When should you use *t* instead of *z*? Look at the values in Appendix B (the *t*-table) and notice that *t*-values converge upon those for z as degrees of freedom increase. At $df = 120$ they are identical; t should only be used when the sample size is less than 120.

One-Tailed versus Two-Tailed Estimates

Looking at Appendix B, you'll notice a distinction between one- and two-tailed tests. To correctly estimate the confidence intervals for the population mean, you'll have to know which chart to use.

One-tailed tests determine how likely it is that an observation is above or below a specific threshold value. For instance, if we used sample data to find the probability of someone living below the poverty line, we would use a one-tailed test. Because we have a hypothesis about the direction of the relationship (that is, the person is expected to be toward the bottom of the income distribution), we can focus our attention on one end of the distribution.

To estimate a population mean from a sample mean, a **two-tailed test** is used. If the sample is a representative sample, there is no reason to hypothesize that the value of the sample mean should be above or below the population mean. If the sample is random, there is an equal likelihood of it being above or below the population mean. There is no way to specify which direction we expect the difference to fall in, so both ends of the distribution have to be included. Two-tailed tests are more common, so they are the only kind shown in the z-table in Appendix A. If you need to conduct a one-tailed test using z, use the t-table with $df = \infty$, which gives you a critical value of 1.645.

The Sample Distribution of Proportions

Much as we are able to make sense of sample means by relating them to the normal curve, we are also able to make sense of proportions. Thanks to the central limit theorem, the **sample distribution of proportions** will also increasingly resemble a normal distribution as the number of trials increases.

To keep things simple, we'll look at proportions with only two categories here. If you were interested in proportions with more than two categories, however (such as voter distribution among four or more political parties in a national election), the logic would need only to be straightforwardly extended. Imagine that you wanted to identify the proportion of the population of Quebec that supported the 2012 student strikes. As you would expect, there is a "real" answer out there, and the only way to find this out would be to ask every single resident of Quebec at one point in time—unlikely to happen. A second possibility would be to take several samples from the population and use the sample proportions to estimate the proportion within the population. Although more likely (you could choose to take random samples of people within certain communities), this would still be a lot of work. Furthermore, since the central limit theorem tells us that one well-chosen sample of sufficient size is all we really need, the easiest and most likely choice we'd make is to take just one sample from the whole population of Quebec. In the section below, we will discuss the techniques for doing this.

Using Degrees of Freedom and the *t*-Distribution to Estimate Population Proportions

Ideally, we would calculate population proportion using data from the population, but rarely do we know population characteristics. To estimate population proportions, a measurement commonly used for opinion polls, we use a slightly different equation for the standard error, where p is the sample proportion and n is the number of cases in the sample, and s_p is the estimated standard error of the sample proportion:

$$s_p = \sqrt{\frac{p(1-p)}{n-1}}$$

The difference is in the numerator, where the proportion in a certain category is multiplied by 1 minus that proportion. The denominator is $n-1$ because the equation deals with a sample.

Once the standard error of the sample proportion has been calculated, the confidence interval is calculated as follows:

$$CI = p \pm (z_{\text{critical}} * s_p)$$

Suppose, for example, that you wanted to gauge Canadian Aboriginal opinion about the 2006 protests in Caledonia, in southwestern Ontario, over land issues. You polled 500 Aboriginal people (composed of First Nations, Métis, and Inuit peoples). In your sample, you found that 62 per cent of all respondents believed that the protests were warranted. As interesting as this percentage is, what you really want to know is the attitude of the *entire* Aboriginal population of Canada, not just the sample. Since there are roughly one million Aboriginal people living in Canada, any claim you make from only 500 respondents could be suspect. Researchers try to put these kinds of doubts to rest by reporting confidence intervals. Using the information we have, let's construct a confidence interval.

First, we need to estimate the standard error of the sample proportion, using the following equation:

$$\begin{aligned} s_p &= \sqrt{\frac{p(1-p)}{n-1}} \\ &= \sqrt{\frac{0.62(1-0.62)}{499}} \\ &= \sqrt{\frac{0.236}{499}} \\ &= 0.022 \end{aligned}$$

To find the 95 per cent confidence interval for the population (also see Box 9.4), we need to know the margin of error (z times the standard error). To find it, we need to find the value of 1.96 from the z chart. Why 1.96? Because it is the cut-point (in standard deviations) within

BOX 9.4

The Steps: How to Estimate Population Proportions by Using Only Sample Characteristics

1. Estimate the standard error of the sample mean:

$$s_p = \sqrt{\frac{p(1-p)}{n-1}}$$

2. Use $df = n - 1$ to calculate the degrees of freedom to find the **critical value of** z, or t, to find the desired confidence interval:

$$\textbf{confidence interval} = p \pm (z_{\text{critical}} * s_p)$$

3. It will be necessary to do this for both the upper and lower bounds (which explains why ± appears in the equation), meaning that you will have to solve the equation for a positive and negative value of z_{critical}.

which 95 per cent of all cases lie. We calculate the degrees of freedom to be n minus 1, or 499. To be 95 per cent confident in our estimate, we need to provide the range within which we are 95 per cent confident that the population value lies. If the sample were smaller, we would need to look up the value of t by using the degrees of freedom instead of z.

Second, multiply the standard error by 1.96 to find the margin of error:

$$\begin{aligned} \textbf{margin of error} &= \mathbf{1.96} * s_p \\ &= \mathbf{1.96 * 0.022} \\ &= \mathbf{0.043} \end{aligned}$$

Third, add and subtract the margin of error from the sample proportion to find the values of the confidence interval:

$$\begin{aligned} \textbf{95\% confidence interval} &= p \pm 1.96 s_p \\ &= \mathbf{0.62 \pm 0.043} \\ &= \mathbf{0.577 \text{ to } 0.663} \end{aligned}$$

Now we can say that we are 95 per cent confident that between 57.7 per cent and 66.3 per cent of all Aboriginal peoples in Canada believe that the protests in Caledon were warranted.

The Binomial Distribution

Technically speaking, the distribution of sample proportions is not a normal distribution. It is instead called the binomial distribution, and it differs from a normal distribution in that there are only two possible outcomes, the sum of which is always one. Furthermore, rather than modelling the dichotomy in a normal distribution, we are modelling the *probability* of the occurrence of an outcome in a binomial distribution. The binomial distribution is beyond the discussion of this text.

Conclusion

As long as we are using samples, we *never* know conclusively what the values of our variables are in the population. The best we can do is provide a confidence interval and state how confident we are that the true population value lies within that interval.

This concludes our discussion of univariate statistics. The coming chapters will explore methods of analysis using more than one variable. Most of what's been covered so far is foundational, but without that knowledge, performing any statistical analysis is nearly impossible. Now the focus will shift to analysis with more than one variable (bivariate and multivariate statistics) and how to measure and hypothesize about relationships between variables.

Glossary Terms

Central limit theorem (p. 82)
Confidence interval (p. 83)
Confidence limits (p. 84)
Critical value of *z* (p. 90)
Degrees of freedom (p. 87)
One-tailed test (p. 88)
Sample distribution of means (p. 81)
Sample distribution of proportions (p. 88)
Standard error of the mean (p. 83)
Student's *t*-distribution (p. 85)
Two-tailed test (p. 88)

Practice Questions

1. A research study was conducted, examining the differences between the perceived life satisfaction of men and women. Ten men and ten women were given a life satisfaction test. Scores on the measure range from 0 to 60, with high scores indicating high life satisfaction and low scores implying the opposite. The data are presented below:

Men	Women
45	34
38	22
52	15
48	27
25	37
39	41
51	24
46	19
55	26
46	36

a. Assuming that this is a sample, calculate the mean, variance, and standard deviation for both men and women.

b. Calculate the standard errors for each mean estimate.

c. Estimate the 95 per cent confidence interval for each sample mean.

2. Professor Smart recently returned a lab assignment to his students. Before doing this, he polled *all of them* on the number of hours they spent on the assignments. There were 24 individuals in the lab, and the data were used to make inferences about subsequent classes. The data are presented below:

4.5	20.0	19.0	9.0
22.0	8.0	7.0	8.5
7.5	2.5	14.5	3.5
9.0	5.0	9.0	8.0
11.0	10.5	9.0	18.0
7.5	15.0	14.0	20.0

Compute and interpret the 95 per cent confidence interval. What does the interval mean?

3. In a sample of 50 individuals, 34 per cent prefer soft drink A to soft drink B. In the population, we could be 99 per cent confident that the real proportion lies between _____ per cent and _____ per cent.

4. What do you think happens to the standard error of a sample mean as the number of observations increases? What about confidence intervals?

5. You are taking a class with 12 other students, and you recently wrote an exam that yields an average score of exactly 78 per cent. When returning your exams, your professor informs you that she is unable to find your exam (all other students receive theirs).

75	44	89	92
92	55	66	86
97	76	83	88

What is your exam score?

6. Suppose that you have 10 pet cats and that you know their average weight is four kilograms (all 10 cats weigh 40 kilograms), with a standard deviation of one kilogram. Recently, two of your cats escaped. Determined to find them, you decide to put posters around your neighbourhood. You record the weights of your remaining eight cats, calculate the standard deviation, and find it to be 0.8. Thinking about the weight of your cats in terms of degrees of freedom, would you be able to identify the weight of the remaining lost cat if you located and weighed the ninth one?

7. Suppose that in a sample of 121 students you find that 40 per cent of students in York Lanes purchase at least one food item per day. What is the 95 per cent confidence interval of the amount in the population?

8. Your friend Clara is very proud of her dog, and over the years she has compared her dog to yours. If you were the type of person who used statistics to assess her various claims, would you use a one-tailed or two-tailed assessment to verify the following statements?

    a. "My dog is bigger than your dog."

    b. "My dog is different from your dog."

    c. "My dog has more hair than your dog."

    d. "My dog wouldn't act like your dog does in public."

    e. "My dog weighs less than your dog."

9. Throughout most of 2011, software designer Rovio sold an average of 40,000 copies of its popular game *Angry Birds* every day, with a standard deviation of 1000 copies. Now, on a random day in early 2012, Rovio has sold 38,000 copies. Can you be 95 per cent confident that this was a regular day?

10. How many texts does the average teen send in a day? Rogers Communications Inc. has asked you to answer this question. Rogers provides you with a random (and anonymous) sample of its client base, yielding the following numbers:

47	11	66	94	77
99	50	10	57	2
100	13	37	93	37
97	79	81	36	78

What range could you give them with a 95 per cent confidence interval?

Answers to the practice questions for Chapter 9 can be found in Appendix H.

PART II

Bivariate Statistics

10 Testing Hypotheses: Comparing Large and Small Samples to a Known Population

LEARNING OBJECTIVES

This chapter will introduce ways to build and test hypotheses. Topics will include:

- null and research hypotheses;
- hypothesis testing with a large sample and a known population; and
- hypothesis testing with a small sample and a known population.

Introduction

So far in this book, we've discussed the gap between sample means and population means and sample proportions and population proportions in terms of sampling error. That is, we've assumed that the only source of discrepancy between a mean calculated from a sample and a mean calculated from a population or a proportion calculated from a sample and a proportion calculated from a population is that the sample does not perfectly represent the population. However, there's another reason to compare the mean or proportion of a sample to a population: hypothesis testing. Here, rather than assume that the difference between the mean of the sample and the mean of the population or the proportion of the sample and the proportion of the population is only due to sampling error, we investigate whether the difference could stem from some other, systematic factors.

In order to make these comparisons, we must have a known population value to which we can compare the sample. Following are examples of the types of questions that can be asked by comparing a sample mean to a known population mean:

- Do test scores from McGill University differ from the national or provincial mean?
- Was the summer of 2011 significantly warmer than average?
- Do business students have a higher mean GPA than other students at a specific university?

Similarly, following are examples of the types of question that can be asked by comparing a sample proportion to a known population proportion:

- Does the percentage of Albertans voting for the NDP in a federal election differ from the overall percentage of Canadians who vote for the NDP?
- Are medical students more or less likely than the entire student body to support a smoking ban on campus?

If we find a difference, some of that difference may still stem from sampling error, but the other factors affecting that difference can now become the matter under investigation. Up to this point we have been thinking of the gap between a population and a sample as sampling error, as something we wanted to get rid of. Now, the gap is what we use to identify whether a sample mean differs from a known population mean in a *systematic* way, or whether a sample proportion differs from a known population proportion in a systematic way.

For now, let's assume we have a sample that is sufficiently large (>120 observations) and that we're interested in making comparisons between that sample and a known population. We must assume that we know the population mean and standard deviation (this is a big assumption, as will be discussed later). Because of the large sample, we get to use the *z*-distribution, rather than the *t*-distribution. We'll look at how to handle comparing a mean from a small sample and a known population by using the *t*-distribution afterwards. But first, let's look at hypotheses.

What's a Hypothesis?

You've probably heard people talk about hypotheses before. Perhaps you've even learned about them in your research methods course, but in case you haven't, a brief description might help.

In essence, a hypothesis is a tentative statement about the relationship between two variables (or potentially more, but we are only looking at two here). A hypothesis is a specific, testable prediction about a relationship that you expect to emerge from your analysis. For example, a study designed to look at the relationship between studying for an exam and grades on the exam might have this as a hypothesis: "People who study for statistics exams will have significantly higher average scores on the exam than the overall average score on the statistics exams."

A hypothesis like the one above (called a **research** or **alternative hypothesis** and denoted by H_a or H_1, H_2, H_3, etc.) is always accompanied by a **null hypothesis.** When casting hypotheses, it's important to remember that the research hypothesis and the null hypothesis must be *mutually exclusive and exhaustive.* A null hypothesis (usually denoted by H_0) is typically a statement of no relationship, that is, the sample mean (the mean exam grade for those who study) will not be statistically significantly different from a population mean (the mean exam grade for everyone who took the exam). There are also instances where a null hypothesis indicates a direction, but these are less common. For example, the research hypothesis about the relationship between studying and statistics grades points to a statistically significantly higher mean grade for those who study. The null hypothesis would be that those who study have mean grades equal to or less than the overall mean grade. The null hypothesis is driven by the research question you are asking.

Typically, a null hypothesis will be directional when the research hypothesis is. So, if our research hypothesis is that the mean for those who study will surpass the overall mean ($\bar{x} > \mu$), we are implying a direction to the relationship. (Recall that $\bar{x}$ refers to a sample mean and μ is the population mean.) Our null hypothesis should also reflect that directionality. For this, we might formulate a null hypothesis that those who study will have a mean that is equal to or less than those who don't study ($\bar{x} \leq \mu$). If we cared only about difference, not directionality, our research hypothesis would need only to state the existence of difference between the sample mean (those who study) and the overall mean ($\bar{x} \neq \mu$), and the null hypothesis would assert no significant difference between the two ($\bar{x} = \mu$). As you can see, these are very different sets of hypotheses, geared to answer different questions. Which of these you choose depends on the information you're interested in obtaining.

Statistically, as you might have gleaned from the discussion above, choosing between null and research hypotheses is often done by comparing a sample mean to a population mean with the help of the normal distribution. Failing to find a statistically significant gap between the sample mean and the population mean implies that there isn't sufficient cause to believe that a sample is statistically significantly different from a population. But if you do find a gap, it means one of two things:

1. A gap exists, and the sample of interest differs from the population; or
2. A gap exists, and it is due to sampling error.

It is impossible to know with 100 per cent certainty which of the two possibilities is true. That is why we must settle for a confidence interval. Similarly, if no gap is found, it could be because there actually isn't a gap or because there's a sampling error. This uncertainty is why we *never* say in the social sciences that we've proven or disproven the null hypothesis. Instead, the best we can say is that we have rejected or failed to reject the null hypothesis. Notice that we are discussing the null, rather than research, hypothesis; out of convention, we typically discuss the null hypothesis when we are rejecting or failing to reject hypotheses. Research hypotheses often don't get much mention.

What this means is that in the social sciences—indeed, in all sciences—hypothesis testing carries with it the possibility of making the wrong decision about rejecting or failing to reject the null hypothesis. If we use a 95 per cent confidence interval, there's a 5 per cent (or 1 in 20) chance that we've made the wrong choice. If we choose to reject the null hypothesis when it is actually true, we've made a **type one error**; if we fail to reject the null hypothesis when it is not true, we've made a **type two error**.

Which of the two error types do you think is more serious? Well, that depends on what you are doing. If, for example, you are looking at the effect of an experimental drug on a serious illness, failing to find an effect when in fact there is one—which would be a type two error—can result in a lost opportunity to develop a potentially life-saving drug. On the other hand, type one errors can be equally problematic because they would lead you to conclude that a treatment or intervention has an effect on an outcome when in fact it doesn't. Can you imagine committing a type one error when you are studying the effect of prison sentences as a deterrent for certain types of crime? The implications of a mistake here can also be serious, underscoring the importance of doing everything possible to minimize both types of errors.

One-Tailed and Two-Tailed Hypothesis Tests

The type of difference you are looking for between your sample and your population will guide you when choosing between a **one-tailed** or **two-tailed test**. If your goal is simply to identify difference without direction (Are Maritimers different from other Canadians in terms of their levels of friendliness?), then you are interested in a two-tailed test. If you believe that there is a direction to the difference (Are Maritimers friendlier than other Canadians?), then you're interested in a one-tailed test. In other words, if you are hypothesizing directionality, then you're likely looking at a one-tailed test; if you're not interested in directionality, then it's probably a two-tailed test that you're after.

Appendix A contains both sets of values of interest. Different disciplines have different conventions, but generally, in the social sciences, we attribute any difference beyond the 95 per cent confidence interval to be statistically significant. In other words, if a sample mean is more than 1.96 standard deviations from the population mean, then we can say with confidence that the difference likely exists in the real world instead of just in the data. As this relates to your null and research hypothesis, you would *reject* the null hypothesis.

Let's further our understanding of hypothesis testing with a hypothetical example. Let's imagine that the federal government conducts tests of the reading ability of all 15-year-olds in Canada every year. Each time, students are shown different kinds of written text, ranging from prose to lists, graphs, and diagrams. The students are set a series of tasks, requiring them to retrieve specific information, to interpret the text, and to reflect on and evaluate what they have read. These texts are from a variety of reading situations, including reading for private use, occupational purposes, education, and public use. Overall, Canadian youth achieved a mean score of 493 ($\sigma = 93$). The Canadian data were calculated from 427,740 students (the population of 15-year-olds in Canada in 2014). Now let's imagine that the province of Alberta wants to know how it compares to the country overall. The mean score in Alberta was 524 ($s = 90$) and the number of students who took the test in Alberta was 246,151. Are Albertan students significantly different from the Canadian mean?

To answer this question, we first need to generate a null (H_0) hypothesis and a research hypothesis (H_1). The hypotheses can be stated in words:

H_0: There is no significant difference between the mean reading scores of Albertan 15-year-olds and the mean reading scores of 15-year-olds in Canada.

H_1: Albertan 15-year-olds' mean reading scores differ significantly from the mean reading scores of 15-year-olds in Canada.

Or, we can articulate our hypotheses with numbers:

$$H_0: \bar{x} = 493$$
$$H_1: \bar{x} \neq 493$$

In this example, we have an abundance of information to compare Albertan 15-year-olds to those in Canada:

- sample size for Alberta and population size for Canada;
- standard deviation for Alberta and Canada; and
- mean for Alberta and for Canada.

We are lucky to have this much information because it allows us to compare the Albertan sample to the Canadian population quite easily. (Remember this is a hypothetical example.) Table 10.1 shows the symbols that we can attach to each number.

TABLE 10.1 | Hypothetical Example of an Albertan Sample of the Canadian Population

The Albertan Sample	The Canadian Population
$n = 246{,}151$	$N = 427{,}740$
$\bar{x} = 524$	$\mu = 493$
$s = 90$	$\sigma = 93$

Our sample of Albertans contains 246,151 people. Since we are dealing with a sample size that exceeds 120, we can use the *z*-statistic.

Recall that the equation for the *z*-statistic is:

$$z = \frac{\bar{x} - \mu}{\sigma_{\bar{x}}}$$

Where:

z = the *z*-score
$\bar{x}$ = the sample mean of individual scores X
μ = the population mean
$\sigma_{\bar{x}}$ = the population standard error

Recall also that to get the standard error of the sample, when you know the population standard deviation, you need to use the following equation:

$$\sigma_{\bar{x}} = \frac{\sigma_x}{\sqrt{n}}$$

This gives us:

$$\sigma_{\bar{x}} = \frac{93}{\sqrt{246{,}151}} = \frac{93}{496} = 0.188$$

Returning to our main equation, we get:

$$z = \frac{524 - 493}{0.188} = 165$$

We're fortunate to have the population standard deviation. It's more common to have only the s_x, the sample standard deviation, which means that we would have to estimate the standard error of the sample while building in some extra uncertainty for using s_x instead of s_x. To do this, we divide the sample standard deviation by the square root of $n - 1$ instead of by the square root of n.

For comparison purposes, let's quickly calculate t:

$$t = \frac{\bar{x} - \mu}{\frac{s_x}{\sqrt{n}}} = \frac{524 - 493}{\frac{90}{\sqrt{246{,}151}}} = \frac{31}{0.181} = 171$$

Where:

t = the t-score
$\bar{x}$ = the mean of individual scores X
μ = the population mean
s_x = the sample standard deviation

Note how close the t-value of 171 is to the z-value of 165. This is because our sample size is so large. As mentioned in Chapter 9, it is only necessary to calculate the t-statistic when the sample size is less than 120. Looking at Appendix A, we see that the $z_{critical}$ for a 95 per cent confidence interval is 1.96, whereas $t_{critical}$ from Appendix B when $df = \infty$ is, you guessed it, 1.96. Our $t_{observed}$ and $z_{observed}$ values both greatly exceed this figure, so we can be 95 per cent confident that Alberta's reading ability scores are significantly different from the Canadian mean. This means that we reject H_0, the null hypothesis.

In this example, we hypothesized the existence of a statistically significant difference between our sample and the population, without making a statement about what that difference would be. This requires a two-tailed test. If we instead hypothesized that Alberta's score is significantly *higher* than the Canadian mean, then we would conduct a one-tailed test. Our hypotheses, stated in words, would be the following:

H_0: The mean reading scores of Albertan 15-year-olds are equal to or lower than the mean reading scores of 15-year-olds in Canada.
H_1: The mean reading scores of Albertan 15-year-olds are significantly higher than the scores of 15-year-olds in Canada.

Notice that both hypotheses are changed to reflect directionality. Furthermore, we'll only use $z_{observed}$, given our sample size of 246,151.

Similarly, stated in numbers we have the following:

$$H_0: \bar{x} \leq 493$$
$$H_1: \bar{x} > 493$$

Statistically, the only difference lies in the value of $z_{critical}$. Rather than 1.96, $z_{critical}$ is now 1.65. The reason for the change is that we no longer need to determine whether or not $z_{observed}$ lies either + or – two standard deviations from the mean. We only need to see if it is beyond one tail, so the 95 per cent confidence z-statistic value is 1.65. Our $z_{observed}$ is also well beyond 1.65, so we can also reject the null hypothesis here. Looking at Appendix A, can you see why we use the value 1.65?

A 95 per cent level of confidence is a rather arbitrary level of certainty, and in certain circumstances it contains a higher level of error in judgment than we want. In this case, we could increase our level of confidence to, say, 99 per cent. For this, we would replace 1.96 and 1.65 with 2.57 and 2.33. Can you see why this is the case? In this example, do we still have a significant finding at this higher level of confidence?

The Return of Gosset: Student's *t*-Distribution

In the reading test example, we used a combination of sample and population characteristics to determine whether the differences witnessed in a sample were likely to exist in the population. This exercise is of little use in the real world, because there isn't usually that much information available. There are rarely values for either σ, the population standard deviation, or μ, the population mean. A value for s and $\bar{x}$, the sample standard deviation, will always be available, or can be calculated. Unfortunately s is a biased estimator of σ; and there's no way to tell how big the difference is.

BOX 10.1

The Steps: Hypothesis Testing with a Large Sample and a Population

1. State the null and the alternative or research hypotheses. Typically, the research hypothesis postulates that there is a significant difference between the sample mean and the population mean, whereas the null hypothesizes that there is no difference.
2. Decide whether a one-tailed or two-tailed test is more appropriate. Pick your confidence level and locate the appropriate critical z-statistic values from Appendix A.
3. Compute z_{observed} with the following equation:

$$z = \frac{\bar{x} - \mu}{\frac{\sigma_x}{\sqrt{n}}}$$

Or, if you prefer to calculate the standard error of the mean before calculating z, compute with this equation:

$$\sigma_{\bar{x}} = \frac{\sigma_x}{\sqrt{n}}$$

then

$$z = \frac{\bar{x} - \mu}{\sigma_{\bar{x}}}$$

4. Compare z_{observed} with the z_{critical} values of 1.96 (at confidence level 95 per cent, for a two-tailed test) or 1.65 (at confidence level 95 per cent, for a one-tailed test). If z_{observed} exceeds z_{critical}, then you must reject H_0; if it does not, then you fail to reject the null hypothesis.

However, once a sample size of about 120 is reached, the differences are negligible. (See for yourself: look at how minimal the differences between z-values and t-values are when $df = 120$ in the z- and t-distributions in Appendices A and B). We can, moreover, consider s to be a decent approximation of σ. Substitute the sample standard deviation for the population standard deviation, and use the techniques above to get the equation:

$$t = \frac{\bar{x} - \mu}{\frac{s_x}{\sqrt{n}}}$$

Or, since

$$s_{\bar{x}} = \frac{s_x}{\sqrt{n}}$$

then

$$t = \frac{\bar{x} - \mu}{s_{\bar{x}}}$$

This looks a lot like the equation for z, except that s replaces σ in the denominator. Like z, t is a standardized unit that tells us how far $\bar{x}$ is from μ in standard deviations. Unlike z, which has only one distribution (the normal distribution), remember that t is a family of distributions, each differing slightly based on sample size. For smaller samples (<120), the *t*-distribution can differ substantially from the *z*-distribution.

Since we need to calculate degrees of freedom to use the *t*-distribution, it would be helpful to once again briefly review the concept. Recall that the degrees of freedom are the number of values in a set of scores that are free to vary. The degrees of freedom will be equal to the number of cases minus one.

To learn about what to do in the absence of information on the population, we'll first look at an example where only the population mean is known in addition to sample characteristics, then in Chapter 11 we'll look at instances where we have information only on samples.

EVERYDAY STATISTICS

Using the *t*-Distribution

In health research, the *t*-distribution is often used to determine whether there is a difference between the birth weights of babies who are born to women who live in poverty and those born to women in the overall population.

Q: Why do you think the *t*-distribution would be useful for comparing the birth weights of babies in a sample of the population?

Hypothesis Testing with One Small Sample and a Population

As we saw in Chapter 7, working with means and standard deviations derived from small samples (<120) requires an additional measure of uncertainty since it is likely that both the $\overline{x}$ and s parameters contain some error.

This part of Chapter 10 will primarily focus on the ***t*-test** to test hypotheses with information from small samples.

The solution requires you to assume that the sample is fairly (but not perfectly) representative of the population and build the uncertainty into our estimates. We acknowledge that our sample is not a perfect representation of our population, and that there's a possibility that any differences we observe are due to chance. However, if there are clear patterns in the sample, they probably exist in the population. To present the differences between sample and population, we acknowledge the uncertainty by reporting the values (e.g., means) for the sample as a confidence interval.

Let's try using the following for a sample mean. Engineering students at Queen's University have been accused of consuming too much alcohol. Their mean number of drinks per week is 12, with a standard deviation of 4.6 (taken from a random sample of 100 engineers), compared to 8 mean number of drinks per week for the entire university population. Obviously, there is a difference between the sample and the population, but is it statistically significant? There are two possibilities:

1. The mean number of drinks per week for the engineers is the same as the university mean (8), and the difference seen in this sample is caused by random chance or sampling error (H_0).
2. The difference is real (statistically significant), and engineers drink more than other students (H_1).

What we've done here is cast the possibilities as a set of competing hypotheses. The first (labelled H_0) states that there is no relationship between being an engineer and alcohol consumption. This is known as the null hypothesis because it predicts that the relationship between two variables is null. The second (labelled H_1) is our research hypothesis, and it asserts that a connection does indeed exist between the two variables. Also note that we are hypothesizing a directional relationship (engineers generally outdrink other students).

Although it would be ideal to be able to prove that the research hypothesis is true, remember that in scientific research the best that we can do is reject or fail to reject the null hypothesis. Remember from the discussion in Chapter 1 that when we have cross-sectional data (data collected at one point in time) and when we are only looking at two variables, we can never be sure whether our relationships are actually causal or if they are instead "merely" correlative. Imagine that males tend to drink more than females and that males are more likely to be engineers. The differences above would be significant, but it wouldn't be because engineers outdrink others. It would instead be because males outdrink females. By rejecting the null hypothesis instead of accepting the research hypothesis, we give ourselves room for such possibilities.

To choose between these possibilities, identify the probability of getting a sample mean of 12 for the engineers with a sample size of 100 when the engineers' true mean is 8. The convention dictates that if we are less than 5 per cent likely to get a sample mean of 12 when the true mean is 8, the sample mean represents an actual difference.

If we treat all the students at the entire university as our population, and engineering students as our sample, we get values of 8 for μ, 12 for $\bar{x}$, 100 for n, and 4.6 for s_x. To calculate t, we use the following equation:

$$
\begin{aligned}
t &= \frac{\bar{x} - \mu}{\frac{s_x}{\sqrt{n}}} \\
&= \frac{12 - 8}{\frac{4.6}{\sqrt{100}}} \\
&= \frac{4}{0.46} \\
&= 8.7
\end{aligned}
$$

The next step is to figure out how big the difference between the mean of the sample and the mean of the population is, using what we know about the normal distribution—particularly the **central limit theorem**. The normal distribution also tells us the probability of finding differences between samples and populations. We want to know if the probability of finding a difference of four drinks per week is less than 0.05 (or 5 per cent).

The equation places the difference between the population mean and sample mean into a common metric (that's why the standard deviation appears in the denominator), so that the value can be assessed with the normal distribution, like any other variable. If we multiply the population mean, sample mean, and standard deviation by 100, 1000, or even 100,000, we could use the same table (*z* or *t*) to assess the significance of the differences. It's often useful to think of *z*- or *t*-statistics as techniques for translating values measured in different units—be they drinks, dollars, or donkey rides—into a common unit of measurement, or standard score, as discussed in Chapter 7.

One characteristic of the equation is that there is a term in the denominator $\left(\sqrt{n-1}\right)$ that "penalizes" t if it's calculated using a small sample. The reason is that sampling error tends to *decrease* as sample size *increases* (and vice versa).

Look at Appendix B to get the critical t-value of 1.66 ($df = 99$ for a one-tailed test), which our observed value of 8.7 greatly exceeds. This tells us that there is less than a 5 per cent chance (actually less than 0.1 per cent chance) that the value we obtained from the sample of engineers is a fluke. Since we set a threshold of 5 per cent, we can be confident (though not completely certain) in rejecting the null hypothesis (that there is no overall differences

BOX 10.2

The Steps: Hypothesis Testing with a Small Sample and a Population

1. State the null and the alternative or research hypotheses. Typically, the research hypothesis postulates that there is a statistically significant difference between the sample mean and the population mean, whereas the null hypothesizes that there is no relationship.
2. Decide whether a one-tailed or two-tailed test is more appropriate, pick your confidence level, state your degrees of freedom ($n - 1$), and locate the appropriate t-statistic values from Appendix B.
3. Compute $t_{observed}$ with the following equation:

$$t = \frac{\bar{x} - \mu}{\frac{s_x}{\sqrt{n}}}$$

Or, if you prefer to calculate the standard error of the mean before calculating z, compute with this equation:

$$s_{\bar{x}} = \frac{s_x}{\sqrt{n}}$$

then

$$t = \frac{\bar{x} - \mu}{s_{\bar{x}}}$$

4. Compare $t_{observed}$ with the $t_{critical}$ values taken from Appendix B at $df = n - 1$. If $t_{observed}$ exceeds $t_{critical}$, then you must reject H_0; if it does not, then you fail to reject the null hypothesis.

between our sample of engineers and population of university students), and we can conclude that we have evidence that a difference exists. In fact, for the differences to be significant, our *t*-value only needed to exceed ±1.66.

Calculating Confidence Intervals in the One-Sample Case

The variables *t* and *z* are also important for approximating the mean for a population when there is only one sample available. Because of sampling error, samples do not perfectly represent populations, so we can *never* be 100 per cent sure that any sample statistic equals the true population parameter. As a partial solution, we express uncertainty about estimates by providing a range of values for the mean.

To analyze a sample from a normal population with an unknown mean (μ), construct the confidence interval around the sample mean ($\bar{x}$), using the following equation:

$$CI = \bar{x} \pm t * \frac{s}{\sqrt{n}}$$

The appropriate value of *t* can be found in the *t*-distribution at the back of the book, at $n - 1$ degrees of freedom.

Suppose that we have a random sample of 15 refugees who moved to Vancouver one year ago, and we want to look at the civic participation rates of refugees living in Vancouver after one year in Canada. To measure civic participation, we ask respondents how many social organizations they belong to. Below are the data:

3	7	4	0	2
1	4	4	5	2
7	6	3	5	7

From these data it is possible to estimate the mean number of organizations that members of the refugee population who have been in Vancouver for one year belong to. First, calculate the mean:

$$\begin{aligned}\bar{x} &= \frac{\sum X_i}{n} \\ &= \frac{60}{15} \\ &= 4\end{aligned}$$

Next, calculate the sample standard deviation, using the following equation:

$$s_x = \sqrt{\frac{\sum(X-\bar{x})^2}{n-1}}$$

$$= \sqrt{\frac{68}{14}}$$

$$= 2.20$$

Use these values for the mean and standard deviation, and the critical value of t, to find the interval where we can be 95 per cent confident the population mean falls. To find the value of t, use the degrees of freedom ($n - 1 = 14$) and a 95 per cent confidence interval to retrieve the critical value of t of 2.145 (this number is found in Appendix B), leaving us with the confidence interval:

$$CI = \bar{x} \pm \left(t * \frac{s}{\sqrt{n}}\right)$$

$$= 4 \pm 2.145 * \frac{2.20}{\sqrt{15}}$$

$$= 4 \pm 1.22$$

$$= 2.78 - 5.22 \textbf{ organizations per individual}$$

We can be 95 per cent confident that the mean number of social organizations for all refugees who have been in Vancouver one year ranges between 2.78 and 5.22. If you wanted to be more confident of the range, say 99 per cent, go back to the t-table and find the critical value for a df of 14 and the 0.01 column. Now the critical value is 2.98, yielding a range of 2.31 to 5.69 social organizations. Notice that now we're more confident, but the range is larger.

Single Sample Proportions

A requirement for calculating and comparing means from samples and populations is that the variables must be measured at the interval/ratio level. However, for nominal and ordinal variables it is possible to assess proportions instead of means in a single sample, using the normal distribution and applying most of the same logic and techniques. For example, you could determine if the same proportion of people with a certain characteristic in a sample is likely to exist in the population. The three things you need to calculate z_{obtained} are the sample proportion, population proportion, and sample size.

Let's illustrate z_{obtained} with an example: a random sample of 120 individuals whose mothers added salt to their food during pregnancy reveals that 50 per cent of those people also add salt to their food. In the whole population, the proportion of people who add salt to their food is about 40 per cent. Are individuals with mothers who added salt to their food different from the rest of the population in this regard?

The proportion for the population is 40 per cent, or 0.4, and the sample proportion is 50 per cent, or 0.5. Using a 95 per cent confidence interval from the z table at the back of the book, we get a critical value of 1.96 (remember that 95 per cent is on both tails, so you will look for $\frac{95}{2}$, or 47.5 per cent, or 0.475 on the table).

To calculate z with proportions, use the following equation:

$$z_{\text{obtained}} = \frac{p_{\text{sample}} - P_{\text{population}}}{\sqrt{\dfrac{P_{\text{population}}\left(1 - P_{\text{population}}\right)}{n}}}$$

This is essentially the same equation as the one we used before, except that the denominator is the standard error for proportions (which is calculated by multiplying the population probability of adding salt, by the probability of not adding salt and dividing by n). Otherwise, the equations are the same.

Inserting our values gives us the following:

$$\begin{aligned} z_{\text{obtained}} &= \frac{p_{\text{sample}} - P_{\text{population}}}{\sqrt{\dfrac{P_{\text{population}}\left(1 - P_{\text{population}}\right)}{n}}} \\ &= \frac{0.5 - 0.4}{\sqrt{\dfrac{0.4(1 - 0.4)}{120}}} \\ &= \frac{0.1}{\sqrt{0.002}} \\ &= 2.24 \end{aligned}$$

Since 2.24 exceeds 1.96 (the critical value of z at 95 per cent confidence interval), we can say that there are differences at the 0.05 level between people whose mothers added salt to their food and the overall population. This suggests that whether or not your mother added salt to her food during pregnancy is a significant predictor of whether you will add salt to your food.

Measuring Association with the Same Group Measured Twice

So far, the focus has been on comparing samples and the populations that they're drawn from. Recently, Canadian researchers have become more interested in following people over a longer period of time. Since the early 1990s, Statistics Canada has launched a series of **longitudinal surveys** (panel surveys are the type of longitudinal survey discussed here). One of the challenges of using this type of data source is comparing the same sample at two points in time. The procedure for doing that has several names: **paired samples *t*-tests**, **repeated measures *t*-tests**, or ***t*-tests for dependent samples**.

To use the same sample at each point in time, we focus only on the *difference* between scores for the first and second times. As with sample means, assume that the difference between means is normally distributed and rely on the *t*-distribution if we have a small sample size.

Despite the introduction of several new equations, the process is almost identical to a one-sample *t*-test. The new equations use new symbols, since they deal with differences between dependent observations.

The first equation calculates the differences between observations:

$$d_i = X_{i1} - X_{i2}$$

Where X_{i1} is the score at time one, X_{i2} is the score at time two, and d_i is equal to the difference between them.

The next equation is for the mean of the differences between time points:

$$\bar{x}_d = \frac{\sum d_i}{n}$$

The standard deviation is defined as

$$s_d = \sqrt{\frac{\sum (d_i - \bar{x}_d)^2}{n-1}}$$

Or, if you want to avoid calculating $\bar{x}_d$

$$s_d = \sqrt{\frac{\sum d_i^2}{n-1} - \left(\bar{X}_1 - \bar{X}_2\right)^2}$$

To calculate the standard error of the difference between means, use this equation:

$$s_{\bar{d}} = \frac{s_d}{\sqrt{n-1}}$$

Finally, the *t*-value for dependent samples is the difference between means, divided by the standard error of that difference:

$$t = \frac{\bar{x}_1 - \bar{x}_2}{s_{\bar{d}}}$$

BOX 10.3

It's Your Turn: *t*-Test for the Same Sample Measured Twice

Joshua surveyed a number of his friends and acquaintances about how many sexual partners they had during the year they were 18 and how many partners they had during the year they were 21. How can he use the data to calculate whether the number of partners people have per year is significantly different between the two time points?

Individual	# of partners per year at age 18	# of partners per year at age 21	$d_i = X_{i1} - X_{i2}$	d^2
1	2	1		
2	0	0		
3	3	2		
4	1	2		
5	8	1		
6	1	1		
7	2	1		
8	0	2		
9	0	4		
10	3	1		
	$\bar{X}_1$	$\bar{X}_2$	Σd_i	Σd^2

1. For each age (column), calculate the mean $\bar{x}_1$ and $\bar{x}_2$.
2. For each individual, calculate the difference between the number of partners at ages 18 and 21 ($d = X_{i1} - X_{i2}$) then sum that value (Σd_i). Square the difference for each person and sum the values.
3. Use the formula $s_d = \sqrt{\frac{\sum d_i^2}{n-1} - (\bar{X}_1 - \bar{X}_2)^2}$ to calculate the standard deviation.
4. Use the standard deviation to calculate the standard error of difference between means $s_{\bar{d}} = \frac{s_d}{\sqrt{n-1}}$
5. Use those values to calculate the *t*-value using $t = \frac{\bar{X}_1 - \bar{X}_2}{s_{\bar{d}}}$
6. Compare the *t*-value to the critical value from the *t*-table for 9 degrees of freedom at a 95 per cent confidence level. Is there a significant difference in the number of sexual partners at the two ages?

The solution for Box 10.3 can be found in Appendix H.

The *t*-value is compared to the critical values, as with one-sample cases, to see if the change within individuals is statistically significant.

Conclusion

In this chapter, you learned how to compare a sample mean with a population mean, and how to compare a sample proportion with a population proportion. You should see some overlap with the material in the last chapter. The difference is that Chapter 9 was about determining how closely a sample mean or proportion resembles a population while taking sampling error into account, while this chapter was about whether a sample mean or proportion *actually* differs significantly from the entire population. In the next chapter, you will learn about how to compare means and proportions from two samples.

Glossary Terms

Alternative hypothesis (p. 97)
Central limit theorem (p. 106)
Longitudinals surveys (p. 109)
Null hypothesis (p. 97)
One-tailed test (p. 99)
Paired samples *t*-test (p. 110)
Repeated measures *t*-test (p. 110)
Research hypothesis (p. 97)
t-test (p. 104)
t-tests for dependent samples (p. 110)
Two-tailed test (p. 99)
Type one error (p. 98)
Type two error (p. 98)

Practice Questions

1. The students of the University of New Brunswick have begun a fundraising initiative to reduce the prevalence of diabetes in the province. Last year, they raised $43,000, and this year the chief fundraiser expects the amount to be higher.
   a. Generate a null and research hypothesis that would allow you to test for a difference without saying anything about the direction of the difference.
   b. Now, do the same thing, except hypothesize a direction in the relationship.
2. For the past few years, Amber's drive to work has taken an average of 75 minutes ($\sigma = 8$ minutes). She thinks she's found a shorter way and wants to see if the mean time for the new route is significantly different from that of the old route. Last week she tried the new route 5 times, and got a mean of 73 minutes. Help her determine whether or not the new way is significantly faster.
   a. Calculate $z_{obtained}$.
   b. Compare this value to $z_{critical}$, using a 95 per cent level of confidence.
3. From 1990 to 2005 (inclusive), Jasper earned approximately $31,000 per year ($\sigma = 27{,}000$). He wants to know if his earnings were significantly different from the Canadian population. Across this period, the mean wage was roughly $29,000/year. Help Jasper determine whether or not his income was significantly different from that of the Canadian population.
4. In a recent Ipsos-Reid poll (www.ipsos-na.com/news/pressrelease.cfm?id=3809) of a random sample of 3219 Canadian adults, 52 per cent thought that the price of food is too high. The 95 per cent confidence interval that Ipsos-Reid provides is 1.7 percentage points. What range is the population mean likely to lie between?
5. Ted loves his basil plants and believes that he has a special relationship with them. In his mind, this causes them to grow more leaves than his neighbour's plants. Ted is a good statistician and feels confident that he can prove his claim statistically. His neighbour is rather indifferent to the entire exercise, but he wants to be a good neighbour so he lets Ted come over and count the leaves on his 10 plants. Ted counts a mean of 57 leaves on his neighbour's plants, with a standard deviation of 8 leaves. On his own 12 plants, he finds 61 leaves, with a standard deviation of 9 leaves. Help Ted determine whether there is any statistical evidence to support his belief in his basil superiority.
6. Scout and Lily are both avid Highland dancers, and they compete annually in the Antigonish Highland Games. Their parents have consistently told the girls that their scores are likely to be significantly higher than the mean because they've gone to so many competitions already and are, therefore, much better prepared than everyone else. Develop a set of null and research hypotheses that would allow you to test the following:

a. if Scout and Lily's parents are right, and the girls are significantly better than their competition; and

b. if Scout and Lily's parents are wrong, and the girls are no different from the other competitors.

7. Roughly 32 per cent of a random sample of 172 individuals in the city of Edmonton feel that the Eskimos are likely to win the next Grey Cup. In the entire Alberta population, the comparable figure is about 27 per cent. Are Edmontonians significantly different from Albertans overall in their love for their home team? Use a 95 per cent confidence interval to draw your conclusion.

8. Sneha has recently become interested in the differences in the amount of personal space required by people of different ethnic groups. Based on her experience, people who were born and raised in Canada require approximately an arm's length of distance between themselves and the people they're talking to. She drafts a set of hypotheses about whether there are differences by group, and after her study she fails to reject the null hypothesis when she should have. Has she committed a type one or type two error?

9. Looking at the z-table in Appendix A, what happens to the probability that you'll reject the null hypothesis as values of $z_{critical}$ increase? Does the probability of rejecting the null hypothesis increase or decrease?

10. If you reject the null hypothesis, are you saying that the sample does or does not differ from the population on your outcome of interest?

Answers to the practice questions for Chapter 10 can be found in Appendix H.

11 Testing Hypotheses: Comparing Two Samples

LEARNING OBJECTIVES

Now that we're working with two samples, there are two sources of sampling error we need to account for when testing hypotheses. This chapter will introduce several ways of dealing with two samples:

- standard error of the difference between means;
- comparing proportions with two samples;
- two-sample *t*-tests; and
- one- and two-tailed tests.

Introduction

In Chapter 10, we looked at comparing means or proportions from a sample to means or proportions from a known population or comparing means or proportions from a sample to itself at different points in time. In this chapter, we will look at comparing means and proportions from two samples. So far, we've assumed that our samples and/or populations had roughly equal variances, and we have needed information about the population. To use the *z*-distribution for one-sample cases, you need to have both the population mean and the standard deviation. To use the *t*-distribution, the only population characteristic you need is the mean. However, both cases require at least some population information. Although the *t*-tests for dependent samples don't require population characteristics, we had to assume that the samples contain the same sampling error. However, we usually do not have data from the population—we just have sample data.

When we examine data from two groups, it is common to find that not just the means but also the variances of the two groups being compared differ. For example, if we wanted to compare men and women in terms of mean BMI, Hondas and Toyotas in terms of mean gas mileage, or even dogs and cats in terms of mean intelligence, it wouldn't be surprising to discover that not just the means but also the variances between the groups are different for most characteristics. Why would they be the same?

This chapter will focus on comparing means and proportions from two distinct groups, or *independent* samples.

The Standard Error of the Difference between Means

Let's start with comparing two sample means. The mean for group one would be calculated as

$$\bar{x}_1 = \frac{\sum x_1}{n_1}$$

For group two, it would be calculated as

$$\bar{x}_2 = \frac{\sum x_2}{n_2}$$

Similarly, the sample standard deviation for group one can be calculated as

$$s_1 = \sqrt{\frac{\sum (x_1 - \bar{x}_1)^2}{n_1 - 1}}$$

For group two, it can be calculated as

$$s_2 = \sqrt{\frac{\sum (x_2 - \bar{x}_2)^2}{n_2 - 1}}$$

To calculate the *z*-value for the difference between sample means we use the following formula:

$$z = \frac{\bar{x}_1 - \bar{x}_2}{s_{\bar{x}_1 - \bar{x}_2}}$$

The *z*-test is used because it's assumed that we have big enough (≥120 combined) samples. (We'll discuss small samples momentarily.) The symbols should be familiar now, except for the denominator term:

$$s_{\bar{x}_1 - \bar{x}_2}$$

This term represents the standard error of the difference between means. It is also referred to as the anticipated level of error between the measurements of the two-sample means and can be found by using the following equation:

$$s_{\bar{x}_1 - \bar{x}_2} = \sqrt{\left(\frac{s_1^2}{n_1} + \frac{s_2^2}{n_2}\right)}$$

The standard error is the degree of certainty about how closely our calculated means (calculated from two samples) resemble the means of the two respective populations. A wider variance (the s^2 values in the numerator) results in higher calculations, as does a reduction in sample size (seen in the denominator).

Once the standard error is determined, the *z*-value can be calculated:

$$z = \frac{\bar{x}_1 - \bar{x}_2}{s_{\bar{x}_1 - \bar{x}_2}}$$

A larger standard error value (the denominator), or a smaller difference between means, will result in a lower *z* figure. This reflects waning confidence in whether the observed difference between the two sample means would also exist between the two populations. This should look familiar, as it is the same equation that you use for calculating *z* for one sample mean, except that now the denominator is the standard error of the difference between means. If you think about this for a moment, it shouldn't surprise you because it is the level of certainty that we attach to our estimates of the means that differs, not the actual estimates of the means. Thus, we need to alter only one component of our calculation, albeit in a much more complicated way.

Using small samples (<120) complicates things further: doing so relies on the *t*-distribution instead of the *z*-distribution and involves a more complicated calculation of the standard error of the difference between means:

$$s_{\bar{x}_1 - \bar{x}_2} = \sqrt{\left(\frac{n_1 s_1^2 + n_2 s_2^2}{n_1 + n_2 - 2}\right)\left(\frac{n_1 + n_2}{n_1 n_2}\right)}$$

Substituting *t* for *z* requires the following equation:

$$t = \frac{\bar{x}_1 - \bar{x}_2}{s_{\bar{x}_1 - \bar{x}_2}}$$

As we saw earlier, the *t*-distribution is actually a family of distributions, making it necessary to identify which distribution to use by calculating the degrees of freedom. The calculation for degrees of freedom is slightly more complicated than in the one-sample case and can be found by entering the two sample sizes in the following equation:

$$df = (n_1 + n_2 - 2)$$

If the calculated *z*- or *t*-statistics exceed the critical value (1.96 for *z*, approaching 1.96 for *t*) for both the *z*- and *t*-distributions, you can be 95 per cent confident that the difference between the two sample means did not occur merely by chance. Since the null hypothesis is *always* to assume that groups are the same, reject the null whenever *z* exceeds $z_{critical}$ or *t* exceeds $t_{critical}$. Once again, as sample size increases, the *t*-statistics and the *t*-distribution increasingly resemble the *z*-statistic and the *z*-distribution.

Let's illustrate with an example. Suppose that we were interested in studying the effects of class size on university learning and that we hypothesized that students in universities with large classes (>75) have lower standardized test scores than students in small-class universities. You take one sample of students from University A, where all the classes are larger than 75. This sample has 100 students in it, and their mean score on the test is 81 ($s = 11$). Next, you take a sample of students from University B, where all the classes are smaller than 75. You have 92 students in this sample, and their mean score on the test is 83 ($s = 10$). You want to know if the difference you observe between your two samples is statistically significant or if you can be 95 per cent confident that a similar difference exists between the two populations. Let's call the large-class students from University A group 1 and the small-class students from University B group 2. For illustration purposes, let's assess the difference by using both *z* and *t*.

Our hypotheses would be as follows:

H_1 = Students from universities with large class sizes (>75) will score lower on the standardized test than students from universities with smaller class sizes (≤75). Or, $\mu_1 < \mu_2$.

Note that because we're dealing with two samples, we are hypothesizing about two population means μ1 and μ2.

H_0 = Students from universities with large class sizes (>75) will not score lower on the standardized test than students from universities with smaller class sizes (≤75). Or, $\mu_1 \geq \mu_2$.

Note that these are directional hypotheses, which point us to a one-tailed test. If we were hypothesizing non-directionality, H_1 would be $\mu_1 \neq \mu_2$ and H_0 would be $\mu_1 = \mu_2$.

Let's put our values above in the appropriate equations. First, *z*. Let's calculate the standard error of the difference between means:

$$
\begin{aligned}
s_{\bar{x}_1-\bar{x}_2} &= \sqrt{\left(\frac{s_1^2}{n_1}+\frac{s_2^2}{n_2}\right)} \\
&= \sqrt{\left(\frac{11^2}{100}+\frac{10^2}{92}\right)} \\
&= \sqrt{\left(\frac{121}{100}+\frac{100}{92}\right)} \\
&= \sqrt{(1.210+1.086)} \\
&= 1.516
\end{aligned}
$$

Now, we simply put this information into our equation for z:

$$z = \frac{\bar{x}_1 - \bar{x}_2}{s_{\bar{x}_1 - \bar{x}_2}}$$
$$= \frac{81 - 83}{1.516}$$
$$= -1.319$$

So, we get a z_{observed} value of −1.319. Now let's calculate t, remembering that our value for the numerator remains the same. This leaves us with the more complicated version of the equation for calculating the standard error of the difference between means.

$$s_{\bar{x}_1 - \bar{x}_2} = \sqrt{\left(\frac{n_1 s_1^2 + n_2 s_2^2}{n_1 + n_2 - 2}\right)\left(\frac{n_1 + n_2}{n_1 n_2}\right)}$$
$$= \sqrt{\left(\frac{100 * 11^2 + 92 * 10^2}{100 + 92 - 2}\right)\left(\frac{100 + 92}{100 * 92}\right)}$$
$$= \sqrt{\left(\frac{12{,}100 + 9200}{190}\right)\left(\frac{192}{9200}\right)}$$
$$= \sqrt{112.1 * 0.02}$$
$$= \sqrt{2.242}$$
$$= 1.50$$

Now, we can calculate t:

$$t = \frac{\bar{x}_1 - \bar{x}_2}{s_{\bar{x}_1 - \bar{x}_2}}$$
$$= \frac{81 - 83}{1.50}$$
$$= -1.333$$

This gives us a t_{observed} value of −1.333. Note how close this number is to z_{observed} above.

Returning to our hypotheses, how do we use these numbers to test our hypothesis? First, let's look at the z-table in Appendix A. For a 95 per cent confidence interval with a one-tailed test, we get a z_{critical} value of 1.65, which is greater than our observed value of 1.313. (Remember that we can drop the negative sign.)

To obtain t_{critical} for a 95 per cent confidence interval, we need to calculate our degrees of freedom for t as $df = (n_1 + n_2 - 2)$, or 190. Since our highest value is for 120, we will use the value for ∞, which is 1.645. Once again, this is greater than our t_{observed} value of 1.336. Regardless of which statistic you calculate, we must draw the same conclusion and fail to reject the null hypothesis. Based on our results from these samples, we cannot say that students from universities with large classes score lower than students from universities with small classes.

In the example above, we calculated both t- and z-statistics for illustrative purposes, even though we really only needed one. When should you use t versus z? There is no broad consensus on this, but a good rule of thumb would be to use the z-statistic when (a) the sum of the two sample sizes is greater than 120, and (b) when each sample has at least 30 observations. Otherwise, opt for the t-statistic. In any instance, you should have at least 10 observations in each group to make meaningful comparisons.

Comparing Proportions with Two Samples

As with the one-sample case, you can compare proportions of a nominal variable across samples. To find the value of $z_{obtained}$ or $t_{obtained}$, several equations, again, are needed. Take a closer look and you'll see that most of these resemble the one-sample case.

Let's revisit the equation for $z_{obtained}$ in the one-sample case:

$$z_{obtained} = \frac{p_{sample} - P_{population}}{\sqrt{\frac{P_{population}\left(1 - P_{population}\right)}{n}}}$$

Now, let's compare it to the formula for $z_{obtained}$ in the two-sample case:

$$z_{obtained} = \frac{p_{s_1} - p_{s_2}}{s_{p_1 - p_2}}$$

The similarities in the numerator are obvious: instead of $P_{population}$, which indicates the proportion of the population in a group or category, we are interested in the proportion of respondents in the second sample p_{1_2}. The numerator subtracts the two sample proportions. So far, so good.

The denominator for a one-sample case is:

$$\sqrt{\frac{P_{population}\ (1 - P_{population})}{n}}$$

Consequently for the one sample case,

$$z_{obtained} = \frac{p_{sample} - P_{population}}{\sqrt{\frac{P_{population}\left(1 - P_{population}\right)}{n}}}$$

Remember that one-sample and two-sample equations are not identical. Calculating the standard deviation of differences between a sample proportion and a population proportion is not the same as calculating the standard deviation between the proportions of

BOX 11.1

It's Your Turn: Difference between Two Sample Means

Sophia is enrolled in a sociology course examining ethnicity in Canada. For her term paper she is interested in the extent to which people who grew up in Canada feel that they "fit into" Canadian society. Using data collected in 2002 for the Ethnic Diversity Survey (Public Use Microdata File), she selected people who were either second generation (i.e., they were born in Canada, but one or both parents were not) or third generation or more (i.e., both the respondents and their parents were born in Canada). The dependent variable was a response to this question "Up until you were age 15, how often did you feel uncomfortable or out of place because of your ethnicity, culture, race, skin colour, language, accent, or religion?" The responses ranged from 1 (all of the time) to 5 (never). Although the response categories were ordinal, Sophia treated the variable as an interval/ratio variable so that she could construct a mean.

Sophia found the following:

	Generation status—2nd or 3rd	N	Mean	Standard deviation	Standard error mean
Felt uncomfortable before age 15	2nd—parents born outside Canada	6799	4.59	.776	.009
	3rd or more—respondent and parents born in Canada	23,237	4.78	.593	.004

Note: Rescaled weights for data have been used (weight/average weight for sample selected). Target population was people aged 15 and older living in private dwellings in the 10 Canadian provinces. Sample excludes those under age 15, people living in collective dwellings, Indian Reserves, people who declared an Aboriginal ethnic origin or identity on the 2001 census, and people living in the territories and remote areas (Statistics Canada User's Guide, p. 4, Catalogue no.89m0019GPE).

Source: Ethnic Diversity Survey Public Microdata File, 2002 accessed at http://www5.statcan.gc.ca/bsolc/olc-cel/olc-cel?catno=89M0019XCB&lang=eng

1. What would Sophia's null hypothesis be?
2. Use the information in the table to calculate the z-score. (This will require you to calculate the standard error of the difference between means.)
3. Comparing the z-score to the critical value for 95 per cent confidence, would you reject or fail to reject the null hypothesis?

The solution for Box 11.1 can be found in Appendix H.

two samples because of differences in the source of error. When comparing a sample to a population, there will be error only in the sample proportion (there cannot be error in the population parameters), so it is sufficient to include n in the denominator to acknowledge that the size of the error is partially a function of sample size.

With two samples, any errors are probably because there are two sources, so the calculation of

$$s_{p_1-p_2}$$

is more complicated than

$$\sigma_{p_1-p_2}$$

To calculate $s_{p_1-p_2}$ we have to find P_u, an estimate of the proportion in the population. In the social sciences, we always assume that the null hypothesis is true and that there are no differences between groups. (For example, if we were comparing smoking rates of men and women, we would assume that the proportion of male and female smokers is the same in the population.) P_u can be calculated as the average of the proportions in samples one *and* two, adjusting for sample size, using the equation:

$$P_u = \frac{n_1 p_{s_1} + n_2 p_{s_2}}{n_1 + n_2}$$

Where p_{s_1} is the proportion in sample one, and p_{s_2} is the proportion in sample two.

On its own, P_u is of little interest (it is basically a weighted average of the two samples), but it is useful for calculating $s_{p_1-p_2}$:

$$s_{p_1-p_2} = \sqrt{p_u * (1 - p_u)\frac{n_1 + n_2}{n_1 n_2}}$$

This number $\left(s_{p_1-p_2}\right)$ is known as the standard deviation of the difference between sample proportions. If the sample size is large enough, the **central limit theorem** tells us that we can assume that the difference between the proportions is also normally distributed.

The value for $s_{p_1-p_2}$ is used in the following equation to calculate the obtained value of z:

$$z_{\text{obtained}} = \frac{p_{s_1} - p_{s_2}}{s_{p_1-p_2}}$$

One- and Two-Tailed Tests, Again

This chapter has only explained how to differentiate one group from another group, without considering the directionality of difference, using a **two-tailed test**. Often, you will be concerned with differences in one direction. For example, instead of studying whether men and women have on average a significantly different number of friends (statistically), you might want to know if males have significantly more friends than females (statistically), or vice versa.

Once again, this introduces the need for a **one-tailed test**, which measures the significance and direction of a relationship. There is no statistical reason for choosing a one-tailed test over a two-tailed test; the choice is theoretically driven. So if you are trying to determine whether or not groups are equal, use a two-tailed test. If you believe that a group has more, or less, of a quality than another group, use a one-tailed test.

To use a one-tailed test, a slight modification of the two-tailed case is needed. The modification is the use of a different value from the z- or t-table. Let's illustrate using the 95 per cent confidence interval from the z-table in Appendix A. As you know, z_{critical} must exceed ±1.96 to be considered significant. In column C of the table, the area beyond z is listed as

BOX 11.2

It's Your Turn: Difference between Two Sample Proportions

Data from the 2001 Aboriginal Peoples Survey (APS), Public Use Microdata File (PUMF), show you that 38 per cent of men and 52 per cent of women engage in prayer. The sample size for men is 114,734, and the sample size for women is 143,013.

Note: APS is a post-censal survey, with selection based on responses to four questions examining Aboriginal identity (e.g., self-report, list of Aboriginal groups in list of ethnic or cultural group(s) they belong to). It includes residents of 10 provinces and 3 territories. Only off-reserve adults are included in the PUMF. Individuals were excluded if they lived in a collective dwelling (Statistics Canada, 2001: 6–9).

1. Calculate the estimate of the proportion of the population who use prayer by using the equation:

$$P_u = \frac{n_1 p_{s_1} + n_2 p_{s_2}}{n_1 + n_2}$$

2. Use this value to calculate the standard deviation of the difference between sample proportions:

$$s_{p_1 - p_2} = \sqrt{p_u * (1 - p_u)\frac{n_1 + n_2}{n_1 n_2}}$$

3. Calculate the value of z_{obtained} using $z_{\text{obtained}} = \frac{p_{s_1} - p_{s_2}}{s_{p_1 - p_2}}$

4. Compare the value of z_{obtained} to the critical value. (Hint: Remember that your hypothesis had a direction.)

The solution for Box 11.2 can be found in Appendix H.

0.025, so when both tails are included we will have a 95 per cent confidence interval. However, the one-tailed case needs to have 5 per cent of all values falling in only one of the two tails, pointing to a z_{critical} value of roughly 1.65. If we hypothesize that one group will have a higher score than another (i.e., that males have more friends than females), we use a value of +1.65. For a lower score (males have fewer friends), we use −1.65. Follow the same process for *t*-scores.

Look at the *z*-table and find the one-tailed values for 90 per cent and 99 per cent confidence intervals.

Conclusion

This chapter has introduced you to comparing means and proportions from two samples. We have examined the logic of assessing whether a difference in means or proportions between two samples is large enough to conclude that this difference did not occur just by chance, and that we would expect to find the difference in the populations from which the samples were derived. Note that all of the examples in this chapter use an independent variable that has just two values (gender, for example) and a dependent variable that is either an interval ratio (so we can calculate a mean) or a dichotomous nominal variable (so that we can calculate a proportion). In the next chapter we will talk about examining the relationship between nominal and ordinal variables that have more than two response categories.

Glossary Terms

Central limit theorem (p. 121)
One-tailed test (p. 121)
Two-tailed test (p. 121)

Practice Questions

1. A research study was conducted to examine the differences between men and women on perceived life satisfaction. In total, 200 men and 200 women were given life satisfaction tests. Scores on the measure range from 0 to 60, with high scores indicating high life satisfaction. Men scored an average of 48 ($s = 7$), and women scored 45 ($s = 5$). Recalling our discussion of hypotheses from Chapter 10, draft a set of non-directional null and research hypotheses that will allow you to identify whether a significant difference exists in the population.

2. From these data, can you conclude that there is a significant difference between the means of these two groups? What is the implication of this for your decision to reject or fail to reject the null hypothesis?

3. In a recent Genworth Financial survey of the housing experiences of recent immigrants to Canada (www.genworth.com), 76 per cent of recent immigrant non-homeowners ($n = 201$) said that distance to work was very important for them, compared with 68 per cent of homeowners ($n = 218$). Generate a set of non-directional null and research hypotheses that will allow you to identify whether a significant difference exists in the population.

4. Are the differences in the proportion of these groups significant at the 95 per cent confidence level?

5. Tess believes that dogs are more intelligent than cats, and she has developed an intelligence test that can be validly administered to both animals. For some reason, she had considerable success convincing dogs to take the exam, and administered 93 tests. She had less luck with cats, however, and could only administer 28 exams so she is unsure about how to proceed. She approaches you and asks for assistance. What would you tell her regarding the following?

   a. Whether a *z*-test or *t*-test is more appropriate, and why

   b. Whether a one-tailed or two-tailed test is more appropriate, and why

6. Assuming that for dogs the mean was 79 and the standard deviation was 12, and for cats the mean was 75 and the standard deviation was 12, what does Tess's test tell you about the comparative intelligence of cats versus dogs?

Answers to the practice questions for Chapter 11 can be found in Appendix H.

12 Bivariate Statistics for Nominal Data

LEARNING OBJECTIVES

Although what's been covered so far in this text is useful, it is foundational information. While you need to know these things to do statistics in the social sciences, what you've learned so far probably won't form the analytical centrepiece of any research project. For that, we usually conduct more sophisticated bivariate (two variables) or multivariate (more than two variables) analysis. This chapter will focus on bivariate analysis by:

- examining some of the reasons that it is useful to study more than one variable at a time;
- learning what independent and dependent variables are;
- studying chi-square tests of statistical significance; and
- learning some popular techniques for measuring the association between two nominal variables.

Introduction

Typically, social scientists are interested in uncovering the **associations** or **relationships** between variables. Some examples of relationships that social scientists might be interested in are

- differences in budgetary spending between majority and minority governments;
- differences in family size, by religion;
- whether members of particular visible minority groups are more susceptible to certain diseases; and
- whether marital status affects the number of hours spent at clubs, discos, malls, churches, synagogues, etc.

In each example, there are two variables of interest. Although we might be interested in the distribution of each particular variable (and, indeed, we should always examine descriptive univariate statistics for all our variables of interest before we move on to bivariate analysis), what is central for the investigation is identifying the relationship *between* the two variables.

First, to simultaneously analyze two variables that cannot be ranked or ordered, you'll need to learn **bivariate analysis** techniques for nominal variables. In this chapter, you will learn how to determine if a relationship exists between two nominal variables and how to measure the strength of the relationship between two nominal variables. Next, you'll learn to deal with ordinal and interval/ratio variables by using contingency tables, which help researchers "visualize" the relationship between two variables. In the following chapters, you will learn how to identify some techniques for assessing the relationship between two continuous variables.

Analysis with Two Nominal Variables

The first step for performing bivariate analysis is organizing the data so that patterns can be easily discerned. Suppose a researcher wants to examine whether the effect of a commercial for a pizza restaurant varies for men and women. She takes a random sample of 1000 people who are watching television and who see the commercial, and examines their behaviour following the commercial. What would be a useful way of identifying and assessing the nature of this relationship?

For nominal data like these, it is useful to create a frequency table, like the one in Table 12.1. In the table, the information is organized into columns. The left column shows the possibilities or response categories. The next columns are divided by sex and then show the percentage distributions. The second column looks at how the women responded, and the fourth looks at how the men responded, with the third and the fifth columns showing the percentage distribution of responses by sex.

We use the columns to show which of the variables is affecting the value of the other. This is often referred to as "determining the order of causality." The variable that we think is affecting the outcome is an **independent variable**, and the outcome of interest is the **dependent variable**. Note: These two terms are important because they apply to all relationships and are used often.

In the example, we hypothesize that the sex of the respondent will affect the reaction to the commercial. Independent variables are conventionally shown as columns; and dependent variables, as rows. These tables are called **contingency tables**, or cross-tabulations.

TABLE 12.1 | Response of 1000 Television Viewers, by Sex, to a Pizza Commercial

	Sex of respondent			
Response of respondent to television commercial	*Female*	*%*	*Male*	*%*
Order a pizza	10	1.6	30	7.9
Grab food from the refrigerator	60	9.7	40	10.5
Change channels	60	9.7	50	3.2
No reaction	490	79.0	260	68.4
Total	620	100.0	380	100.0

Table 12.1 shows us that women are more likely to have no reaction to the pizza commercial than men. This is determined by looking at the percentage of women who had no reaction to the pizza commercial (490 women out of 620, or roughly 79 per cent). On the other hand, of the 380 men in the study, only 260, or 68.4 per cent, of them did not react to the commercial ($\frac{260}{380}$ = 68.4 per cent).

What is more interesting (at least to the pizza company) is not who does *not* respond to the pizza commercial, but who actually does. Only 10 women out of 620 (or 1.6 per cent) ordered a pizza after viewing the commercial, compared with 30 out of 380 men (or 7.9 per cent). There are several ways to explain these results: (1) men like pizza more than women do; (2) men and women like pizza equally, but the pizza commercial resonates with men more than women; or (3) women like pizza more than men do, but TV is not an effective way to target women.

We don't know if any one of these explanations is true. Before we can explore the alternatives, we need to determine whether the results, which are based on a sample of 1000 men and women, indicate that we can expect to find this difference in the population from which our sample was drawn. To assess the statistical significance of any bivariate association, we need to learn about the **chi-square test** of statistical significance.

BOX 12.1

Cross Tabulations: History of a Term

In the second half of the seventeenth century, the German empire was fractured, suffering from balkanization and the resulting social ills of poverty and civil disruption. In response, the state bureaucracies started defining and cataloguing the micro-states, seeking not only to describe but also to organize, hierarchize, and classify them. Bureaucrats began using an early version of a cross-tabulation. The countries constituting the empire were put into the rows, and state characteristics were used as columns. In this way, rulers could compare states based on the presence of a particular characteristic—from art and culture to agriculture—and the total character of any state could be read across the rows. Using the table, it was possible to sum up a state and compare it to other states by using nominal features.

This new method of comparative taxonomy was met with resistance. Critics saw the tables as "vulgar" statistics that allowed qualitative equivalence and conceptual reduction. This was at odds with what the critics valued as "subtle and distinguished" statistical analysis—an analysis that did not employ tables at all. Neither form of schematic qualitative analysis is used any longer, having been replaced by quantitative analysis, which historically has also been driven by nations' desires to organize and understand their populations (Desrosiers, 1998).

Statistician Karl Pearson (1857–1936), following in the footsteps of Francis Galton (1822–1911), argued that contingency tables allow one to schematize "the *partial relationship* between two phenomena, midway between two limits—absolute independence and absolute dependence—or *correlation*, synonymous with association" (Desrosiers, 1998: 110). The emphasis on "contingence" is important. Pearson believed absolute unilinear causation to be rare to non-existent. Philosophical understandings of the nature of what was being described lie behind the table's schematization. Pearson described them as contingency tables because he considered each event in the observable world to be unique; the human process of observation and classification is what allows for predictability. Regularity is a consequence of human conceptualization and abstraction, not an inherent feature of an external world.

EVERYDAY STATISTICS

The Utility of Contingency Tables

During an election period, news stations will often report the percentage of the population that is planning on voting for a particular political party. This information is collected from a sample of the population and is sometimes presented with data on the age or sex composition of those surveyed. A contingency table is often presented to give the public an overall picture of the expected outcome of the election.

Q: Why is a contingency table a good way to represent the information collected during an election period?

The Chi-Square Test of Statistical Significance

The chi-square test, introduced by Karl Pearson in 1900, is a widely used method for determining the level of agreement between the frequencies in a distribution of observed data and the frequencies calculated on the assumption of a random distribution. Pearson wanted to determine the frequency with which patterns seen in a sample would also be seen in the population. He developed the chi-square to serve that purpose. Since then, it has become an important part of statistical theory and practice.

The chi-square can be used for all levels of measurement, not just nominal data, and for combinations of levels—nominal/ordinal, ordinal/interval, nominal/ratio, etc.—making it extremely useful. It is a **non-parametric test**, so you don't have to make assumptions about the shape or distribution of a sample and/or population. (They do not need to be normally distributed, for example.) Using chi-square makes it easy to identify statistically significant relationships between variables without worrying about whether the variables are normally distributed until after a relationship has been identified.

Perhaps the most compelling aspect of the chi-square is that it can be directly calculated from bivariate tables, such as Table 12.1, by comparing the discrepancies between observed values and expected values. It determines whether the observed differences in data come from random sampling error or if they are large enough for us to conclude that they exist in the population.

To calculate the chi-square, you need to do the following:

1. Compute row and column totals, or **marginals**.
2. Divide the marginals by the total number of possible variable values to determine the distribution of highest probability. These are the expected values.

For example, imagine there are 400 students registering for 10 statistics classes. If none of the variables that distinguish the statistics classes from each other (the professor, the time of day the class is taught, etc.) influence registration, (that is, there are no independent variables predicting allocation of students across classes), we would expect each class to have 40 students. However, if there were discrepancies between the expected class sizes and the observed class sizes, we might suspect that there are factors other than random allocation (popularity of teachers, location of friends, etc.) behind class selection. Chi-square measures the magnitude of that discrepancy and tells us the likelihood of observing such a discrepancy by chance. We can then decide whether we should pursue further analysis.

To further illustrate the calculation of chi-square, let's imagine that we were looking at the sex composition of players on local soccer teams. Imagine we randomly sampled two soccer teams, A and B, each made up of 11 people (there are 22 people in total). Of these, 12 are female, and 10 are male. If we guessed at the distribution of men and women across teams A and B, assuming that there were no factors influencing their distribution, we would guess that each team would be expected to have 6 females and 5 males.

Table 12.2 shows the actual (observed) frequencies of men and women on the two sampled teams, with the expected frequencies shown in parentheses, surrounded by row and column totals. The expected frequency is the number of observations we would expect to see in each cell, if the independent variable had no relationship with the dependent variable. It can be calculated using the equation:

$$f_e = \frac{\sum \text{column} * \sum \text{row}}{n}$$

The expected frequency, f_e, is equal to the column marginal, multiplied by the row marginal, divided by the total number of observations. There are a total of 12 women and 10 men, with 11 people on each soccer team. Let's calculate the expected frequency for each of the four cells:

$$f_e = \frac{\sum \text{column} * \sum \text{row}}{n}$$

Upper left:

$$f_e = \frac{12 * 11}{22} = 6$$

TABLE 12.2 | Sex Composition of Two Hypothetical Soccer Teams

	Female	Male	Total
A	7 (6)	4 (5)	11
B	5 (6)	6 (5)	11
Total	12	10	22

Upper right:

$$f_e = \frac{10 * 11}{22} = 5$$

Lower left:

$$f_e = \frac{12 * 11}{22} = 6$$

Lower right:

$$f_e = \frac{10 * 11}{22} = 5$$

These numbers are the most likely distribution of men and women across the two teams, assuming a random distribution with no other explanatory factors or independent variables. In the event of a **non-integer**, it is acceptable to have an expected frequency that is not a whole number (e.g., 5.3) even though it is not likely to occur in reality. (It is not possible to have part-persons on soccer teams.)

The numbers that represent the expected distribution (typically expressed as f_e for expected frequencies) are used with the numbers you have observed (f_o for observed frequencies) to calculate chi-square χ^2 by using the following formula:

$$\chi^2 = \sum \frac{(f_o - f_e)^2}{f_e}$$

To reduce the possibility of calculation errors, it is useful to organize the data in a table like Table 12.3. In the far left column, each of the four cells is identified by its team name (A or B), and whether the cell refers to females or males (F or M). "AF" refers to the cell containing the frequency of females on team A. The observed frequencies, f_o, appear in the second column, followed by expected frequencies, f_e, in column three. Column four is $(f_o - f_e)^2$, the solution for the numerator of the chi-square equation, and column five solves the equation for each

TABLE 12.3 | Sex Composition of Two Hypothetical Soccer Teams

Group	f_o	f_e	$(f_o - f_e)^2$	$\frac{(f_o - f_e)^2}{f_e}$
AF	7	6	1	0.17
AM	4	5	1	0.20
BF	5	6	1	0.17
BM	6	5	1	0.20
Total	22	22		$\chi^2 = 0.74$

row. The summation of values appears in the bottom right cell and represents the value of chi-square in this example.

Now that we have the chi-square value, we need to assess it by using a chi-square table. Like other distributions (normal distribution, *t*-distribution, etc.), chi-square has a known distribution (the table is shown in Appendix C), which means that we can determine if there are significant differences between observed and expected values. To do this, we need to revisit **degrees of freedom**, but with a slightly different calculation to reflect that two variables are being used.

When we looked at the *t*-distribution to assess sample means, we calculated the degrees of freedom to be equal to sample size minus one. We subtracted one from the sample size because the individual values of any particular set of numbers can be determined using $n - 1$ values. Since there was one variable, only one value was determined by the others.

Now we are looking at two variables, but the same logic applies. Remember that it is possible to determine the value of any cell by knowing the table total and the value of all other cells. Now there are two values that are determined entirely by other values. In this case, degrees of freedom can be defined as the number of rows minus one, times the number of columns minus one:

$$df = (\text{rows} - 1)(\text{columns} - 1)$$

In Table 12.3, there are two rows and two columns (we never count the totals columns), so the degrees of freedom are $(2 - 1)\ (2 - 1) = (1)\ (1) = 1$. Using Appendix C, we can choose a level of significance (also often called α or the *p*-value) of either 0.05 or 0.01. Next, find the correct number for the degrees of freedom that we calculated (1). For our example, with a level of significance of 0.05, the critical chi-square value is 3.841. To determine whether our observed values differ significantly from the expected values, we need to compare the calculated chi-square value of 0.74 to 3.841. If the calculated value exceeds the critical value, we can conclude that the observed values differ significantly from expected values. Note

BOX 12.2

Chi-Square: The Steps

1. Compute row and column totals, or marginals.
2. For each cell, multiply the row and column marginal together. Then divide that product by the total number of people in the sample (*n*) to determine the distribution of highest probability. These are the expected values.
3. Subtract expected cell frequencies from observed cell frequencies.
4. Square that number and divide it by the expected frequency.
5. Sum the product of these calculations. This is your observed chi-square value.
6. Assess the statistical significance of this number using the chi-square chart in Appendix C; $df = (\text{row} - 1)(\text{column} - 1)$.

that this is a hypothetical example, with a small sample and only two possible categories on the independent variable (two teams). Normally, we use chi-square on much larger samples.

Our calculated value is well *below* the critical value, suggesting that a respondent's sex is not a significant factor in determining which team he or she will be on.

Further analysis of this association is probably unnecessary since chi-square suggests that the observed trends could have easily appeared at random. An important point to make about the chi-square is that it depends on both the strength of the relationship *and* the sample size. The difference in sex composition across teams may be too slight for us to be certain about its existence. It is also possible that the relationship is not statistically significant because our sample of 22 people is too small to determine if a significant association exists.

Measures of Association for Nominal Data

As you know by now, researchers are interested in identifying not only the existence of relationships between variables but also the strength of those relationships. Before the pizza company revamps its advertising strategy to more effectively target women, it would be useful to know if the findings we saw in the sample of 1000 people are statistically significant.

(Remember we are assuming that our sample is random and represents all people seeing the commercial.) We need a test of **statistical significance** to assess the relationship between sex and the reactions of the study members. This measure will allow us to determine whether the observed patterns actually exist in the population or if they were seen in the sample by chance. Chi-square will tell us whether this relationship is statistically significant. What we don't have, but need, is a measure of the *strength* of the association.

The level of measurement of a particular variable determines which measures of association are suitable. For nominal data, we'll learn about *phi*, *Cramer's V*, and *lambda*.

Phi

Phi is a chi-square-based measure of association used only for tables where both the independent variable and the dependent variable have exactly two response categories (two by two tables). The chi-square coefficient depends on the strength of the relationship and sample size, but phi eliminates the influence of sample size by dividing chi-square by the sample size, n, and taking the square root.

$$\phi = \sqrt{\frac{\chi^2}{n}}$$

Since phi is a **symmetrical** measure, it doesn't matter which of the two variables you believe is the independent variable (although it's good practice to keep them straight). Phi is the percentage of difference between a product of the diagonal cells and the product of the non-diagonal cells. It is the magnitude of difference between observed and expected values, adjusting for sample size. Phi defines a perfect association as one with complete statistical dependence, for example, if all members of a certain category of the independent variable are also all members of one response of the dependent variable.

Let's suppose that you were interested in the relationship between owning a gun to protect personal property and being a victim of a crime. You want to determine if people who have been a victim of a crime are more likely to own guns. Since both gun ownership and victim data are nominal variables, with only two possible categories (yes/no), using phi is a good way of determining whether an association exists. Let's use the 2004 Victimization Survey (GSS cycle 18) as a data source. (See Table 12.4.)

The first thing we need to do is create a table to calculate chi-square, as we did in Table 12.3 (see Table 12.5).

Looking at Appendix C, we can see that chi-square is significant. Now that we have chi-square, the calculation of phi is relatively straightforward:

$$\phi = \sqrt{\frac{\chi^2}{n}} = \sqrt{\frac{\mathbf{700.34}}{\mathbf{25{,}848{,}700}}} = \mathbf{0.005}$$

There is a statistically significant association between gun ownership and victimization (700.34 is much higher than the critical value of 3.841 found in Appendix C). Turning to

TABLE 12.4 | Gun Ownership and Victims of Crimes

	Have you been a victim of a crime in the past 12 months?		
Owns gun	Yes	No	Total
Yes	21,400	234,900	256,300
	(17,994)	(238,306)	
No	1,793,400	23,799,000	25,592,400
	(1,796,806)	(23,795,594)	
Total	1,814,800	24,033,900	25,848,700

Source: Statistics Canada. 2004. General Social Survey on Victimization, Cycle 18: An Overview of Findings. Catalogue no. 85-565-XIE

TABLE 12.5 | Computing Chi-Square for Gun Ownership and Victims of Crimes

Group	f_o	f_e	$(f_o - f_e)^2$	$\frac{(f_o - f_e)^2}{f_e}$
YY	21,400	17,994	11,600,836	644.71
YN	234,900	238,306	11,600,836	48.68
NY	1,793,400	1,796,806	11,600,836	6.46
NN	23,799,000	23,795,594	11,600,836	0.49
Total	25,848,700	25,848,700		$\chi^2 = 700.34$

Source: 2001 Census of Canada

BOX 12.3

It's Your Turn: Phi—Drinking and Daily Exercise

A student was thinking about the different men he was friends with: he felt that some men were always exercising and other men were always drinking. It seemed that none of them liked to both exercise and drink, but he thought that maybe this just happened among his friends. Using data from the Canadian Community Health Survey (CCHS, wave 2.1), he randomly selected 20 cases to test his null hypothesis that men who exercise at least 15 minutes a day are as likely as those who exercise less to regularly consume 12 or more drinks in a week. (These cases were selected only from male respondents who provided an answer for both questions.) If the student were to reject his null hypothesis, he would be able to conclude that there is a significant relationship between exercise and drinking for males.

TABLE 12.6 | Drinking and Exercise, 20 Cases from the Canadian Community Health Survey, Wave 2.1

Case	Regularly has more than 12 drinks a week (1 = yes, 0 = no)	Participates in at least 15 minutes of exercise daily (1 = yes, 0 = no)	Case	Regularly has more than 12 drinks a week (1 = yes, 0 = no)	Participates in at least 15 minutes of exercise daily (1 = yes, 0 = no)
1	1	0	11	0	1
2	0	0	12	0	0
3	1	0	13	0	0
4	1	1	14	1	1
5	0	1	15	0	0
6	0	0	16	1	0
7	1	1	17	0	0
8	1	0	18	0	0
9	0	0	19	1	0
10	0	0	20	1	1

1. Create a table (like Table 12.4) that compares your dependent (as rows) and independent (as columns) variables. Include column and row totals. (In this example, because we are not testing a causal relationship, either variable can be the independent or dependent variable.)
2. Compute the expected values for each cell and add values to the table in question 1.

$$f_e = \frac{\sum \text{column} * \sum \text{row}}{n}$$

3. Create a chi-square table (like Table 12.5). To do this, you will (1) subtract the expected from the observed value in each cell, and square the number, and (2) divide it by the expected value $\frac{(f_o - f_e)^2}{f_e}$. Once that is done for each cell, compute the totals to find the χ^2 value.
4. Calculate phi $\phi = \sqrt{\frac{\chi^2}{n}}$. Is there a strong or weak association between the variables?

The solution for Box 12.3 can be found in Appendix H.

the phi value of 0.005, researchers typically regard values between zero and 0.10 as a weak association, 0.10 and 0.30 as a moderate association, and values between 0.30 and 1.0 as a strong association. At the extremes, when two variables are completely dependent on each other, phi will take the value of one. When there is no dependence, phi will take the value of zero. These results (phi = 0.005) point to a weak association.

Phi is a popular measure for two by two tables. However, phi is only useful for demonstrating the existence and measuring the strength of a relationship. It is not possible to analyze the pattern of the relationship without looking at the contingency table. You should always examine the contingency table first and understand which values of the independent variable go along with which values of the dependent variable. When describing a relationship, you can use phi to show the existence and strength of the relationship, and use the contingency table to describe the pattern of the relationship.

As explained above, phi is only used with two by two tables. For tables larger than two by two, the appropriate chi-square-based measure is Cramer's *V*.

Cramer's *V*

Cramer's *V* is an elaboration of phi, except that it always assumes a value between zero and one. Calculating Cramer's *V* is simple, using the following equation:

$$V = \sqrt{\frac{\chi^2}{(n)(min(\mathbf{row} - \mathbf{1})|(\mathbf{column} - \mathbf{1}))}}$$

Although the equation seems more daunting than the one for phi, the two are actually similar. The difference is in the denominator, which is the number of observations multiplied by the lesser of the number of rows – 1, or the number of columns – 1.

If we calculated Cramer's *V* for city of residence and favourite Canadian National Hockey League (NHL) hockey team from a sample of 1000 people, we would have 26 different cities and 6 hockey teams. The denominator for the Cramer's *V* equation would be 1000 * (6 – 1), or 5000, because the hockey teams variable has 6 values, considerably fewer than the 26 values for city of residence. For a two by two table, the denominator would be *n* (2 – 1), which equals *n*, and reduces the equation for Cramer's *V* to that of phi.

Since Cramer's *V* is so similar to phi, they share almost all of the same properties and weaknesses. (See Box 12.4.)

The Proportional Reduction of Error: Lambda

The final measure of association for nominal data that we will study is lambda. Like phi and Cramer's *V*, lambda is used to measure the strength of a relationship between two nominal variables. It is non-directional, and since it relies on a **proportional reduction of error**, its numbers are interpreted more straightforwardly than either phi or Cramer's *V*. Lambda

BOX 12.4

Phi and Cramer's *V*: The Steps

1. Calculate the observed chi-square value (Box 12.2).
2. For phi, divide chi-square by the number of observations, then take the square root.
3. For Cramer's *V*, divide chi-square by the number of observations, multiplied *either* by the number of rows minus 1 *or* by the number of columns minus 1 (use the smaller of the two values), then take the square root.

BOX 12.5

It's Your Turn: Cramer's *V*

Based on readings that Jeff had done for his Sociology of the Family course, he wondered if women were more or less likely to have worked full-time rather than part-time in the last year if they were married, divorced, or single. Using data from the 2001 Census of Canada, he selected only women who answered both the questions on full- vs part-time work and marital status, and recoded marital status to combine divorced and separated. (People who were widowed were excluded.) He found the following observed and expected (in brackets) values for the Canadian population:

TABLE 12.7 | Chi-Square for Women's Relationship between Work Status and Marital Status in 2000

Work status in 2000	Marital status			
	Married	Divorced/separated	Single	Total
Worked mainly full-time weeks	3,485,748 (3,321,615.99)	604,011 (525,916.09)	1,193,175 (1,435,401.92)	2,582,934
Worked mainly part-time weeks	1,291,201 (1,455,333.0)	152,330 (230,424.9)	871,134 (628,907.1)	2,314,665
Total	4,776,949	756,341	2,064,309	7,597,599

Source: 2001 Census of Canada

1. Using the observed and expected frequencies for each cell, compute chi-square.
2. Instead of computing phi, compute Cramer's *V* (because the table is bigger than 2 * 2) using this equation:

$$V = \sqrt{\frac{\chi^2}{(n)(min(\text{row} - 1) \mid (\text{column} - 1))}}$$

The solution for Box 12.5 can be found in Appendix H.

allows us to answer the following question: How much is our ability to predict one variable improved by taking another variable into account?

To get lambda, first you calculate the extreme possibilities, then you calculate the total classification error by comparing the extreme predictions with the actual distribution of observations across response categories.

Like the other measures, the mechanics of calculating lambda are better illustrated with an example. Suppose we wanted to determine if there is a relationship between smoking and the sex of the respondent, and saw the patterns found in Table 12.8 in our hypothetical data.

From Table 12.8, we see that a higher proportion of females in our sample smoke. From this, we might conclude that knowing the sex of a respondent increases the accuracy of our prediction about whether or not a respondent is a smoker. However, what we don't know yet is how big the difference is. To determine if that is the case, we need to make two predictions. For our first prediction, we ignore the influence of the independent variable (sex). The first prediction is made without the independent variable. Our first prediction is always the overall mode for the whole sample. So in this case we would predict that everyone is a smoker (S = 103, NS = 0) Then we calculate the classification error, which is the number of observations minus the number of correctly classified cases.

For smokers, it would be equal to

103 (the total number of observations) – 63 (the number of correctly classified cases) = a total of 40 misclassifications

We will call this number E_1. For this example, $E_1 = 40$. E_1 is the amount of error made of the dependent variable, with no information on the independent variable.

TABLE 12.8 | Smoking and Sex of Respondent (Fictional Data)

	Female	Male	Row total
Smokes	42	21	63
Does not smoke	10	30	40
Column total	52	51	103

BOX 12.6

Why Does Non-Directionality Matter?

Because variables that are measured at the nominal level cannot be ranked or ordered, it's not surprising that these measures are non-directional. A non-directional measure is the best we can hope for.

The next step is to determine how much better we can do by using the values of our independent variable, sex of respondent. This time, we will make a prediction within each category of the independent variable. First we will make a prediction for females and then for males. Looking at the female column we see that the modal category is "smoke," so we will predict that all females smoke. Predicting that all females smoke yields a classification error of 10 (52 – 42 = 10). For males, the modal category is "don't smoke," so we will predict that all males don't smoke. Predicting that all males do not smoke misclassifies 21 people (51 – 30 = 21).

We sum the errors for each group (10 + 21 = 31) to produce E_2, which is the total classification error that can be made with the information provided for the independent variable. Here, when we take into account sex, we get a classification error of 31. This is the second error, and it represents the best possible prediction that can be made by including the independent variable, sex.

Finally, we define lambda (λ) as the percentage improvement in error from knowing a person's sex, using the following equation:

$$\begin{aligned} \lambda &= \frac{E_1 - E_2}{E_1} \\ &= \frac{40 - 31}{40} \\ &= 0.225 \end{aligned}$$

The number tells us that the change in the number of correct predictions between knowing the value for the independent variable and not knowing that value. The accuracy of the smoking prediction improves by 22.5 per cent when the value of the independent variable (sex of respondent) is known. Note that lambda is a relatively crude measure. If the mode does not change across the categories of the independent variable, lambda will be 0, even though there may be changes in other categories of the independent variable. (See Box 12.7.)

BOX 12.7

Calculating Lambda: The Steps

1. Predict the mode for the whole sample. Subtract *n* from the number of observations in the modal categories for the dependent variable. The lowest number gives you E_1.
2. For each category of the independent variable, predict the mode. Subtract *n* from the number of observations in the modal category for each category of the independent variable. Sum these amounts. This is E_2.
3. Calculate lambda as $\lambda = \frac{E_1 - E_2}{E_1}$

BOX 12.8

It's Your Turn: Lambda

Jose and his friend Vanessa have been arguing about whether women are more safety-conscious than men when they play sports. Vanessa is sure that they are, but Jose is sure they are just as likely as men to take risks. To test whether there is any difference between the two, Vanessa and Jose randomly select 100 individuals from the second wave of the CCHS who had gone in-line skating during the previous three months and who had responded to a question about whether they use protective gear when skating. They found the following:

Wears all protective equipment for in-line skating	Female	Male	Row total
Yes	9	2	11
No	42	47	89
Column total	51	49	100

1. Make the two extreme predictions for the dependent variable. The lower prediction will be E_1.
2. Calculate the extreme predictions by using the independent variable to determine E_2.
3. Calculate lambda (or the percentage increase in predictive accuracy): $\lambda = \frac{E_1 - E_2}{E_1}$

The solution for Box 12.8 can be found in Appendix H.

Conclusion

This chapter has covered statistics used to examine the association between two variables measured at the nominal level of measurement. First, we introduced contingency tables which enable you to identify whether there appears to be a pattern between the independent and dependent variables. Next, we discussed chi-square, a measure of statistical significance, which allows you to discuss whether a **bivariate relationship** seen in a sample is likely to also exist in the population from which the sample is drawn. Finally, we covered three measures of association, which allow you to examine the existence and strength of bivariate relationships: phi, Cramer's *V* and lambda. The next chapter will cover measures of association between two variables measured at the ordinal level of measurement.

Glossary Terms

Associations (p. 124)
Bivariate analysis (p. 125)
Bivariate relationship (p. 138)
Chi-square test (p. 126)
Contingency tables (p. 125)
Cramer's *V* (p. 134)
Degrees of freedom (p. 130)
Dependent variable (p. 125)
Independent variable (p. 125)
Marginals (p. 127)
Non-integer (p. 129)
Non-parametric test (p. 127)

Phi (p. 131)
Proportional reduction of error (p. 134)
Relationships (p. 124)
Statistical significance (p. 131)
Symmetrical (p. 131)

Practice Questions

We're interested in the relationship between religion and city of residence. Questions 1–5 refer to the results of the data from a selected sample from the 1901 census:

	City		
Religion	Saint John	Toronto	Quebec City
Roman Catholic	98	329	641
Church of England	73	552	30
Methodist	26	414	8
Presbyterian	55	403	5

Source: Fourth Census of Canada, 1901, Ottawa, Census Office, 1902 (AMICUS 7196327).

1. Why are these variables nominal?
2. Calculate chi-square for this table.
3. How many degrees of freedom are there in the table?
4. Calculate phi and Cramer's *V*. Which of the two measures would you use? Why?
5. Imagine that we had data only for Roman Catholics and the Church of England for Saint John and Quebec City. Choose the appropriate statistic to measure the association and calculate it.
    a. Looking back at our discussion of hypotheses in Chapter 10, draft a set of directional null and alternative hypotheses about a potential difference between Saint John and Quebec City. (For example, is one city significantly more Roman Catholic than the other?)

On 25 July 2011, *The Globe and Mail* polled its readers to gauge their opinion on whether facial-recognition technology constituted an infringement of privacy. Questions 6–8 refer to the results of this poll by the sex of the respondent, listed below:

Does facial-recognition technology constitute an infringement of privacy?	Female	Male	Row total
Yes	699	376	1075
No	1000	842	1842
Column total	1699	1218	2917

Source: *Globe and Mail*, 25 July 2011, readers' poll.

6. Draft null and research hypotheses for the table above.
7. Calculate lambda for the table above.

8. Based on your results in Question 7, do you find any reason to reject or fail to reject the null hypothesis you postulated in Question 6?

9. Bo and Jonah both like to study how people behave in society. One day, they decide to observe people in the public library for a morning, to see how many people could work in a public space without communicating with anyone. They hypothesize that those over the age of 30 will have different patterns of communication than those 30 and under. Here are the data they generate:

	Age 30 and under	Over age 30
Doesn't speak to anyone	17	15
Speaks with at least one person	12	22

Calculate chi-square for the above table. (Since we are only interested in the statistical significance of the differences, we can use chi-square even though other measures like lambda would give us more information.) Then, identify whether you have amassed any support for Bo and Jonah's hypothesis.

Answers to the practice questions for Chapter 12 can be found in Appendix H.

13 Bivariate Statistics for Ordinal Data

LEARNING OBJECTIVES

This chapter will continue our overview of the measures of bivariate association for variables at different levels of measurement. We'll cover several measures for ordinal data, specifically:

- Kruskal's gamma (γ);
- Somers' *d*;
- Kendall's tau-*b* and tau-*c*; and
- Spearman's *rho* (ρ_s).

Introduction

In Chapter 12, we looked at some of the tests of significance and measures of association frequently used to test for relationships between nominal variables. Some of the measures (phi, Cramer's *V*) are based on chi-square, whereas others (lambda) rely on a "proportional reduction in error." Since nominal data cannot be ranked or ordered, we did not discuss any statistics that assess the *direction* of a relationship between two variables. When we are assessing the relationship between nominal variables we must look at the **contingency tables** to determine what happens to the value of the dependent variable as the value of the independent variable changes. Relationships between nominal variables do not have a "direction" per se because the values of the variables themselves are not ordered.

In this chapter, we will look at tests of significance and measures of association for ordinal variables. Since ordinal variables have more desirable statistical properties than nominal variables (notably, the ability to rank response categories), after completing Chapter 13, not only will you be able to assess the existence and statistical significance of a relationship between two variables, you will also be able to identify the **direction** of the relationship. You will thus also be able to generate directional hypotheses.

There are at least five popular measures of association for variables measured at the ordinal level: Kruskal's gamma (γ), Somers' *d*, Kendall's tau-*b* and tau-*c*, and Spearman's *rho* (ρ_s). As with nominal variables, one of the most useful ways to understand the relationships between ordinal variables is to look at a contingency table.

Contingency Tables/Cross-Tabulations

To illustrate measures of association between ordinal variables, consider an example from mental health research. Suppose you want to determine if there are differences in self-perceived mental health by level of education. You hypothesize that people with higher education levels are likely to report higher levels of mental health than people with lower levels of education. What would you need to determine whether that relationship actually exists? In Canada, this inquiry can be made using the Canadian Community Health Survey (CCHS). We'll use wave 2.1 of the Canadian Community Health Survey, which provides data from 2003 on a broad range of health topics.

The first thing to do is to look at the distribution of each of the variables. For the purposes of this example, we have removed all the respondents who did not answer either the education or the self-rated mental health question in the CCHS. First we look at educational attainment levels of the Canadian population, as shown in Table 13.1.

Then we must look at self-perceived mental health, as shown in Table 13.2.

In Table 13.3, we show the bivariate distribution of self-rated mental health contingent on education. The cells in Table 13.3 contain the number of observations. In Table 13.4, the

TABLE 13.1 | Educational Attainment Levels of the Canadian Population, Canadian Community Health Survey, Wave 2.1, Valid Observations Only

Highest level – respond. 4 levels (*D*)		Frequency	Per cent	Valid per cent	Cumulative per cent
Valid	Less than secondary	6,923,116	26.6	26.6	26.6
	Secondary graduate	4,726,816	18.2	18.2	44.8
	Other post-secondary	1,996,934	7.7	7.7	52.5
	Post-secondary graduate	12,333,484	47.5	47.5	100.0
	Total	25,980,349	100.0	100.0	

Source: Canadian Community Health Survey Public Microdata File (82M0013X)

TABLE 13.2 | Self-Perceived Mental Health of the Canadian Population, Canadian Community Health Survey, Wave 2.1, Valid Observations Only

Self-perceived mental health		Frequency	Per cent	Valid per cent	Cumulative per cent
Valid	Excellent	9,923,648	38.4	38.4	38.4
	Very good	9,055,679	35.0	35.0	73.5
	Good	5,652,134	21.9	21.9	95.3
	Fair	994,779	3.9	3.9	99.2
	Poor	210,764	0.8	0.8	100.0
	Total	25,837,004	100.0	100.0	

Source: Canadian Community Health Survey Public Use Microdata File (82M0013X)

TABLE 13.3 | Self-Perceived Mental Health Status by Education, Canadian Community Health Survey, Wave 2.1, Valid Observations Only

	Highest level respond. 4 levels (*D*)				
Self-perceived mental health	Less than secondary	Secondary graduate	Other post-secondary	Post-secondary graduate	Total
Excellent	2,158,456	1,725,723	768,166	5,103,471	9,755,816
Very good	2,237,899	1,639,759	685,646	4,327,080	8,890,384
Good	1,789,087	1,028,018	413,825	2,299,402	5,530,332
Fair	331,363	181,424	87,898	373,042	973,727
Poor	70,770	41,202	18,494	76,288	206,754
Total	6,587,575	4,616,126	1,974,029	12,179,283	25,357,013

Source: Canadian Community Health Survey Public Use Microdata File (82M0013X)

same information exists but is presented in percentages. Pay particular attention to how the data are "clustered" in certain cells. Note that in Table 13.3, every cell has a fairly high number of observations, so we needn't worry about **sparsity** affecting our results. Other than that, it is difficult to glean any useful information from Table 13.3. For this reason, when working with large data sets it is easier to identify patterns by looking at the percentages of observations in each cell, rather than the raw number of observations.

Looking at Table 13.4, we start to see that a relationship exists between education and perceived mental health. Taking a look at the entire table, note that the percentage of people reporting better perceived mental health gradually increases as education levels get higher. Beyond identifying the direction of the relationship between the two variables, it is difficult to identify the strength of the relationship just by looking at the table. We need to rely on measures of association partly for this reason. Some of the leading measures for ordinal data are discussed below.

TABLE 13.4 | Self-Perceived Mental Health Status by Education Presented as Percentages, Canadian Community Health Survey, Wave 2.1, Valid Observations Only

	Highest level respond. 4 levels (*D*)				
Self-perceived Mental health	Less than secondary (%)	Secondary graduate (%)	Other post-secondary (%)	Post-secondary graduate (%)	Total (%)
Poor	1.1	0.9	0.9	0.6	0.8
Fair	5.0	3.9	4.5	3.1	3.8
Good	27.2	22.3	21.0	18.9	21.8
Very good	34.0	35.5	34.7	35.5	35.1
Excellent	32.8	37.4	38.9	41.9	38.5
Total	100.0	100.0	100.0	100.0	100.0

Source: Canadian Community Health Survey Public Use Microdata File (82M0013X)

Kruskal's Gamma (γ)

Kruskal's gamma (also called **Kruskal and Goodman's gamma**) is a measure of association that can be used with ordinal variables. Values of gamma range between −1.00 and +1.00, because the measure is directional. (If the value is positive, the two variables are positively related to each other; if the value is negative, the two variables are negatively related to each other.)

To compute gamma, two quantities are necessary:

1. N_{same} is the number of case pairs that are ranked in the *same* order on both variables, also known as **concordant pairs** or concordant observations.
2. $N_{different}$ is the number of case pairs that are ranked in a *different* order on each variable, also known as **discordant pairs** or discordant observations.

To understand what we mean by N_{same} and $N_{different}$, consider an example. Although we could continue with the mental health example above, it is a fairly large cross-tabulation for calculating gamma. So let's look at something simpler.

Suppose that we are interested in examining the relationship between educational attainment and physical activity among working-age Canadians. To do this, we can use the 2012 Canadian Community Health Survey. For this analysis, we are restricting our sample to respondents age 25–64 who answered questions about their educational attainment and their physical activity (n = 33,383). Although there is more detailed information in the survey, pretend for the purpose of this example that we only have two dichotomous, ordinal measures of education and physical activity. The measure of educational attainment has two response categories: "high school diploma or below" and "greater than high school." The measure of engagement in physical activity also has two response categories: "occasional or infrequent physical activity" and "regular physical activity."

Looking at Table 13.5, you can see that Canadians who have at least some post-secondary education are more likely to engage in regular physical activity (72 per cent) than those who have a high school diploma or less (63 per cent).

TABLE 13.5 | The Relationship between Educational Attainment and Physical Activity for Working-Age Canadians, CCHS, 2012

		Educational attainment		
		HS or below	Greater than HS	Total
Physical activity level	Occasional or infrequent	3623 (37%)	6592 (28%)	10,215
	Regular	6098 (63%)	17,070 (72%)	23,168
	Total	9721	23,662	33,383

Source: Statistics Canada. Canadian Community Health Survey: Public Use Microdata File (82M0013X).)

Table 13.5 suggests that there is a relationship between educational attainment and physical activity. What we do not know at this point is whether the relationship is strong, moderate, or weak. Our first step is to generate a research and null hypothesis:

> **H_1: Those with higher levels of education are more likely to report engaging in regular physical activity than those with lower levels of education.**
> **H_0: Education is not related to physical activity.**

We can use gamma to test these hypotheses. To calculate gamma we need two numbers. The first, N_{same}, is calculated by multiplying the number of people who have higher values on both variables (education = greater than high school; and physical activity = regular) by the number of people who have lower values on both variables (education = high school or below; and physical activity = occasional or infrequent). This gives us:

$$\mathbf{3623 * 17{,}070 = 61{,}844{,}610}$$

Next, calculate, $N_{different}$, which is equal to the number of observations with different values on each variable. In a two by two table, these are the two opposite cells of the ones used to calculate N_{same} yielding: 6592 * 6098 = 40,198,016

Now that we have N_{same} and $N_{different}$, we calculate gamma by using the following equation:

$$\lambda = \frac{N_{same} - N_{different}}{N_{same} + N_{different}}$$

Inserting our numbers of interest, then making the necessary calculations, yields the following:

$$\lambda = \frac{61{,}844{,}610 - 40{,}198{,}016}{61{,}844{,}610 + 40{,}198{,}016}$$

$$= \frac{21{,}646{,}594}{102{,}042{,}626}$$

$$= 0.212$$

The equation gives us a gamma value of 0.212.

The equation may look daunting, but that's only because of the big numbers. Using smaller samples would make these numbers less intimidating.

If you look at Table 13.6, you can see from the rough guidelines for interpreting gamma that this value counts as a moderate relationship, so now we know that there is a moderate relationship between educational attainment and physical activity among working-age Canadians.

TABLE 13.6 | Rough Guidelines for the Interpretation of Gamma Values

Value	Strength
Between 0.0 and 0.10	Weak
Higher than 0.10 and less than 0.30	Moderate
0.30 or higher	Strong

TABLE 13.7 | Support for Smoking Ban in the Workplace, by Level of Education (in a Two by Two Table)

	No university training	At least some university training	Total
No support for ban	18	10	28
Full support for ban	10	14	24
Total	28	24	52

Gamma ranges in value from −1.00 to +1.00. A value of −1.00 indicates that all untied pairs are discordant, which implies a perfect negative relationship: that is, knowing a person's score on an independent variable improves your ability to predict the score on the dependent variable, and the relationship is negative. A value of +1.00 indicates the opposite; all untied pairs are concordant and the relationship between independent variable and dependent variable is perfect and positive. A discordant pair is defined as any value where values on each variable run in a different "direction"—meaning that somebody who has a high score on the independent variable has a low score on the dependent variable, and vice versa. A concordant pair refers to any observation with scores that run in the same direction; here, a high score on the independent variable would be matched by a high score on the dependent variable. Let's use another example: the relationship between support for a smoking ban in the workplace and an individual's level of education. (See Table 13.7.)

In a two by two table like Table 13.7, calculating concordant and discordant pairs is easy. When your contingency table is arranged properly (so that values consistently go from high to low, or low to high, on both axes), concordant pairs in a two by two table are defined as those in the top left and bottom right cells, and discordant pairs are those in the top right and bottom left cells. The top left cell of frequencies (where individuals do not support smoking bans and hold less than a university degree), represents one of the two concordant cells; 18 people with both low support for smoking bans and lower levels of educational attainment are placed there. The other concordant table is the bottom right cell of frequencies; 14 people with high support for smoking bans and higher levels of education are found there.

In a table that is larger than two by two, it is more difficult to calculate concordant and discordant pairs. This is because determining whether a pair is concordant or discordant depends on the location of a particular cell. Imagine Table 13.7 having three categories for each variable, instead of two.

A concordant pair is defined as any positive diagonal of a particular cell. A positive cell is one where responses to one variable are in the same direction as responses to another variable. (Ties are ignored for gamma in a table larger than two by two.) For example, in Table 13.8 the cells concordant with the top left cell, *a* (less than high school diploma, no support for smoking ban), are cells *e*, *f*, *h*, and *i*. Each one is both below and to the right of cell *a*. Other concordant cells in Table 13.8 are *f* and *i* with *b*; *h* and *i* with *d*; and *i* with *e*. In a three by three table, these form the four concordant sets.

The next step is to sum the number of observations that are concordant to each cell, and multiply that sum by the cell frequency, giving us N_{same}. To help with that, it is useful to organize the information in the manner seen in Table 13.9.

Now that we have N_{same}, we must calculate the number of discordant cells and observations ($N_{different}$). The logic is the same as for calculating concordant cells, except that we tally the number of observations operating in the *opposite* direction (to the left and below). For cell *c*, the discordant cells would be *e*, *d*, *g*, and *h*; for cell *b* it is *d* and *g*; for *f* it is *g* and *h*; and for *e* it is *g*. Table 13.10 organizes information in the same way as Table 13.11, but this time for the discordant cells.

TABLE 13.8 | Support for Smoking Ban in the Workplace, by Level of Education (in a Table Larger than Two by Two)

	Less than high school diploma	High school diploma, no university	At least some university training	Total
No support for ban	18(a)	10(b)	10(c)	38
Some support for ban	14(d)	12(e)	11(f)	37
Full support for ban	10(g)	14(h)	14(i)	38
Total	42	36	35	113

TABLE 13.9 | Concordant Cells of Table 13.8

Cell	# of concordant cells	# of concordant observations	Contribution to N_s
A	4 (e, f, h, i)	12 + 11 + 14 + 14 = 51	18 * 51 = 918
B	2 (f, I)	11 + 14 = 25	10 * 25 = 250
C	0		
D	2 (h, I)	14 + 14 = 28	14 * 28 = 392
E	1 (I)	14	12 * 14 = 168
F	0		
G	0		
H	0		
I	0		
			N_s = 1,728

TABLE 13.10 | Discordant Cells of Table 13.8

Cell	# of discordant cells	# of discordant observations	Contribution to N_d
a	0		
b	2 (d, g)	14 + 10 = 24	10 * 24 = 240
c	4 (d, e, g, h)	14 + 12 + 10 + 14 = 50	10 * 50 = 500
d	0		
e	1 (g)	10	12 * 10 = 120
f	2 (g, h)	10 + 14 = 24	11 * 24 = 264
g	0		
h	0		
i	0		
			N_d = **1,124**

Tables 13.9 and 13.10 give us N_{same} and $N_{different}$, so we can calculate gamma using the same formula as before:

$$\lambda = \frac{N_{same} - N_{different}}{N_{same} + N_{different}}$$

$$= \frac{1728 - 1124}{1728 + 1124}$$

$$= 0.212$$

The value of 0.212 implies a moderately strong and positive relationship. If the number were negative, the relationship would be moderate but negative.

Although gamma is widely used, it has the limitation of ignoring tied pairs, or values that are neither concordant nor discordant. A significant proportion of all observations are often not used in the calculation of a relationship, suggesting that gamma tends to overestimate the relationship between variables, especially when there are a lot of tied cases.

BOX 13.1

Gamma: The Steps

1. Calculate N_{same} (or the number of concordant pairs).
2. Calculate $N_{different}$ (or the number of discordant pairs).
3. Calculate gamma by using the following equation:

$$\lambda = \frac{N_{same} - N_{different}}{N_{same} + N_{different}}$$

4. As a rough guideline, consider values between 0 and 0.10 to be weak, higher than 0.10 and less than 0.30 to be moderate, and 0.30 or higher to be strong.

BOX 13.2

It's Your Turn: Calculating Gamma

Zane has recently noticed a lot of media attention about how people are afraid of walking alone in his city at night. He wonders if this fear is related to beliefs about the Canadian justice system. To find out if this is true, he uses data from the 2004 General Social Survey (GSS18), which focused on victimization. Since the GSS is a huge survey and Zane knows he would have to compute gamma by hand (his computer is broken), he decides to select a sample of 500 people. Once the missing cases are removed, Zane has 413 observations. The contingency table is presented below.

Relationship between Views of Justice and Frequency of Walking Alone at Night

		Walk alone at night			
		At least once a week	Up to once a month	Never	Total
Courts do a good job of quick justice	Good	40	18	16	74
	Average	77	61	26	164
	Poor	86	51	38	175
	Total	203	130	80	413

General Social Survey, Cycle 18: Victimization (2004): Public Use Microdata File and Documentation, 2004 (12M0018X)

1. Compute N_{same} and $N_{different}$. (Hint: use a table to make this easier.)

2. Calculate gamma: $\lambda = \dfrac{N_{same} - N_{different}}{N_{same} + N_{different}}$

3. Determine if there is a weak, moderate, or strong relationship between people's views on the efficiency of the courts and walking alone in neighbourhoods after dark.

   The solution for Box 13.2 can be found in Appendix H.

Somers' *d*

Somers' *d* is one of several alternatives to gamma. It is quite similar to gamma, except that it adjusts for tied ranks on the dependent variable. Ties occur when two pairs of scores are both either concordant or discordant for the independent variable and tied on the dependent variable. The following equation is used to calculate Somers' *d*:

$$d = \frac{N_{same} - N_{different}}{N_{same} + N_{different} + Ties_y}$$

As you can see, this equation is very similar to gamma except for the addition of a term in the denominator. Calculating the number of ties can be problematic; see Table 13.11 for help.

TABLE 13.11 | Levels of Education and Support for Smoking Ban

	Less than high school diploma	High school diploma, no university	At least some university training	Total
No support for ban	18(a)	10(b)	10(c)	38
Some support for ban	14(d)	12(e)	11(f)	37
Full support for ban	10(g)	14(h)	14(i)	38
Total	42	36	35	113

Let us go back to the example of levels of education and support for a smoking ban. Table 13.8 is repeated here, renumbered as Table 13.11. In Table 13.11, cells *b* and *c* are tied with *a*; *c* is tied with *b*; *e* and *f* are tied with *d*; *f* is tied with *e*; *h* and *i* are tied with *g*; and *i* is tied with *h*, yielding the following value for $Ties_y$:

$$\begin{aligned} Ties_y &= 18(10+10)+10(10)+14(12+11)+12(11)+10(14+14)+14(14) \\ &= 360+100+322+132+280+196 \\ &= 1390 \end{aligned}$$

N_{same} is 1728 and $N_{different}$ is 1124, so we can find *d*:

$$\begin{aligned} d &= \frac{N_{same} - N_{different}}{N_{same} + N_{different} + Ties_y} \\ &= \frac{1728 - 1124}{1728 + 1124 + 1390} \\ &= 0.142 \end{aligned}$$

This is a lot lower than the gamma value of 0.212 because there is an additional term in the denominator. Like gamma, values for Somers' *d* range between −1 and +1 with an approximately similar interpretation. Hypothesis testing and formulation is also similar in that it is directional.

BOX 13.3

Somers' *d*: The Steps

1. Calculate N_{same}, or the number of concordant pairs.
2. Calculate $N_{different}$, or the number of discordant pairs.
3. Calculate $Ties_y$, or the number of ties.
4. Calculate Somers' *d* using the following equation:

$$d = \frac{N_{same} - N_{different}}{N_{same} + N_{different} + Ties_y}$$

Just as when you're using gamma, consider values between 0 and 0.10 to be weak, higher than 0.10 and less than 0.30, to be moderate, and 0.30 or higher to be strong.

BOX 13.4

It's Your Turn: Calculating Somers' *d*

Isa is convinced that people who study full-time have dirty homes, but she wants to test her theory using Somers' *d*. Using data from the 2001 Census of Canada Individual File, Isa randomly selects 250 people who had answered questions about their educational status and the number of hours they spent on housework. She recodes educational status, going from not being a student to being in school full-time. She also recodes hours of unpaid housework into three categories.

The Relationship between Educational Status and Hours Spent on Unpaid Labour per Week

	Educational status			
Hours on unpaid household labour/week	Not studying	Part-time student	Full-time student	Total
Less than 5	52	1	20	73
5 to 14	59	2	5	66
15 or more	106	2	3	111
Total	217	5	28	250

Source: 2001 Census of Canada

1. Compute N_{same} and $N_{different}$ and $Ties_y$.
2. Calculate Somers' *d*: $d = \dfrac{N_{same} - N_{different}}{N_{same} + N_{different} + Ties_y}$
3. Based on your calculation, what can you say about the relationship between hours of unpaid household work and being a student, using Somers' *d*? Do you find support for Isa's theory?

The solution for Box 13.4 can be found in Appendix H.

Kendall's Tau-*b*

Kendall's tau-*b* is conceptually similar to gamma and Somers' *d*, but it goes one step further than Somers' *d* and corrects for tied pairs on both the dependent variable and the independent variable. Its equation is as follows:

$$\text{tau-}b = \frac{N_{same} - N_{different}}{\sqrt{(N_{same} + N_{different} + Ties_y)(N_{same} + N_{different} + Ties_x)}}$$

Calculating $Ties_x$ is similar to calculating $Ties_y$, except instead of looking for ties on the dependent variable, you look for ties on the independent variable. So, for Table 13.11 the calculation is as follows:

$$\begin{aligned} Ties_x &= 18(14+10)+14(10)+10(12+14)+12(14)+10(11+14)+11(14) \\ &= 432+140+260+168+250+154 \\ &= 1404 \end{aligned}$$

And so the equation is:

$$\text{tau-}b = \frac{N_{\text{same}} - N_{\text{different}}}{\sqrt{(N_{\text{same}} + N_{\text{different}} + Ties_y)(N_{\text{same}} + N_{\text{different}} + Ties_x)}}$$

$$= \frac{1728 - 1124}{\sqrt{(1728 + 1124 + 1390)(1728 + 1124 + 1404)}}$$

$$= \frac{604}{\sqrt{(4242)(4256)}}$$

$$= 0.142$$

In our example, Somers' *d* and tau-*b* are identical, because 1390 and 1404 are so close that multiplying essentially equal numbers (4242 and 4256) and then taking the square root has almost no effect on the denominator.

Note: When the independent and dependent variables do not have the same number of categories, we use **Kendall's tau-*c***. This is beyond the scope of this text.

Spearman's *rho*

As you can imagine, calculating gamma isn't easy using any table larger than two by two. Also, given the limitations of gamma, it is often useful to consider other options.

Spearman's *rho*, or Spearman's rank order correlation coefficient, is one such option. The traditional formula for calculating the *rho* is:

$$\rho_s = 1 - \frac{6 * \sum D^2}{N(N^2 - 1)}$$

Spearman's *rho* relies exclusively on the *rank* of observations, rather than on the value, making it a prime candidate for calculations using ordinal data. Since it is not possible to measure distances between values with ordinal data, Spearman's *rho* takes the approach of comparing the level of concordance between the rank order of one variable and the rank order of another.

To illustrate, consider an example. Suppose we suspect that there is a relationship between the number of races a cyclist wins and the number of endorsements he or she receives. To keep it simple, let's look at five cyclists from different countries: Japan, US, Canada, Russia, and Venezuela. Each has won a number of races, and we believe that winning a race makes a particular candidate more attractive to sponsors, leading to an increase in the number of endorsements received.

Since *rho* is only concerned with ranks, it is necessary to sort and rank each of the variables.

TABLE 13.12 | Spearman's *rho* and the Relationship between Races Won and Endorsement Participation

	Japan	US	Canada	Russia	Venezuela	Totals
# of races won	5	2	3	6	7	
Rank	3	5	4	2	1	
# of endorsements	4	3	5	7	6	
Rank	4	5	3	1	2	
D (Rank for wins − rank for endorsements)	−1	0	1	1	−1	0
D^2	1	0	1	1	1	4

The easiest way to calculate Spearman's *rho* is by organizing all of the information in a table, like Table 13.12. It is also helpful to have a null and research hypothesis:

> **H_1: As the number of races that a cyclist wins increases, so too does the number of endorsements he or she receives.**
> **H_0: There is no relationship between race wins and endorsements.**

Once we have the information organized, calculating *rho* is easy because we can straightforwardly measure the discrepancy for each observation between the rank on one variable and the rank on the other. *D* is the difference between these two rankings—in this case, the difference between the rank on number of races won and the rank on number of endorsements. D^2 is $D * D$. By summing D^2, we get one of the numbers that we need to calculate *rho*.

$$\begin{aligned}\rho_s &= 1 - \frac{6 * \sum D^2}{N(N^2 - 1)} \\ &= 1 - \frac{6 * 4}{5(25 - 1)} \\ &= 1 - \frac{24}{120} \\ &= 0.80\end{aligned}$$

The sign of *rho* indicates the direction of the relationship. Our value here indicates that we have a strong and positive relationship and that we have found considerable reason to reject the null hypothesis.

Squaring *rho* allows for a PRE-type interpretation (Proportional Reduction in Error), which means that we can calculate the reduction in our errors in predicting the value on the dependent variable by knowing the value on the independent variable. In this case it's $0.8^2 = 0.64$, so our errors of prediction for the cyclists' endorsement rank will be reduced by 64 per cent once we know their rank on the number of races won.

BOX 13.5

Spearman's *rho*

Psychologist Charles Spearman (1863–1945) first proposed the measure of correlation we call Spearman's *rho* in a 1904 paper titled "The Proof and Measurement of Association between Two Things." In that paper, he did not call the value he arrived at *rho*, nor did he give it the Greek letter ρ, which is used today. In that early paper, the value was similar to, but not the same as, *rho* as we know it today. Calling it a "method of rank differences," Spearman listed some disadvantages:

1. It can be done with ranks only, not measurements.
2. The probable error we get tells us what correlation we could expect from independent variables. It doesn't tell us the error we can expect to attribute to measurement (as Pearson's probable error does).
3. Because the values of Somers' *d* do not follow a normal, or even symmetrical, distribution, there are problems when one takes negative values as inverse correlation.
4. This value ρ_s is not the same as the *rho* of other methods of correlation.

Spearman continued working and reflecting on formulae for assessing correlation and took up *r* again in a 1906 paper titled "Foot Rule for Measuring Correlation," where he advocated it as a quick rule for determining correlation, while still being comparable to **Pearson's *r*.**

Spearman's motivation apparently was to establish a suitable rule for correlation in psychology that is both pragmatic and statistically meaningful. One of his supporting arguments for this measure of correlation is that it can be done in less than a minute by hand, in some cases entirely in one's head, and can provide a guideline (or foot rule, as Spearman calls it) for what sort of correlation, or lack thereof, one is looking at. It is up to the reader whether the standard formula for *rho* fulfills Spearman's desire for a foot rule requiring only a "trifling expenditure" on the part of the researcher. Perhaps, just as electronic calculation has supplanted the extensive need for logarithmic tables, the ease of use of statistical formulae is no longer a determining factor in their development.

BOX 13.6

Spearman's *rho*: The Steps

1. Organize the observations into a chart, with columns for score on each variable of interest, their respective ranks, the differences between the two (*D*), and the differences between the two squared (D^2). D^2 will be the most useful for calculating *rho*.

1	2	3	4	5	6	7
Case	Raw score of variable 1	Rank of variable 1 (1st, 2nd, etc.)	Raw score of variable 2	Rank of variable 2 (1st, 2nd, etc.)	*D* (column 3 − column 5)	D^2
n						

2. Calculate *rho* as

$$\rho_s = 1 - \frac{6 * \sum D^2}{N(N^2 - 1)}$$

BOX 13.7

It's Your Turn: Calculating Spearman's *rho*

Andy feels that he's been doing pretty well in his statistics class. His friend Marianne is surprised—they always study together, but she doesn't do as well. What could be the difference? "Andy does go out drinking every weekend," she thinks to herself. To figure out if drinking more might improve her grades, Marianne conducts a small study. She asks Andy and seven other friends what grade they received on the mid-term, and how many drinks they usually have in a week. These are the data she collects:

Case	Number of drinks a week	Grade
1	10	65
2	2	75
3	2	52
4	0	98
5	1	45
6	5	55
7	20	70
8	3	60

At first glance, there doesn't appear to be much of a pattern, but she realizes that she needs to make a ranking table.

1. Complete the ranking table below.
   Note that there are two people who have the same number of drinks (two drinks per week). When this happens, the ranking is the average of the two ranks. If they take up the places for ranks three and four, they will each rank 3.5. When in doubt about how to handle ties, you'll know you've made the right choice when the sum of *D* equals zero.

Case	Number of drinks a week	Drinks rank	Grade	Grade rank	D	D^2
1	10		65			
2	2		75			
3	2		52			
4	0		98			
5	1		45			
6	5		55			
7	20		70			
8	3		60			

2. Calculate *rho*: $\rho_s = 1 - \frac{6 * \sum D^2}{N(N^2 - 1)}$
3. Square the value of *rho*.
4. How much does knowing the number of drinks a person has in a week improve your ability to estimate how well Marianne and Andy will do in their statistics class?

The solution for Box 13.7 can be found in Appendix H.

What about Statistical Significance?

You may have noticed that we haven't been discussing statistical significance a whole lot in this chapter. This exclusion was intentional, and although the considerations covered in earlier chapters also apply here, calculating significance for the relationship between two ordinal variables is the same process as that described in Box 12.2: Chi-Square: The Steps. Generally, however, ordinal data have three or more categories for each variable, making the calculation of chi-square cumbersome, though certainly possible.

It is often helpful to rely on chi-square to identify whether an association is statistically significant. Indeed, it can be quite difficult to test hypotheses without a measure of significance. However, chi-square treats both your independent and dependent variables as nominal, forcing you to disregard some of the information inherent to ordinal variables, such as the direction of the relationship. The upside of chi-square is that it can be calculated with relative ease for smaller tables. In larger tables, where the calculations become rather cumbersome, it is still possible to understand the underlying principle even if you are relying on a computer for the calculations. For this reason, chi square is often used to test hypotheses with ordinal variables, even though, strictly speaking, it is a rather blunt instrument for doing so. As we discussed in Chapter 12, the degrees of freedom are equal to $df = (\text{row} - 1) * (\text{column} - 1)$. When chi-square is below the critical value, we say that our variables of interest are independent and that we fail to reject the null hypothesis. If the calculated value exceeds the critical value, though, we can reject the null hypothesis and say that the variables are likely not independent in the population.

Thinking about this in terms of association, the null hypothesis would be that there is no association between the two variables, or that the value of gamma, tau-*b*, Somers' *d*, or Spearman's *rho* is 0. The test you are performing is to determine whether the measure of association you have calculated for your sample is significantly different from 0. You can determine this with chi-square, although you must implicitly treat the variables as nominal to do so.

A superior, though perhaps more complicated, method for identifying the statistical significance of a measure of association for gamma, tau-*b*, and Somers' *d* can be taken from the normal curve values of *z*. To calculate z_{obtained}, use the following equation (shown using gamma, although also suitable for tau-*b* and Somers' *d*):

$$z_{\text{obtained}} = \sqrt{\frac{N_{\text{same}} + N_{\text{different}}}{N\,(1-\gamma^2)}}$$

You should have at least 120 observations to use this test. Otherwise, use the *t*-value and define the degrees of freedom as $df = n - 1$. In either instance, values that fall below the critical value are insufficient to reject the null; values that exceed the critical value lead you to reject the null.

For testing the significance of Spearman's *rho*, use the following equation:

$$t_{\text{obtained}} = \sqrt{\frac{N-2}{1-\rho_s^2}}$$

where degrees of freedom are defined as $df = n - 2$. (We subtract 2 because we can determine a person's rank on each variable once we know all the other values.)

EVERYDAY STATISTICS

Choosing the Right Significance Test

In the early 1950s, Cyril Burt, a famous British psychologist, concluded that genetic factors are more important than environmental factors in determining IQ. Burt based this conclusion on a study that he conducted with 42 pairs of identical and non-identical twins who were reared apart. Burt then measured the difference in the IQ scores of the twins, and found that the IQ scores of identical twins were closer than those of the non-identical twins.

Q: What type of significance test do you think Burt conducted to obtain his findings?

Conclusion: Which One to Use?

In this chapter, we have covered four of five different measures of association for ordinal data. How do you choose which of these measures to use? Unfortunately, that is a hard question to answer.

There are occasions when one measure is preferable to another. For example, if there are a lot of ties on either the independent or the dependent variable, it is desirable to use something other than gamma. Spearman's *rho* is often easiest to understand when you're more interested in ranks than scores. Some of the other measures can be computationally intense and might be undesirable in certain situations, such as when you must calculate by hand. It's

BOX 13.8

Kendall's Tau-*b*: The Steps

1. Calculate N_{same}, or the number of concordant pairs.
2. Calculate $N_{different}$, or the number of discordant pairs.
3. Calculate $Ties_y$, or the number of ties in the dependent variable.
4. Calculate $Ties_x$, or the number of ties in the independent variable.
5. Calculate Kendall's tau-*b* by using the following equation:

$$\text{tau-}b = \frac{N_{same} - N_{different}}{\sqrt{(N_{same} + N_{different} + Ties_y)(N_{same} + N_{different} + Ties_x)}}$$

As with gamma and Somers' *d*, values between 0 and 0.10 are weak, higher than 0.10 and less than 0.30 are moderate, and 0.30 or higher are strong.

BOX 13.9

It's Your Turn: Kendall's Tau-*b*

Using the same data that Isa collected in Box 13.4, can you see any changes in the relationship when you calculate Kendall's tau-*b*?

1. Compute the value for $Ties_x$.
2. Calculate Kendall's tau-*b*:

$$\text{tau-}b = \frac{N_{\text{same}} - N_{\text{different}}}{\sqrt{(N_{\text{same}} + N_{\text{different}} + Ties_y)(N_{\text{same}} + N_{\text{different}} + Ties_x)}}$$

3. Does the inclusion of the ties on the independent variable change what Isa can say about the relationship between students and hours spent on housework?

The solution to Box 13.9 can be found in Appendix H.

important to remember that although they won't give you exactly the same numbers, most measures will yield similar results in most circumstances (except for gamma, which will often be higher because it ignores the tied pairs). The next chapter will introduce you to measures of association for variables measured at the interval and ratio levels of measurement.

Glossary Terms

Concordant pairs (p. 144)
Contingency tables (p. 141)
Direction (p. 141)
Discordant pairs (p. 144)
Kendall's tau-*b* (p. 151)
Kendall's tau-*c* (p. 152)
Kruskal's gamma (p. 144)
Kruskal and Goodman's gamma (p. 144)
Pearson's *r* (p. 154)
Somers' *d* (p. 149)
Sparsity (p. 143)
Spearman's *rho* (p. 152)

Practice Questions

1. Jerry is interested in the relationship between exercising every week and improvements in health. He thinks it might be part of why older people care for themselves better than young people do. Here are the data he is working with:

	Changes in self-rated health	
Exercises every week	Poor	Good
No	13	11
Yes	17	25

Formulate a null and research hypothesis to test Jerry's hunch.

2. Calculate gamma for the table above, then determine its statistical significance by using the *t*-distribution. Does this support Jerry's hypothesis?

3. Universities across Canada are working hard to increase their research revenues. It may be difficult for some universities to increase their profiles, however, because funding agencies might be more likely to continue to give money to the universities that they have given to in the past, creating a pattern of path dependency. To test this prospect, you are hired by a consortium of universities.

Universities and Research Revenue in 2000 and 2009

Case	University	2000 research revenue, $000	2009 research revenue, $000
1	University of Toronto	372,119	858,182
2	University of British Columbia	165,992	524,569
3	University of Alberta	206,667	507,613
4	Université de Montréal	253,099	486,179
5	McGill University	234,340	432,118
6	McMaster University	106,892	377,732
7	Université Laval	168,382	282,657
8	University of Calgary	134,507	264,358

Note: Figures not adjusted for inflation

Source: Research Infosource (http://www.researchinfosource.com/top50.shtml)

What null and research hypothesis would allow you to identify the presence of path dependence (assume that the funding comes from sources in the same ratios across years)?

4. Calculate Spearman's *rho* for the ranking of the eight universities above. Do you find support for path dependence? Is your value of *rho* statistically significant? How do you know?

5. Suppose you are interested in squirrels and nuts (who isn't?), and you hypothesize that the heavier a squirrel is, the more food that squirrel will hide away for the winter. Since the nuts vary dramatically in size, you can only treat the data as ordinal. Formulate a set of hypotheses to test whether this is true.

6. Calculate Somers' *d* to identify the relationship between squirrel mass and the number of stashed nuts. Do your results support or fail to support your null hypothesis from #5 above?

	Weight of squirrel			
Number of nuts stored for winter	<0.5 kg	0.5 kg–1 kg	>1 kg	Total
Fewer than 25	12	11	5	28
25 to 44	6	7	9	22
45 or more	3	2	12	17
Total	21	20	26	67

7. Now, calculate Kendall's tau-*b* for the same data. Does tau-*b* lead you to make a different decision regarding your hypotheses?

Answers to the practice questions for Chapter 13 can be found in Appendix H.

14 Bivariate Statistics for Interval/Ratio Data

LEARNING OBJECTIVES

Chapter 14 will continue to examine techniques for assessing associations between two variables. Now that we've introduced measures for relationships between nominal and interval/ratio variables, nominal and nominal variables, and ordinal and ordinal variables, we will look at measures of association for relationships between two interval/ratio variables. Specifically, this chapter will focus on:

- what r tells us about explained variance;
- the correlation matrix; and
- what to do when your independent and dependent variables are measured at different levels of measurement.

Introduction

By now, you know that when we examine associations between variables, we are primarily interested in how the categories of one variable relate to categories of another variable. Similar to ordinal data, when we have interval/ratio data, we can determine whether a relationship exists between two variables, and we can also determine the *direction* and *strength* of that relationship. Unlike the measures of association we learned for nominal and ordinal levels of measurement, though, measures of association for interval/ratio variables can also be used to represent relationships graphically, so we can determine the approximate *rate of change* in the dependent variable across the values of the independent variable.

Pearson's r: The Correlation Coefficient

The primary measure of association used for interval/ratio variables is another contribution of statistician Karl Pearson. (See Box 14.1 for a history of the term.) **Pearson's r** measures the amount of change in Y (the dependent variable) produced by a unit change in X (the independent variable), where the units are expressed as standard deviations. Like some of the ordinal measures covered in the last chapter, r ranges between -1 and $+1$, with 0 representing no relationship between variables. A value of -1 denotes a perfectly negative relationship: as X rises/declines in value, Y always moves by the same amount (as measured in standard

BOX 14.1

Pearson's Correlation Coefficient: History of a Term

Pearson's *r* is an indicator of how closely two variables are related and designates the strength and direction of a linear relationship. Pearson called his method for calculating this coefficient the product, which he initially called the "moment method" (Porter, 1989). He developed it while working on developing regression analysis and by taking up Galton's work, for it was Galton who first suggested the indicator, in the context of heredity. Although Pearson published his method in an 1895 paper, it was not received with much fanfare. Francis Galton himself was one of the paper's reviewers, and he thought it should be published rather than read aloud in a lecture. In Galton's words, "It would be too dull to *read*" (Stigler, 1986: 344).

The attribution for this coefficient brings up important issues about who gets credited with developing statistical methods. Although Pearson can be credited with much of the work, he did not operate in a vacuum. As we mentioned before, *r* also stems from Galton's research. Further, Pearson and statistician Francis Edgeworth corresponded and reviewed each other's work. Although Pearson did credit Edgeworth early on, referring to "Edgeworth's theorem," in his later work he did not do so.

deviations) in the opposite direction. A perfectly positive relationship of +1 indicates the opposite: as *X* rises/declines in value, *Y* always moves by the same amount in the same direction.

There are a few different formulas for Pearson's *r*. The first is considered an "illustrational" formula, because it illustrates the process, but it can be cumbersome to use:

$$r = \frac{\sum\left[(X-\bar{x})(Y-\bar{y})\right]}{\sqrt{\left[\sum(X-\bar{x})^2 * \sum(Y-\bar{y})^2\right]}}$$

The problem with this equation is that it requires you to subtract the average from each *X* and *Y* value line by line, which would take forever and a day in large data sets. Fortunately, the equation can be modified into what is called a "computational" formula, so that that task is unnecessary:

$$r = \frac{N\sum XY - \left(\sum X\right)\left(\sum Y\right)}{\sqrt{\left[N\sum X^2 - \left(\sum X\right)^2\right]\left[N\sum Y^2 - \left(\sum Y\right)^2\right]}}$$

Most of the time, you'll want to use the second equation. Even though it looks more difficult, it is actually much easier to use.

Take note of the $(\Sigma X)^2$ and $(\Sigma Y)^2$ in the denominator. According to BEDMAS, the order of operations that we discussed in Chapter 2 (Brackets-Exponents-Division-Multiplication-Addition-Subtraction), the calculations within the parentheses must be performed first, then the ones outside them. This means that the *X* and *Y* values must be summed *first*, before raising them to the power of two. People often forget to do this the first time they calculate *r*, and they calculate it incorrectly as a result.

With a little care, you won't be one of those people. To make sure of that, we'll demonstrate the calculation of *r* with an example: Suppose we are interested in the relationship between a person's age and the age of their spouse in colonial Canada (or New France as it was then known) in 1665 and 1666. These data are available because Jean Talon, Intendant of Justice, Police, and Finance in 1665–8 and 1670–2, decided that the best way to learn about the colony was to take a numerical inventory of its inhabitants. In North America's first census, Talon enumerated the 3278 residents of New France, a critical step in establishing New France as a self-sufficient settlement.

These records, an interesting and important part of Canada's history, can be used to demonstrate Pearson's *r*. As with most other measures of association, it is helpful to arrange your data in a table first so that you can calculate the values of interest easily. Table 14.1 shows these data for 5 selected cases from the 513 suitable for analysis. (Since we are comparing the ages of husbands and wives, our unit of analysis is the household, not the individual.) (The first four cases and the last case in the full data set are shown.) In the table, *X* represents the age of the husband; and *Y*, the age of his wife. As we've done so many times already, it is useful to hypothesize a relationship between the variables. Since Pearson's *r* is a directional measure, we should capitalize on the ability to indicate the direction of the relationship.

> **H_1: The age of the husband is related to the age of his wife; the older a husband is, the older his wife will be and vice versa.**
>
> **H_0: There is no relationship between the age of a husband and the age of his wife.**

Once these data are organized and each of the columns is summed, it is easy to calculate *r* by inserting the appropriate values into the formula. Note that the sum at the bottom of each column pertains to *all* the 513 observations in the data set, not just those shown in the table.

A Rough Interpretation of *r*

This may surprise you, but we're not going to worry about the calculations for a moment (it is +0.686, and Box 14.2 contains the steps you need to follow), and focus on the meaning of

TABLE 14.1 | Age of Married Couples in 1666 New France

Observation #	X	Y	X^2	Y^2	XY
1	41	40	1681	1600	1640
2	36	47	1296	2209	1692
3	40	30	1600	900	1200
4	27	18	729	324	486
...	...	...	...	...	...
513	42	28	1764	784	1176
Sum	19,021	14,808	764,779	497,544	593,371

Note: There is only one observation per household, and only couples for whom both ages are listed are included. Sums reflect all observations, not only those shown above.

Source: 1881 Census of Canada 100% sample file

BOX 14.2

Pearson's *r*: The Steps

1. Using a spreadsheet program, square each value of the independent variable *X*.
2. Square each value of the dependent variable *Y*.
3. Find $X * Y$ for each observation.
4. Sum *X*, *Y*, the squared values of *X* and *Y*, and the *XY* product.
5. Pearson's *r* can be solved by using the following equation:

$$r = \frac{N\sum XY - \left(\sum X\right)\left(\sum Y\right)}{\sqrt{\left[N\sum X^2 - \left(\sum X\right)^2\right]\left[N\sum Y^2 - \left(\sum Y\right)^2\right]}}$$

Values between ±0.01 and ±0.30 are considered weak correlations; from ±0.31 to ±0.70, moderate correlations; and from ±0.7001 to ±1, strong correlations.

the statistic. The calculated value of +0.686 has little meaning, unless we read it with some guidelines in mind:

- a value between ±0.01 and ±0.30 is considered a weak correlation;
- ±0.31 to ±0.70 is a moderate correlation; and
- ±0.7001 to ±1 suggests that the variables are strongly correlated

Here, the value of 0.686 suggests a conclusion that in New France a husband's age was moderately and positively correlated with the age of his spouse. Usually, we'd have to make sure that *r* is statistically significant before we can assess our hypotheses (we'll look at that in a moment), but here we have the entire population in our data set (rather than just a sample), so we can say that we have cause to reject the null hypothesis without assessing significance.

You should note that *r* is important because it tells you what happens to one variable, *Y*, as the value of the other variable, *X*, changes. In our example, the value is positive, so we can say that, on average, as the husband's age increases so does the age of his spouse. If the value were negative, we would say the opposite: on average, as the husband's age increases, the age of his spouse *decreases*.

A Visual Representation of *r*

To further illustrate how *r* describes bivariate relationships between two interval/ratio variables, let's look at a couple of graphs, each with a different set of data.

Let's start with Figure 14.1. In this figure, the independent variable *X* is positively related to *Y*, with a Pearson's *r* value of 0.25. The dots represent (simulated) values of *X* plotted against *Y*. The straight line represents fitted, or expected, values. Although a slight relationship between Y and X is detectable, the fit line doesn't do a very good job of representing the relationship. Note that if Pearson's r were negative, for example −0.25, the line would slope downward.

With a stronger positive association (as r increases), we expect the average distance between a particular observation and the line to shrink, indicating a reduction in estimation error. This can be seen in Figure 14.2, where a Pearson's r value of 0.50 is illustrated.

A close examination of Figure 14.2 suggests that the observations are closer to the line than in Figure 14.1. The convergence becomes more obvious when Pearson's r increases to 0.75, which happens in Figure 14.3. In Figure 14.4, variable X finally has a perfect positive relationship to Y. This demonstrates that by knowing a person's score on an independent variable (X), the score on the dependent variable (Y) can be predicted. Thus, *all* of the observed values lie perfectly on the line of best fit.

Note that scatter plots can be hard to read when n is large. In such circumstances, it is sometimes helpful to draw a random sample of 5 per cent from your data and plot just the 5 per cent.

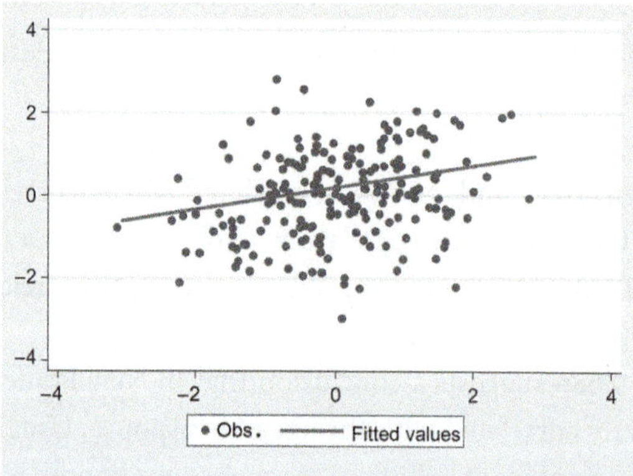

FIGURE 14.1 | An Illustration of a Pearson's r Value of 0.25 Using Simulated Data

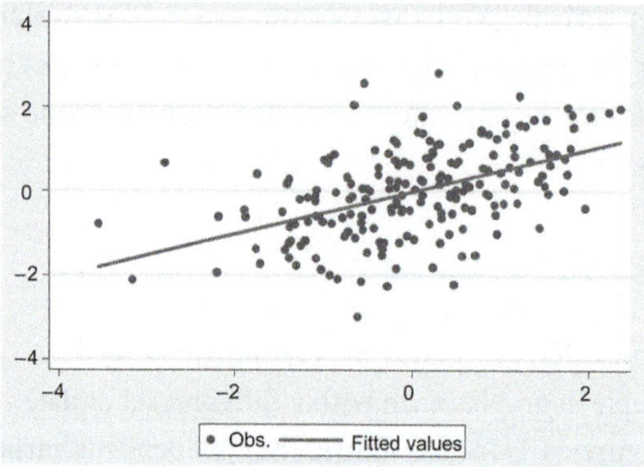

FIGURE 14.2 | An Illustration of a Pearson's r Value of 0.50 Using Simulated Data

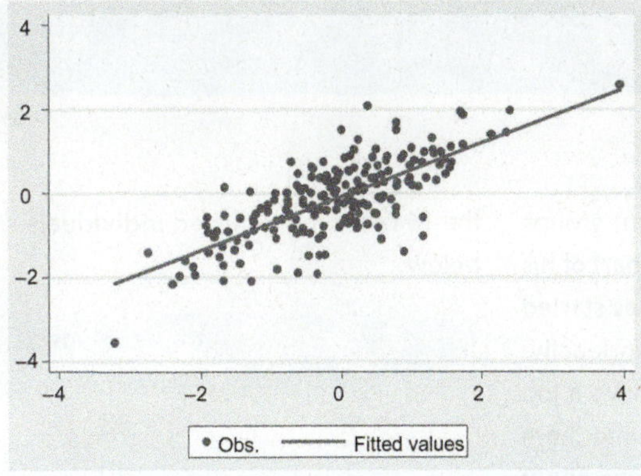

FIGURE 14.3 | An Illustration of a Pearson's *r* Value of 0.75 Using Simulated Data

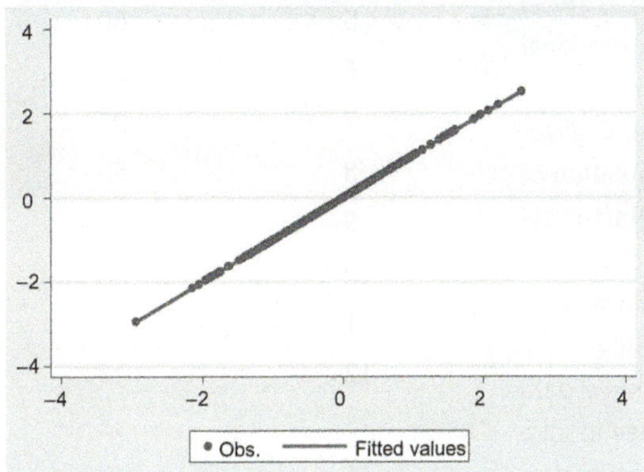

FIGURE 14.4 | An Illustration of a Pearson's *r* Value of 1.0 Using Simulated Data

What *r* Tells Us about Explained Variance

Using Pearson's *r*, it is possible to determine how much of the variation or variance in a variable can be explained with the use of other variables. The value of zero does not reflect a relationship between two variables, but a value of +1 indicates a perfect positive relationship, and a value of −1 indicates a perfect negative relationship. Values between −1 and +1 do not have a direct interpretation, unless you're comfortable with interpreting values in terms of standard deviations.

To determine the variance explained, we find the coefficient of determination by squaring *r*, yielding r^2. The coefficient of determination is the percentage of all variation in the dependent variable that can be explained by the values of the independent variable. The variable r^2 is the percentage by which errors are reduced when the information found in the independent variable is incorporated into the prediction of the dependent variable.

This introduces **explained variation** and **unexplained variation** (or explained variance and unexplained variance) in the dependent variable. Explained variation is how much more

BOX 14.3

It's Your Turn: Calculating Pearson's *r*

Marcus has been involved with a number of student groups during his undergraduate career and has loved that part of his life. As he nears the end of his degree program, he has started to wonder if he will be able to keep up with non-work activities while working, especially since he expects to have a job that will demand a lot of his time. It seems to him that there won't be enough hours in the day, but he has often read about people who manage to juggle organizational involvement and successful careers. Are these people exceptional, or are they the norm? As a research question, he wonders if an increase in work hours has a negative correlation with organizational involvement.

To study this, Marcus uses the Survey on Social Engagement (GSS 17) Public Use Microdata File (target population of Canadians over 15, excluding those in Yukon, the Northwest Territories, and Nunavut, and full-time residents of institutions). Since he is concerned about his own prospects, Marcus has decided to limit the sample to men, and only those who have completed high school. Since this is a huge data set and Marcus doesn't have a computer, he decides to take a random sample of 15 people. He is careful to make sure that none of the 15 are missing information on the variables he is interested in. To measure work hours, he examines the variable "hours of work/week." To measure organizational involvement, he examines the variable "number of civic groups." Only individuals who were involved in at least one civic group are included as Marcus wants to look at only people with some likelihood of involvement. He also limits the number of hours worked to 74 or less (individuals who worked over 75 hours were grouped together). The data for the 15 randomly selected individuals are shown in the table below:

Respondent	Hours of work/ week	Number of civic groups
1	38	2
2	40	2
3	60	2
4	50	2
5	60	2
6	50	4
7	35	1
8	50	2
9	36	2
10	30	3
11	65	1
12	45	1
13	48	3
14	40	2
15	55	3

Source: General Social Survey—Social Engagement (GSS). Detailed information for 2003 (Cycle 17)

1. Construct and complete a table like Table 14.1.
2. Calculate the value of *r*.
3. What can you say about the relationship between the variables based on your calculation?

The solution for Box 14.3 can be found in Appendix H.

accurate a prediction becomes when the independent variable, *X*, is taken into account. Unsurprisingly, unexplained variation is the remaining prediction error, which could be due to variables that weren't included as predictors, measurement error, or random error. The sum of the explained variation and unexplained variation in a dependent variable is equal to its **total variation.**

At this point it is important to remember that just because an independent variable "explains variation" in a dependent variable, that does not mean that the independent variable has "caused" the dependent variable, especially when we are examining cross-sectional data. Remember that in order to prove causation we must be able to show that the independent variable precedes the dependent variable in time and that there can be no other reason the two variables are related. (Nothing else is explaining the variation in *both* the independent and the dependent variables.)

A More Precise Interpretation of *r*

Looking at what the calculation of r achieves makes it possible to move beyond examining rough relationships between variables. The calculation for r standardizes the values of each variable, allowing correlation coefficients to be compared without worrying about how many values each variable has or what the standard deviation of each variable is. When each variable is standardized, r refers to the degree of correspondence between a person's z-score on one variable (X) and his or her z-score on another variable (Y), measuring the relationship between the average person's score in standard deviation units on one variable and the standard deviation score on a second variable. In an instance where $r = 1$, a person with a variable that has a change in the z-score from zero to two on one variable of interest will see a corresponding increase of a z-score of two on the other variable of interest. Conversely, an r value of -1 will translate to a z-score of $+1$ on one variable and a score of -1 on another. Remember, r is a measure of what happens to one variable as the value of another variable changes.

The Correlation Matrix

A **correlation matrix** is used to show correlations between variables and all of the possible relationships between variables on a grid.

Suppose that we wanted to study the relationship between incidents of domestic violence and unwanted sexual acts perpetrated by an ex-spouse/domestic partner because we want to know if non-sexual violent acts and unwanted sexual acts are interchangeable techniques for asserting dominance or if they are qualitatively different. To determine this, we could use the eighteenth wave of the General Social Survey (GSS), Public-Use Microdata File (the Victimization Survey), to construct a correlation matrix between the occurrence and frequency of non-sexual violent acts and unwanted sexual acts (see Table 14.2).

A typical correlation matrix presents a series of numbers in a triangular pattern. Each cell represents the correlation between the variables that are listed above and to the left of the number. For example, the number 0.2391 refers to the correlation between the number of unwanted sexual acts and the number of non-sexual violent acts. In a correlation matrix, the diagonal that runs from the top left cell to the bottom right cell, which refers to the correlation between a variable and itself, will always be one. Using Table 14.2, we can see that there is a weak relationship between non-sexual violent acts and unwanted sexual acts committed by and against domestic partners. That weak relationship suggests that there are other factors involved with both those phenomena.

A correlation matrix can contain as many relevant variables as you can imagine. It shows the bivariate relationships among each set of two variables.

TABLE 14.2 | Correlation between Non-Sexual Violent Acts and Unwanted Sexual Acts between Spouses/Partners

	# Violent acts	# Unwanted sexual acts
# Violent acts	1.0000	
# Unwanted sexual acts	0.2391	1.0000

Note: Includes persons aged 15 or older in Canada, excluding residents of the three territories and full-time residents of institutions (Statistics Canada, 2005).

Source: Statistics Canada, General Social Survey, 1985 to 2013.

Using a *t*-Test to Assess the Significance of *r*

When using *r*, it is usually necessary to determine if we can assume that the relationships we observe in the sample also exist in the population. (Remember that our 1666 Census example included the entire population of New France, so we did not need to examine statistical significance.)

By mathematically manipulating *r*, we can use *t*-distributions to assess its representativeness. Remember that *t*-distributions are closely related to the *z*-distribution, except that *t*-distributions make adjustments (largely to the tails) for sample size. The first step is to select the appropriate *t*-distribution by using the degrees of freedom. Because there are two known parameters (two variables we know the values for), the degrees of freedom are equal to $(n - 2)$, where n is equal to the sample size. If, for example, we had a sample of 22 people and we wanted to be 95 per cent confident of the generalizability of our results, we would need a *t*-value of at least 2.086. To get this value, we look at the row for $df = 20$ (remember to look at $n - 2$ degrees of freedom) in Appendix B, and look for a 0.05 level of significance in a two-tailed test. (Remember that levels of significance are usually stated as alpha values in a *t*-distribution, and can be calculated as one minus your desired confidence interval). This value is known as t_{critical}.

The other value we need to calculate is t_{observed}, which is done using this formula:

$$t_{\text{observed}} = r\sqrt{\frac{n-2}{1-r^2}}$$

Using the same example we used to calculate the correlation matrix (Table 14.2) and a hypothetical sample size of 22 observations, here is the calculation:

$$\begin{aligned} t_{\text{observed}} &= r\sqrt{\frac{n-2}{1-r^2}} \\ &= 0.2391\sqrt{\frac{22-2}{1-0.2391^2}} \\ &= 0.2391\sqrt{\frac{20}{1-0.057}} \\ &= 0.2391 * 4.6063 \\ &= 1.101 \end{aligned}$$

Since the $t_{observed}$ value of 1.101 is below the critical value of 2.086, we cannot be 95 per cent confident that the sample r value of 0.2391 did not occur by chance. Based on our sample of 22 respondents, we cannot be certain that there is a correlation between violent and unwanted sexual acts perpetrated by ex-spouses/domestic partners.

What to Do When Your Independent and Dependent Variables Are Measured at Different Levels of Measurement

Pearson's r provides a rough and ready measure of the relationship between two interval/ratio variables. The variable r assumes that the relationship between two variables is the same, no matter what level of the variables you are at, and this can be problematic at times. Let's see why this is the case with an example.

Imagine that we wanted to identify the relationship between age and amount of time spent on entertainment. We expect that teenagers have more time than their parents do to go to movies, dance clubs, or malls. We also assume that retirees, too, have a high proportion of spare time. This suggests that the relationship between age and time spent on entertainment is not linear; the rate at which available time to spend for entertainment increases is not consistent across the range of ages. There could be a very strong and positive relationship across the teenage years, after which the relationship becomes negative as many Canadians buy houses, get married, have children, develop their careers, etc., and thus have less time to spend on entertainment. Finally, as people get older, their children might leave home, they might begin to work less, etc., and they have more spare time to spend on entertainment again, affecting the relationship between entertainment and age and causing it to become positive again. To consider how this relationship might look, let's examine Figure 14.5.

The curvilinear pattern of circles represents the actual relationship between age and time spent on entertainment, and the straight line represents the relationship described by r.

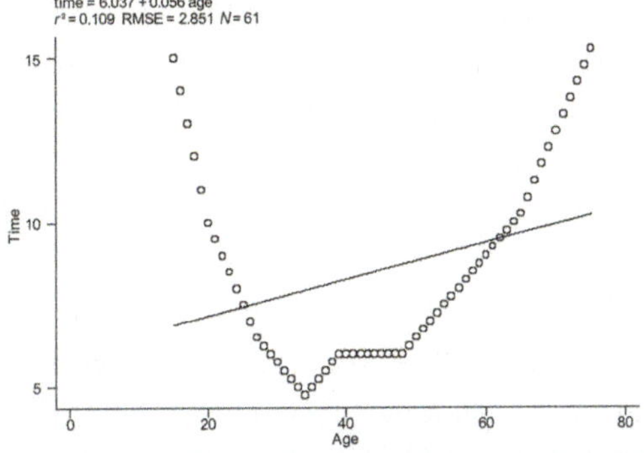

FIGURE 14.5 | A Graphical Representation of the Relationship between Age and the Proportion of Time Spent on Entertainment

(Actually, this is known as the least squares regression line, but we'll talk about that in Chapter 16.) In this case, the relationship is not linear and thus using regression analysis is not appropriate for this relationship. Before embarking on a deeper discussion of the fit line, a brief discussion about using variables measured at different levels is necessary. Most measures of association assume that both variables are measured at the same level and that you are working with two nominal, or two ordinal, or two interval/ratio variables. However, this is often not the case. When that happens, you'll find yourself in a position where you need to assess the association between variables by using different levels of measurement. Don't worry, there are a few short guidelines that we can apply to help us through this added layer of complexity.

Measuring Association between Interval/Ratio and Nominal or Ordinal Variables: Using the Lowest Common Measure of Association

Although previous chapters presented the measures of association as being specifically for one level of measurement, it is possible, under certain circumstances, to use the measures for other levels of measurement.

Recall that there are essentially three levels of measurement: nominal, ordinal, and interval/ratio. If we were working with two nominal variables, we could choose between phi, Cramer's *V*, or **Kruskal and Goodman's lambda**, depending on the nature of the variables and the information we seek. However, if we are looking at a nominal by ordinal relationship, we could use any of those three measures. (Remember that we only use phi for a two by two relationship.) Since these measures are non-directional, they can be used even if one of the two variables can be ranked. By using the measure of association for the variable at a lower level of measurement, we "discard" the additional information in the ordinal variable.

However, the opposite is not possible—a nominal variable cannot be "infused" with information that would allow it to be ranked. If we wanted to look for a relationship between visible minority status and attitudes about the seriousness of climate change (coded as extremely serious, very serious, not very serious, and not at all serious), we would use a nominal level of measurement because ranking information for the ordinal variable, climate change, could be discarded to make it a nominal variable. However, it is not possible to rank visible minority categories, so we would choose a measure of association for nominal variables.

Demoting the level of measurement for particular variables is a technique that can be used for several combinations of variables. For ordinal and interval/ratio variables, it is necessary to discard the information on differences between response categories in the interval/ratio variables and treat that variable as ordinal. To measure the association between attitudes toward the seriousness of climate change and income in dollars, we would have to choose from gamma, *rho*, Somers' *d*, or Kendall's tau-*b*.

Conclusion

This chapter has introduced you to a measure of association for two interval/ratio variables: Pearson's *r*. The following chapter will expand on this chapter by teaching you how to calculate a *regression equation*, which can measure the association between more than two interval ratio variables.

Glossary Terms

Correlation matrix (p. 167)
Explained variation (p. 165)
Kruskal and Goodman's lambda (p. 170)
Pearson's *r* (p. 160)
Total variation (p. 166)
Unexplained variation (p. 165)

Practice Questions

1. The following data from the 1971 Census of Canada describe two pieces of information about 10 families.

Family number	# People in family	Total income ($)
1	2	1330
2	4	800
3	5	1200
4	4	1600
5	6	900
6	4	6144
7	2	3490
8	4	1310
9	3	5670
10	2	1330

Source: 1971 Census of Canada. STC cat. 12–540.

a. Develop a set of research hypotheses about the relationship between family income and family size.
b. Calculate *r*.
c. Assess the statistical significance of *r* by using a *t*-test.

2. Consider the following plot for the values of total income and family size. Do you think that a line summarizing the association will slope upwards or downwards?

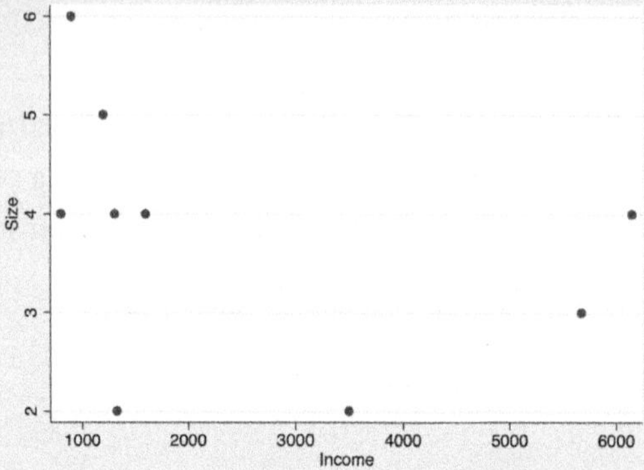

3. Chris and Josie like to eat out at restaurants, and they are worried that they are spending too much, in part because they tend to follow each other's lead in the price of the entrée they order. They are both aspiring statisticians and are therefore keen to apply what they just learned about Pearson's *r*, even though they will only be able to draw conclusions about whether they tend to spend as much as each other. (We don't know if they actually follow each other or if they independently choose similarly priced entrées.) Help them determine if this concern is warranted by drafting a set of research hypotheses, keeping in mind the limitations of Pearson's *r* in this instance.

4. To help you test the hypotheses you composed in #3 above, Chris and Josie eat at a restaurant eight times and decide to keep track of their expenditures. Calculate the value of Pearson's *r* for their expenses.

Chris	Josie
4	5
12	20
13	8
14	6
8	11
3	10
5	6
6	9

5. Determine the statistical significance of your calculation. Does this lead you to reject or fail to reject your null hypothesis above?

6. Now you must report back to Chris and Josie. What would you tell them about their concerns based on your hypothesis test? Does each of them tend to spend as much as the other?

Answers to the practice questions for Chapter 14 can be found in Appendix H.

15 One-Way Analysis of Variance

LEARNING OBJECTIVES

Now that many of the techniques for measuring bivariate relationships have been covered, this chapter will discuss the final technique for identifying relationships between two variables in this text: one-way analysis of variance (ANOVA). Topics will include:

- what a one-way ANOVA is and when it should be used;
- how to calculate within-group sums of squares, between-group sums of squares, and total sums of squares, the three main components of one-way ANOVAs;
- how to use the *F*-distribution to assess significance with ANOVA; and
- limitations of ANOVA.

Introduction

The last few chapters have been about strategies for identifying the existence and strength of a relationship between two variables. This chapter will build on what we've learned by discussing another method for identifying associations between variables measured at different levels. We'll focus on interval/ratio dependent variables, and nominal or ordinal independent variables that have more than two categories by discussing a procedure known as ANOVA, or the **ANalysis Of VAriance.***

What Is ANOVA?

ANOVA is a little like a *t*-test. The *t*-test can be used in any situation where a comparison between two groups on a continuous variable is desired. The *t*-test is useful, but what if we want to compare more than two groups?

For example, say we want to compare the level of control an individual feels he or she has over his or her own life (sometimes called "mastery"), checking to see if it differs by religion, by using Statistics Canada's 2003 Social Engagement Survey. Specifically, we want

* Throughout this chapter, we'll discuss only one-way ANOVA, even though there are multiple ANOVA techniques. For brevity, here ANOVA will be used as short-hand to refer to one-way ANOVA.

to compare the differences across Roman Catholics, Protestants, members of the United Church, and people with no religious affiliation.

Table 15.1 shows the levels of mastery for five respondents in each group. Higher scores indicate superior mastery. A glance at the columns suggests that there are differences across religious groups, but we do not know if the differences are the result of chance, sampling error (most of the time, we can't distinguish between the two), or differences that exist across groups. How can we find this out?

One way is to conduct a series of *t*-tests, comparing two groups at a time. How many *t*-tests would we have to do?

- For *No Religion*, there would be three: NR-RC, NR-UC, and NR-P.
- For *Roman Catholics*, there would be two: RC-UC and RC-P. (There are only two comparisons because Roman Catholics have already been compared with no religion.)
- For the *United Church*, there would be one: UC-P (the only remaining comparison to be made).

That's six *t*-tests. What's wrong with that? First, it's a computational nightmare. Can you imagine wrestling with the results of six *t*-tests? With that many calculations, the chance of making an error increases dramatically—or it does for us, anyway.

Even if you were precise with your calculator, the bad news is that conducting multiple *t*-tests (or any other analysis conducted repeatedly) increases the probability of committing a **type one error**—as we discussed earlier, this means rejecting the null hypothesis when it should be retained—increasing the risk of making a bad judgment about your hypothesis.

Recall that a 0.05 level of significance is used to assess hypotheses, meaning that we accept that 5 per cent of the time we will mistakenly conclude that the differences we observe in our sample actually exist in the population. If each of the *t*-tests in the sample containing six *t*-tests has a 5 per cent chance of yielding significant results by mistake, our 95 per cent significance level decreases to about 73.5 per cent. (To get this number, multiply the independent probabilities: $0.95 * 0.95 * 0.95 * 0.95 * 0.95 * 0.95$.)

TABLE 15.1 | Levels of Mastery by Religious Affiliation for 20 Respondents, 2003 Social Engagement Survey

Levels of mastery			
No religion	Roman Catholic	United Church	Protestant
21	17	22	21
25	21	14	23
26	19	16	22
24	20	13	24
22	20	14	21

Source: General Social Survey—Social Engagement (GSS). Detailed information for 2003 (Cycle 17)

Clearly, a single statistical test to assess differences between more than two groups is necessary. That test is ANOVA, which can be thought of as an elaborated *t*-test. It compares three things:

1. Differences between means
2. Differences in values within samples
3. Differences in values across samples

Like a *t*-test, there is an internal logic to ANOVA. Each observation is different from the grand mean (the overall mean for the whole sample) by some amount. There are two sources of this difference:

1. The independent variable
2. Random unexplained error

ANOVA compares the variation around the mean within groups to the variation across or between groups. Is a Roman Catholic more likely to be similar to another Roman Catholic in terms of mastery than to a person in another category? What about a Protestant's level of mastery compared to another Protestant's level of mastery, etc.? If the answer to these questions is yes, then we could conclude that there are religious differences in levels of personal mastery.

Although that description is oversimplified, it nicely captures what is going on with ANOVA. Before conducting ANOVA it is important to compare the variances across groups. ANOVA will not work if the variances are too different. Most statistical software packages include tests to compare the variances.

We need to go over several equations before we can compare variation within groups to variation between groups.

The first equation is the **sum of squares**, which is defined as the total variation of all observations from a **grand mean** (or **overall sample mean**). To determine whether there is more variation within groups than across groups, we need to identify the total variation available to be distributed between and across groups. This is the **total sum of squares**, or SS_{total}, and equals the total variation of all observations from the grand mean. The equation for SS_{total} is this:

$$SS_{total} = \sum (X - \bar{x})^2$$

Stated in words, the equation is a sum of the differences between individual observations (X) and the grand mean ($\bar{x}$) squared. The differences are squared to remove the negative numbers that would result from subtracting the means from a value that is below the mean. For example, if a person had a mastery score of 15, subtracting the grand mean of 18.85 would yield −3.85. On their own, negative numbers would not be a problem, but here they are because they are summed and therefore cancelled out by positive values. Remember that we take the sum of all deviations; since the mean is the midpoint of all values, the negative numbers resulting from values below the mean would cancel out all values above the mean. If deviations aren't squared, then the sum of the deviations will always equal zero.

Once the total sum of squares has been calculated, equations are used to calculate the differences within groups and the differences across groups. These terms are known as the sum of squares within groups (SS_{within}) and the sum of squares between groups (SS_{between}).

The equation for the sum of squares within groups is

$$SS_{\text{within}} = \sum (X - \bar{x}_{\text{group}})^2$$

This equation is strikingly similar to the equation for the total sum of squares. The major difference between the **within-groups sum of squares** and the total sum of squares is that instead of trying to calculate the total variation of *all* observations, we are trying to calculate the variation that exists *within* each group. (We calculate deviations of each group member from their group mean rather than from the grand mean.) For the equation, the difference is that the grand mean $\bar{x}$ is replaced by the group mean $\bar{x}_{\text{group}}$.

The final equation is a little different. Since we're interested in the variation across groups, each group is treated as an observation:

$$SS_{\text{between}} = \sum n_{\text{group}} (\bar{x}_{\text{group}} - \bar{x})^2$$

To find the **between-groups sum of squares**, subtract the grand mean from each group mean, square the difference, multiply it by the number of observations in each group, and add the results across all groups.

EVERYDAY STATISTICS

When to Use ANOVA

ANOVA is often used in cases where the researcher is attempting to examine the outcome of a particular treatment on a subgroup of the population. In these types of experiments, there are often several subgroups that receive a varying degree of treatment. Usually, one group does not receive the treatment. Once the treatment is over, the groups are compared to see which ones have changed noticeably.

Q: Why do you think ANOVA would be a good method to employ in the above situation?

The Sum of Squares: An Easier Way

While it's possible to use those equations, they are a lot of work. You have to subtract the mean from each observation, which involves going through your data set line by line and calculating deviations. In a sample of 100,000 observations, that would be incredibly tedious and time-consuming, even with a spreadsheet program.

Luckily, there's an easier way: summing variables *before* taking differences. Since the mean is the arithmetic midpoint of all values of a particular variable, it can be multiplied by the number of observations and squared, speeding up the calculation of the sums of squares. Unfortunately, you'll still need to calculate the sum of all observations and square it, but the following equations are still fairly easy to work with:

$$SS_{total} = \sum X_{total}^2 - n_{total}\bar{x}_{total}^2$$

$$SS_{within} = \sum X_{total}^2 - \sum n_{group}\bar{x}_{group}^2$$

$$SS_{between} = \sum n_{group}\bar{x}_{group}^2 - n_{total}\bar{x}_{total}^2$$

A few words of caution: First, for the total sum of squares and within-groups sum of squares, it is important to square the value of the observations (in each case, this first term in the equation) *before* summing them. The mean value in both of the second terms needs to be squared before being multiplied by either n_{total} or n_{group}. Second, when finding the sum of squares between groups, the group average and the grand mean need to be squared *before* being multiplied by n_{group} or n_{total}. Skipping either of these steps will result in the wrong value.

So what exactly is the sum of squares, and why is it useful? The sum of squares is defined as the squared and summed measures of variance that exist between observations. It is a standardized measure of deviance from a measure of central tendency, which is why it periodically pops up in statistics. When two samples of the same size are being compared, a higher sum of squares indicates that one sample has greater variation between observations.

The sum of squares is useful in certain circumstances, but, like the *t*-statistic, larger summed deviations depend in part on sample size. A higher sum of squares is more likely in an analysis of 100,000 observations than with a sample size of 10, suggesting that the sum of squares (total, within, and between) increases with either greater variation between scores or sample size. When measuring variation between scores, the sum of squares must be standardized so that values can be compared regardless of sample size. This creates a standardized measure called the **mean square**.

The equation for the mean square within groups is as follows:

$$MS_{within} = \frac{SS_{within}}{df_{within}}$$

The equation for the mean square between groups is this:

$$MS_{between} = \frac{SS_{between}}{df_{between}}$$

You should recognize the sum of squares within groups and between groups in the numerator of the equation. The denominator is new. The symbols refer to the degrees of freedom within groups (df_{within}) and between groups (df_{between}). Calculating these values is fairly simple:

$$df_{\text{within}} = n_{\text{total}} - k$$

and

$$df_{\text{between}} = k - 1$$

where k = the number of groups you are comparing.

Remember that for the t-distribution, the degrees of freedom are defined as the number of values that are free to vary. Although the equations for degrees of freedom differ slightly, the logic is the same.

First, since we're estimating group means within the population by using sample means, only the number of observations in each group minus one can assume any value. Consequently, in each group there is one predetermined value. We define df_{within} as $n_{\text{total}} - k$, since one value of each group is perfectly determined by all other values within that group, when the group mean is known.

Second, since we're calculating the observed variation between groups, we don't need to know the means for every group. If the grand mean is known, then the values for $k - 1$ groups will determine the value of the last group. Therefore, only $k - 1$ values for df_{between} can vary freely.

The *F*-Distribution

By now, you should know that when statistical equations seek standardized numbers, it means that they're going to be assessed against a pre-established benchmark. These equations are no different, but now, instead of using the t- or z-distribution, we're going to use the ***F*-distribution**.

To assess values against the F-distribution, a test statistic must be calculated first. This test statistic is known as the F-ratio and is calculated using the following equation:

$$F_{\text{observed}} = \frac{MS_{\text{between}}}{MS_{\text{within}}}$$

Since MS_{between} and MS_{within} are the only values needed, we won't bother with MS_{total}. (We're sure you'll agree that we have already covered enough equations in this chapter!) Please note that sometime F_{observed} is called $F_{\text{calculated}}$. They mean the same thing.

Finally, we have enough information to assess our observed test statistic against the F-distribution, our predetermined benchmark. Like the t-distribution, the shape of the F-distribution depends on sample size. Unlike the t-distribution, the F-distribution depends on the number of groups. You need to look at both the within-group and the between-group degrees of freedom. Table 15.2 is an abbreviated version of a table that lists the critical values of the F-distribution for the 95 per cent level of significance. (The same table can also be found in Appendix D.)

TABLE 15.2 | The *F*-Distribution for 95 Per Cent Level of Significance

$df_{between}$	2	3	4	5	6	7	8
df_{within}							
1	199.50	215.71	224.58	230.16	233.99	236.77	238.88
2	19.00	19.16	19.25	19.30	19.33	19.35	19.37
3	9.55	9.28	9.12	9.01	8.94	8.89	8.85
4	6.94	6.59	6.39	6.26	6.16	6.09	6.04
5	5.79	5.41	5.19	5.05	4.95	4.88	4.82
6	5.14	4.76	4.53	4.39	4.28	4.21	4.15
7	4.74	4.35	4.12	3.97	3.87	3.79	3.73
8	4.46	4.07	3.84	3.69	3.58	3.50	3.44
9	4.26	3.86	3.63	3.48	3.37	3.29	3.23
10	4.10	3.71	3.48	3.33	3.22	3.14	3.07
11	3.98	3.59	3.36	3.20	3.10	3.01	2.95
12	3.89	3.49	3.26	3.11	3.00	2.91	2.85
13	3.81	3.41	3.18	3.03	2.92	2.83	2.77
14	3.74	3.34	3.11	2.96	2.85	2.76	2.70
15	3.68	3.29	3.06	2.90	2.79	2.71	2.64
16	3.63	3.24	3.01	2.85	2.74	2.66	2.59
17	3.59	3.20	2.97	2.81	2.70	2.61	2.55
18	3.56	3.16	2.93	2.77	2.66	2.58	2.51
19	3.52	3.13	2.90	2.74	2.63	2.54	2.48
20	3.49	3.10	2.87	2.71	2.60	2.51	2.45
30	3.32	2.92	2.69	2.53	2.42	2.33	2.27
40	3.23	2.84	2.61	2.45	2.34	2.25	2.18
50	3.18	2.79	2.56	2.40	2.29	2.20	2.13
60	3.15	2.76	2.53	2.37	2.25	2.17	2.10
70	3.13	2.74	2.50	2.35	2.23	2.14	2.07
80	3.11	2.72	2.49	2.33	2.21	2.13	2.06
90	3.10	2.71	2.47	2.32	2.20	2.11	2.04
100	3.09	2.70	2.46	2.31	2.19	2.10	2.03
120	3.07	2.68	2.45	2.29	2.18	2.09	2.02
∞	3.00	2.61	2.37	2.22	2.10	2.01	1.94

If you encounter degrees of freedom values that do not perfectly coincide with the provided values, always use the lower value. This will give you a more conservative basis for assessing your calculated or observed *F*-statistic. For example, if you have a $df_{between}$ of 2 and a df_{within} of 118 (reflecting an independent variable with three categories, the equivalent of an independent samples *t*-test), choose a $df_{between}$ of 2 and a df_{within} of 100.

To help this sink in, let's keep using the example from before. Remember that we were comparing levels of mastery by four religious groups. For ease of reference, Table 15.1 is replicated below as Table 15.3. Each group has five people in it, so there are 20 people total in this sample.

TABLE 15.3 | Levels of Mastery by Religious Affiliation for 20 Respondents, 2003 Social Engagement Survey

Levels of mastery			
No religion	Roman Catholic	United Church	Protestant
21	17	22	21
25	21	14	23
26	19	16	22
24	20	13	24
22	20	14	21

Source: General Social Survey—Social Engagement (GSS). Detailed information for 2003 (Cycle 17)

To work through the necessary equations, the first thing we do is square the individual X values and find the sums; see Tables 15.4 and 15.5.

Here are the equations again:

$$SS_{total} = \sum X^2_{total} - n_{total}\bar{x}^2_{total}$$

$$SS_{within} = \sum X^2_{total} - \sum n_{group}\bar{x}^2_{group}$$

$$SS_{between} = \sum n_{group}\bar{x}^2_{group} - n_{total}\bar{x}^2_{total}$$

The X^2 values in Table 15.4 are useful for calculating the within-groups sum of squares (the sum of these values represents the first part of each equation) and the total sum of squares (the sum of the within-group sum of squares). We also need the grand mean and the group means, so we'll have to calculate a few more numbers before using the equations.

Now that we have all of the information we need, let's calculate the sums of squares, starting first with the total sum of squares.

$$\begin{aligned} SS_{total} &= \sum X^2_{total} - n_{total}\bar{x}^2_{total} \\ &= (2802 + 1891 + 1301 + 2471) - 20(20.25)^2 \\ &= 8465 - 20(410.06) \\ &= 8465 - 8201.2 \\ &= 263.8 \end{aligned}$$

This calculation is fairly straightforward. The only thing you need to watch for is the order of operations in the second part of the equation. Remember to square the group mean before multiplying it by the number of observations. Failing to do so will result in a calculation error. The total sum of squares represents the total amount of variation within our sample that can be explained by the sample's characteristics.

The next calculation is the within-group sum of squares. (Since we already calculated the first term in this equation when calculating the total sums of squares, we can take a shortcut here; the first term is 8465):

$$\begin{aligned} SS_{\text{within}} &= \sum X^2_{\text{total}} - \sum n_{\text{group}} \bar{x}^2_{\text{group}} \\ &= 8465 - 5\left[(23.6)^2 + (19.4)^2 + (15.8)^2 + (22.2)^2\right] \\ &= 8465 - 5 * 1675.8 \\ &= 86 \end{aligned}$$

For the between-group sum of squares, we can either use the equation, or we can subtract the within-group sum of squares from the total sum of squares. Recall that the total sum of squares represents the total amount of variation within the sample, and the within-group

TABLE 15.4 | Levels of Mastery by Religious Affiliation for 20 Respondents, 2003 Social Engagement Survey

Levels of mastery							
No religion		Roman Catholic		United Church		Protestant	
X	X^2	X	X^2	X	X^2	X	X^2
21	441	17	289	22	484	21	441
25	625	21	441	14	196	23	529
26	676	19	361	16	256	22	484
24	576	20	400	13	169	24	576
22	484	20	400	14	196	21	441

Source: General Social Survey—Social Engagement (GSS). Detailed information for 2003 (Cycle 17)

TABLE 15.5 | Average Levels of Mastery by Religious Affiliation, 2003 Social Engagement Survey

	Levels of mastery							
Observation #	No religion		Roman Catholic		United Church		Protestant	
	X	X^2	X	X^2	X	X^2	X	X^2
1	21	441	17	289	22	484	21	441
2	25	625	21	441	14	196	23	529
3	26	676	19	361	16	256	22	484
4	24	576	20	400	13	169	24	576
5	22	484	20	400	14	196	21	441
Σ	118	2802	97	1891	79	1301	111	2471
$\bar{x}_{\text{group}}$	23.6		19.4		15.8		22.2	
n	20		Grand Mean $\bar{x}$		20.25			

Source: General Social Survey—Social Engagement (GSS). Detailed information for 2003 (Cycle 17)

sum of squares represents the total variation within each group. The between-group sum of squares should be the difference between the two, but let's make sure.

The first part of the equation for the between-group sum of squares is the same as the last part of the equation for the within-group sum of squares. The second part is the same as the second part of the equation for total sum of squares. We can insert those values.

$$\begin{aligned} SS_{\text{between}} &= \sum n_{\text{group}} \bar{x}^2_{\text{group}} - n_{\text{total}} \bar{x}^2_{\text{total}} \\ &= 8379 - 8201.12 \\ &= 177.8 \end{aligned}$$

The answer should be the same as what we would get by subtracting the within-group sum of squares from the total sum of squares:

$$\begin{aligned} SS_{\text{between}} &= SS_{\text{total}} - SS_{\text{within}} \\ &= 263.8 - 86 \\ &= 177.8 \end{aligned}$$

It is, so we can be fairly certain that our calculations are correct. (To be even more certain, we should choose not to borrow $\sum n_{\text{group}} \bar{x}^2_{\text{group}}$ and $n_{\text{total}} \bar{x}^2_{\text{total}}$ from our previous equations, since any errors we made before will be embedded in them.)

Now that we have our sums of squares, the next thing to do is calculate the mean squares. To do so, we need the two degrees of freedom values:

$$\begin{aligned} df_{\text{within}} &= n_{\text{total}} - k \\ &= 20 - 4 \\ &= 16 \end{aligned}$$

and

$$\begin{aligned} df_{\text{between}} &= k - 1 \\ &= 4 - 1 \\ &= 3 \end{aligned}$$

We can put these values into our equations for the mean square. First, for within-groups:

$$\begin{aligned} MS_{\text{within}} &= \frac{SS_{\text{within}}}{df_{\text{within}}} \\ &= \frac{86}{16} \\ &= 5.38 \end{aligned}$$

And then for between-groups:

$$MS_{between} = \frac{SS_{between}}{df_{between}}$$
$$= \frac{177.8}{3}$$
$$= 59.27$$

Finally, we can calculate our F-statistic:

$$F = \frac{MS_{between}}{MS_{within}}$$
$$= \frac{59.27}{5.38}$$
$$= 11.02$$

Next, we need to compare that value with the $F_{critical}$ value on the F-table. For degrees of freedom, 3 and 16 return the critical value of 3.24, and since 11.02 greatly exceeds that number, we can be 95 per cent confident that there is at least one significant difference in levels of mastery across religions in the population. In other words, we now know with 95 per cent certainty that at least one of our groups differs significantly from the overall mean in terms of mastery.

Is This New?

Although a lot of material has been covered in this chapter—especially equations—there are striking similarities between the techniques we've been using and the techniques for *t*-tests. We have already compared variances between samples with variances within samples to determine where the greater differences lay. ANOVA is an elaboration of this technique because it makes it possible to compare more than two groups. Think of ANOVA as a *t*-test for more than two groups, with a lower probability of type one error.

Limitations of ANOVA

Of course, ANOVA has limitations. Two of them in particular are of concern to us.

First, ANOVA requires the assumption of equal variances across groups within the population. This assumption suggests that each of the groups within a population has approximately the same distribution of values around the mean. Slight differences between groups are acceptable, but ANOVA becomes increasingly unusable when there are large differences. Therefore, it is essential to compare the variances prior to conducting an ANOVA.

The second limitation of ANOVA is the determination of what a significant F-value means. When $F_{observed}$ exceeds $F_{critical}$, the only information provided is that the mean of at least one group in the sample is significantly different from that of the population as a whole. If we were interested in determining *which* group was different, we'd have to conduct a series of complicated **post-hoc tests**. These can be calculated by using statistical software.

Using ANOVA alone it is not possible to determine which group mean is different or whether that group mean is significantly higher or lower than the other group means on an outcome of interest. ANOVA by itself can only test the hypothesis that "at least one of the group means is different from the overall/grand mean."

There are two additional steps to take in order to best interpret ANOVA. First, you should examine the group means. You should be able to say which group means are higher, which group means are lower, and which are similar to each other. Next, you should run post-hoc tests using your statistical software. The most common post-hoc test to run is the Bonferroni (but this is beyond our focus here). This test will tell you which groups have means that are statistically significantly different from each other. Running an ANOVA with post-hoc tests allows you to state which groups differ from each other.

Conclusion

This chapter has covered ANOVA, a method for comparing group means across more than two groups. Ordinary least squares regression, covered in the next chapter, is superior to ANOVA with post-hoc tests because you can assess the statistical significance of differences while holding other characteristics constant. For example, the differences in levels of mastery across religious groups that we saw could be due to differences in the average age or level of education of each group. Suppose that mastery and religion are heavily correlated with age and education. It's possible then that the differences observed in an ANOVA are the result of one or both of the third or fourth variables. In a regression framework, we can determine if that is the case. In a one-way ANOVA, however, we cannot.

On the other hand, many social science disciplines use ANOVA frequently, testifying to its utility. There have been numerous elaborations on the simple technique described in this chapter that address the concerns, such as MANOVA (multivariate analysis of variance). However, these are beyond what is necessary for an introductory statistics course.

In Part III, which follows this chapter, we will move on to methods for analyzing relationships between more than two variables.

BOX 15.1

ANOVA: The Steps

1. Find the grand mean and the mean for each group.
2. Find group sums, sum of squared scores.
3. Find SS_{total}, SS_{within}, $SS_{between}$.
4. Find $df_{between}$, and df_{within}.
5. Find $MS_{between}$ and MS_{within}.
6. Obtain the F-ratio.
7. Compare $F_{observed}$ to $F_{critical}$.

BOX 15.2

One-Way ANOVA: History of a Term

In the mid-nineteenth century, the agriculturist James F.W. Johnston recognized a problem with agricultural analysis. The importance of practical agricultural knowledge to individual and national economic development, was acknowledged, but unlike fields such as physics, where researchers can control conditions, agriculture is not easily translatable to the laboratory. For example, if there are two crops with differing methods of production, how can the cause of differences in yield be determined? Influences can't be isolated, so there is no way to discern whether the treatment of a crop has a significant effect on its behaviour and yield.

Gigerenzer et al. (1991) relate the story of how statistician and geneticist Ronald Fisher (1890–1962) attempted to solve the problem. He aimed "to ascertain whether a difference in means between treated population and controls indicates the causal efficacy of the treatment" (Gigerenzer et al., 1991: 73). Specifically, Fisher wanted to know how to evaluate the application of manure or bone meal when the yield from plots is not constant but has a particular distribution. When there is variation among plots *not* treated with bone meal or manure, how can the distribution of yield of certain plots be accounted for when an independent variable is introduced (e.g., fertilizer)? Fisher developed the statistical and systematic means for comparing within-group variance and the variance between groups to solve this problem—the *F* in *F*-ratios and the *F*-test comes from him.

Glossary Terms

Analysis of variance (ANOVA) (p. 173)
Between-groups sum of squares (p. 176)
F-distribution (p. 178)
Grand mean or overall sample mean (p. 175)
Mean square (p. 177)
Post-hoc tests (p. 184)
Sum of squares (p. 175)
Total sum of squares (p. 175)
Type one error (p. 174)
Within-groups sum of squares (p. 176)

Practice Questions

1. Javier wants to join ballet and, consequently, plans to start exercising so that he can lose weight. He wonders whether he will burn the same number of calories no matter what type of activity he does. Javier chooses five different exercises and asks five different people the number of calories they burn by doing each activity for a period of one hour. Develop a research and null hypothesis for Javier's study.
2. Consider the following data from Javier's data collection technique.

Number of Calories Burned per Hour				
Bicycling	House cleaning	Health club exercise	Yoga	Tennis
236	207	325	236	413
321	249	401	312	599
345	292	452	345	604
292	302	474	281	434
301	222	353	301	477

a. Construct and complete a table containing X^2, n, the average time spent on each activity, the sum of all the times, and the grand mean for all activities.

b. Calculate SS_{total}, SS_{within}, and $SS_{between}$.

c. Calculate df_{within} and $df_{between}$.

d. Calculate MS_{within} and $MS_{between}$.

e. Obtain the F-ratio.

f. Compare $F_{calculated}$ to $F_{critical}$. What are the implications for your hypotheses?

3. Ann's friend Jen always tells her that she should buy more expensive jeans because they last longer. Ann asked five people how many days they wore the four types of jeans shown in the table before they were no longer able to wear them, to see if there are significant differences between the types of jeans and their durability. What are the research and null hypotheses that would allow you to test this prospect?

4. Here are the data from Ann's observations:

		Durability of Jeans		
Observation #	Levi's	People's Liberty	Silvers	Guess
1	182	209	1,040	260
2	130	225	780	624
3	91	156	520	416
4	200	260	340	222
5	154	101	416	85

a. Construct and complete a table that contains the necessary information for finding $F_{calculated}$.

b. Compare $F_{calculated}$ to $F_{critical}$.

c. Do you find evidence to reject or fail to reject the null hypothesis?

5. Edmonton's three major malls (other than the West Edmonton Mall) are Kingsway, City Centre, and Southgate Mall. Here is a fictitious list of the number of daily visitors to each of the malls for four Saturdays in a row, measured in thousands. Calculate a one-way ANOVA to see if at least one of the malls is significantly different from the others in the number of visitors.

Date	Kingsway	City Centre	Southgate
Saturday, August 6, 2011	45	32	51
Saturday, August 13, 2011	32	47	55
Saturday, August 20, 2011	44	55	31
Saturday, August 27, 2011	43	41	30

Answers to the practice questions for Chapter 15 can be found in Appendix H.

PART III

Multivariate Techniques

16 Regression 1—Modelling Continuous Outcomes

LEARNING OBJECTIVES

In this chapter, we're moving beyond bivariate techniques and focusing on analysis with more than two variables. Specifically, we'll look at:

- ordinary least squares (OLS) regression;
- why and when multivariate analysis might be necessary;
- how to calculate and interpret OLS coefficients and how to identify statistical significance;
- standardized partial slopes;
- dummy variables; and
- inference and regression.

Introduction

You probably expect your income to increase as you age, but do you know how to determine whether that is a reasonable expectation? Can you calculate the rate at which it will increase? You could run a bivariate correlation between age and income, but what if you believe that other variables, such as education, will also have an effect? What about being male or female? Black or white? A resident of Winnipeg or Halifax? How can you look at the relationship between two variables and still acknowledge that other variables matter?

So far, a lot of this text has been about levels of measurement, variance, and correlations between variables. In a way, this has all been "build-up" to this chapter and the next. In this chapter, many of these things will come together with ordinary least squares regression, the multivariate technique of choice for continuous dependent variables. Chapter 17 will then cover logistic regression, which is often used with binary outcomes.

Ordinary Least-Squares Regression: The Idea

When two sets of numbers are plotted on a graph (say, age and income), and you think that you can see a general pattern or relationship between the two sets of numbers, you might be tempted to draw a "trend line" (like the one we used in Chapter 14 to illustrate Pearson's r) to describe the relationship that you think you see. Although this can be done by eye, some quick calculations will probably result in a more accurate line. If you drew several lines, how would you know which line was best? Your goal is to find the line that best represents the relationship between individual scores on an independent variable and a dependent variable. You want the sum of the distances between the data points and the line to be as small as possible. That distance is known as the estimation error.

If all of the points were on a straight line, we could trace that line, and the sum of the distances between the data points and the line would be 0. This would indicate a perfect relationship between your independent and your dependent variable. This is rarely the case, though, so we must decide where to draw the line. To do that, we use regression. The shorthand term *regression* refers to a set of techniques that allow relationships between two or more variables to be identified and generalized (or summarized). For now, we'll consider only cases where the relationship between the variables is linear, although linearity isn't necessary with more complicated techniques. In Chapter 17, we'll look at one such technique for examining a non-linear relationship, the logistic curve.

Social scientists use computers to help them with regression. However, the best practitioners are also familiar with the "behind-the-scenes" calculations. In this chapter, we'll focus on calculating regression coefficients by hand, so that when you use Excel, SAS, STATA, SPSS, or any other program to calculate regression, you'll understand what the numbers in the output mean and how they were calculated. Although it is possible to compute multiple regression with many independent variables, because of the complexity of the calculations for regression, we'll use only examples with one or two independent variables for our hand calculations.

Before we proceed, let's quickly review independent and dependent variables. Independent variables are those that elicit an effect on an outcome, whereas dependent variables are the outcomes you wish to explain. So, if you're looking at age and income, age is the independent variable and income is the dependent variable. How do we know this? Largely because it makes more sense to hypothesize that age has an effect on a person's income than it does to think that income affects a person's age.

If you are ever stuck trying to figure out which is which, think about temporal ordering: which occurs first? If you can answer this question, you can be pretty sure that you have distinguished independent variables from the dependent variable. Every person has an age long before he or she has an income; your age for the rest of your life was determined at birth. However, many things in your life will determine your income. To calculate a regression equation, you must be able to identify independent and dependent variables.

Onward from Bivariate Correlation: Multivariate Analysis

To understand ordinary least squares regression, think of it as an extension of a bivariate correlation between two interval/ratio variables. Like a bivariate correlation, Pearson's correlation coefficient measures the strength of the association between two variables.

(Let's call them *X* and *Y*.) However, in multivariate analysis it is possible to examine the strength of an association between *Y* and more than one independent variable at once (such as *X and Z*).

To illustrate, suppose that you wanted to look at the relationship between age and hours of sleep. You suspect that years of schooling also affects hours of sleep, but you aren't actually interested in this second relationship. Perhaps the most important feature of regression is that each relationship can be assessed while accounting for the impact of other potential explanatory variables. If you were interested in the relationship between age and hours of sleep, but acknowledge that years of schooling probably has an effect on that relationship, you can "control" for years of schooling by entering it into the regression equation as a second independent variable. If you still observe a relationship between age and hours of sleep, it now represents the relationship after the effect of years of schooling has been removed. Regression assesses the effect of each independent variable while holding the value of all other independent variables in the model at a constant value.

Think of it as trying to summarize the relationship between independent variable(s) and a dependent variable with a line (called the **least squares regression line** or **line of best fit**). It is the characteristics of that line that you are deriving with a regression.

BOX 16.1

Why Is Regression Called Regression?

The term "*regression*" is a bit odd. Its origin is an interesting one that brings us back to Sir Francis Galton and his pea seeds.

Francis Galton (1822–1911) had a problem. He wanted to know the answer to this question: "How is it possible for a whole population to remain alike in its features, as a whole, during many successive generations, if the *average* produce of each couple resembles their parents?" (Walker, 1929:103).

To answer the question, Galton took several hundred pea seeds and sent them to friends. He asked them to grow the seeds, keeping their soil constant, and to class the offspring according to parental size. Then Galton asked his friends to return those peas. He weighed the parents of each pea seed and the pea seeds themselves, then divided them into classes:

> The measurements . . . led him to his first statement of the law of regression: the mean of every batch of progeny was displaced from the general mean in an amount proportional to the displacement of their parents. The mean displacement of the offspring, however, was always less than that of their parents; they had, on the average, reverted part way back to the mean for the entire race." (Porter, 1986: 286–7)

The idea that developed into what we know now as regression was described as a regression, or reversion, of attributes (such as height) toward the mean. Even if height is hereditary, heights of successive generations continue to observe a normal distribution and appear to be "pulled back" toward a new mean value, away from radical deviation.

Regression: The Formula

A little bit of basic geometry (the relationship between points and lines) should give you a better understanding of regression. Regression takes the following form:

$$Y = a + b_1X_{1i} + b_2X_{2i} + \cdots + b_nX_{ni} + e_i$$

Where:

Y = the dependent variable
a = the Y intercept
b_1 = the partial slope of X_1 on Y
b_2 = the partial slope of X_2 on Y
b_n = the partial slope of X_n on Y
X_{1i} = the first independent variable for individual i
X_{2i} = the second independent variable for individual i
X_{ni} = the nth independent variable for individual i
e_i = error for individual i

The equation represents a formalized hypothesis about the relationship between the independent variables, X, and one dependent variable, Y, acknowledging that each prediction is likely to carry at least some error, meaning that the combination of the variables will "miss" the actual score of the dependent variable by some margin. This margin is represented by e_i, an error term that each individual will have a distinct value for. Error is defined as the gap between predicted and observed values, the amount by which an estimate "misses" its mark. The addition sign between each partial slope coefficient/independent variable couplet implies an additive relationship, which means that each independent or explanatory variable has equal potential to affect the dependent variable.

For now, we assume that the relationship between independent variables and the dependent variable is linear; therefore, it can be plotted as a straight line. The values of the independent variables are multiplied by their slopes (the *b*'s) and summed to determine the predicted value of the dependent variable. The *b*'s are referred to as **partial slope coefficients** because they are the slope assigned to any given independent variable holding all the other independent variables constant. The term *coefficient* refers to a number or quantity that expresses the nature of the relationship between an independent variable and the dependent variable.

Let's look at a simple example: the relationship between age and income, using individuals of all ages. Since this is probably your first time using regression, we'll look at only one independent variable, age, even though it would be easy to think of numerous other characteristics that shape income (gender, occupation, number of hours worked, years of education, etc.). The basic equation looks like this:

$$\textbf{income} = a + b_1\textbf{age}_i + e_i$$

A regression takes the score of the independent variable (age) for each individual and determines how that value relates to the dependent variable, income.

The equation identifies three values: the *y*-intercept, the slope coefficient for age, and the individual error. We've already discussed slopes and errors, and you may remember from high school geometry that the *y*-intercept represents the value where the regression line (you may have called it the best fit line) crosses the *y*-axis. (Note: now you finally know why you had to learn geometry!) The *y*-intercept refers to the point on the *y*-axis where the regression line crosses. This is the point where *X* is equal to 0.

The interpretation in this example is more nuanced because it also represents our prediction of a person's income when his or her age is 0. We probably do not have anyone who is 0 years of age in our data, so the actual value of the *y*-intercept is not meaningful in this case. (We do not need to estimate the income of 0-year-olds!) It is important to think about the ranges for all variables when estimating regression models, because they are critical to describing and understanding patterns in your data. The slope coefficient (there is only one slope coefficient in this case), represents the effect of an additional year of age on our income prediction, starting at the lowest value for that variable.

Usually, the slopes and *y*-intercepts are the values of greatest interest. Occasionally, it may be interesting to look at the error, or the disparity between predicted values (our "guess" of a person's income) and actual values (what their income actually is).

Now that we know what each term in the regression equation means, let's estimate the relationship between age and income by using the 2006 Census of Canada. (See Table 16.1.) We'll restrict the sample so that a person must have valid values on both variables. For now, we'll use a computer to do the calculations. In the next section, we'll learn how to calculate intercepts and coefficients.

TABLE 16.1 | Regression Output for a Single Independent Variable Model

Source	SS	df	MS		
Model	9.8500e+12	1	9.8500e+12	Number of obs	= 645,961
Residual	4.6279e+14,645,959		716,439,688	F(1645959)	= 13,748.61
				Prob > F	= 0.0000
Total	4.7264e+14,645,960		731,687,274	R-squared	= 0.0208
				Adj R-squared	= 0.0208
				Root MSE	= 26,766

Totincp	Coef.	Std. Err.	T	p>|t|	[95% Conf.	Interval]
agep	218.9251	1.867095	117.25	0.000	215.2657	222.5846
_cons	17,795.57	88.38844	201.33	0.000	17,622.33	17,968.81

Source: Statistics Canada. 2001 Census. Catalogue 92–377–X.

Each computer program presents results differently. The regression in Table 16.1 was calculated with STATA. For now, focus on the two values in the column with the heading "Coef." These are the coefficients for age (called agep in the 2006 Public Use Census File) and the y-intercept (called _cons in STATA), respectively. The value of 17,795.57 tells us that our regression predicts that a person with 0 years of age will be earning $17,795.57 (since nobody under age 15 has valid income values on the Census, this is not really meaningful), and the agep coefficient tells us that the value is expected to increase by $218.93 for every additional year of age. Inserting these values into our equation above, a now carries a value of 17,795.57 and b is 218.93. For now, ignore the other columns, even though some of them (Std. Err., t, 95% Conf. Interval) may look familiar.

Graphing the relationship with a scatterplot can make it easier to visualize the relationship. Although it may be difficult to see, Figure 16.1 shows that our estimate of the y-intercept (17,795.57) corresponds with where the line crosses the y-axis. Equally difficult to see is how the pitch of the regression line corresponds with this slope coefficient value (218.93) in our equation. These numbers should be identical. The equation describes the characteristics of the regression line and that the distance between the line and any given observation is the error.

You can probably see some similarities between regression and correlation by now. In some ways, regression is a correlation with many variables.

Regression allows you to predict the value of the dependent variable by using the information provided in independent variables. For example we could estimate the income of a 25-year-old as follows:

$$
\begin{aligned}
\textbf{income} &= \mathbf{17{,}795.57 + 218.93age}_i \\
&= \mathbf{17{,}795.57 + 218.93 * 25} \\
&= \mathbf{17{,}795.57 + 5473.25} \\
&= \mathbf{\$23{,}268.82}
\end{aligned}
$$

Based on our regression equation, we can predict that a person who is 25 years old will earn $23,268.82.

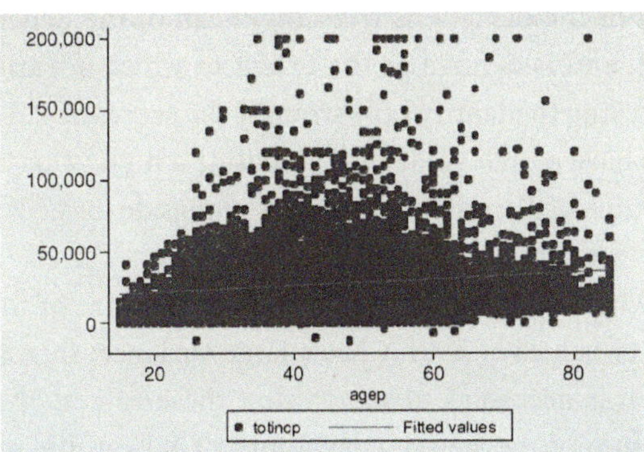

FIGURE 16.1 | Regression of Age and Income, 2006 Canada

Source: 2001 Census of Canada

How accurate is that estimate? According to the 2006 Census file, the actual average income for a 25-year-old is \$21,663.40. Although our guess is pretty good—we're only off by \$1605.42—there are things we could do to get a more accurate estimate. One is to add more explanatory variables (education, gender, occupation, etc.). For now, let's just work with what we have and develop our understanding of the one independent variable equation.

Now that we know how to estimate *Y* by using slopes and intercepts, how are the *y*-intercept, the slope coefficient, and the error terms obtained? We are kind of working backwards, but with some careful thought we can imagine what some of the necessary information might be. Since the regression line is supposed to minimize errors (the average gap between observed values and predicted values should be as small as possible), we need an equation that will create a line of best fit where average distance between observations above the line are the same as, or as close as possible to, those below the line, and where the total sum of these distances is minimized.

That makes sense: with the best fitting line, which is what the regression line is supposed to be, we want the average level of *under*estimation to be approximately equal to the average level of *over*estimation. If positive errors (those above the line) surpass negative errors, then our line will not cut midway through the data.

The line of best fit can also be stated as the sum of cross-product deviations from their mean of independent variable *X* and dependent variable *Y*, divided by the sum of squared deviations from its mean of independent variable *X*. This is the equation for the line of best fit:

$$b = \frac{\sum (X - \bar{x})(Y - \bar{y})}{\sum (X - \bar{x})^2}$$

Where:

$\bar{x}$ = the mean value of variable X
$\bar{y}$ = the mean value of variable Y

The numerator is the sum of the deviations from the mean of the independent variable, multiplied by the sum of the deviations from the mean of the dependent variable (this is called the **covariance**), and is defined as the extent to which *Y* varies with changes in *X*. Remember that we're trying to identify how strongly the scores of *Y* depend on the scores of *X*. When observations have a high value of *X* alongside a high value of *Y*, a middle value of *X* alongside a middle value of *Y*, and lower *X* values alongside lower *Y* values, the numerator will be high, resulting in a larger coefficient.

What's going on with the denominator? Given that the size of the numerator partially depends on the range of values of *X* (if *X* has a large variance, then the numerator will be large, and vice versa), it is necessary to standardize these values. This is done by dividing the numerator by the sum of the squared deviations of *X* from the mean. These values are squared because if they are not, then the sum of deviations from the mean will be equal to 0. You might remember this from when we calculated variance and standard deviation.

Once the coefficient b is found, calculating the y-intercept and the error is easy. We know that the regression line goes through the point where the mean of X and the mean of Y intersect. Therefore, we can substitute this point into our regression equation, and the y-intercept a is defined as follows:

$$a = \bar{y} - b\bar{x}$$

Error e is defined as:

$$e = Y - \hat{y}$$

Where: $\hat{y}$ is the predicted value (the value of Y on the regression line at a particular value of X).

The goal is to ensure that the total estimation errors above the line (which would occur when our estimated value is lower than the actual value) are as close as possible to the total estimation errors below the line (when our estimated values are too high). We can do this by minimizing the sum of all errors. The problem with doing this is that when our line of best fit is optimally placed (the errors above the line equal those below it), the sum of all errors will be 0. For this reason, rather than looking at the sum of errors, we typically compare the sum of squared errors because squaring values will eliminate any negative values.

Let's illustrate this with an example. Using the 1971 Census of Canada Public Use File, we can use age at first marriage to predict family size, focusing on people who were 35 years old in 1971. As with other statistical techniques, the calculations are cumbersome in large samples, so we'll restrict our analysis to 12 randomly selected observations. These observations appear in Table 16.2.

TABLE 16.2 | Family Size and Age of First Marriage among 12 Randomly Selected Cases, 1971 Census of Canada

Age at first marriage (X)	Family size (Y)
21	4
32	2
21	3
25	5
24	1
17	3
22	5
21	4
22	2
22	5
28	2
26	3

Source: 1971 Census of Canada. STC cat. 12–540.

Let's revisit our equation for b:

$$b = \frac{\sum(X - \bar{x})(Y - \bar{y})}{\sum(X - \bar{x})^2}$$

The first things we'll need to find, then, are the means of family size (Y, the dependent variable) and age at first marriage (X, the independent variable).

For family size we get:

$$\frac{4 + 2 + 3 + 5 + 1 + 3 + 5 + 4 + 2 + 5 + 2 + 3}{12} = 3.25$$

For age at first marriage we get:

$$\frac{21 + 32 + 21 + 25 + 24 + 17 + 22 + 21 + 22 + 22 + 28 + 26}{12} = 23.42$$

We'll also need the sum of cross-product deviations of independent variable X and dependent variable y from their means. This is a little harder to calculate, but a chart like the one in Table 16.3 will make things easier.

We are primarily interested in two pieces of information in Table 16.3, $\sum(X - \bar{x})(Y - \bar{y})$, which is often referred to as the sum of products (*SP*), and $\sum(X - \bar{x})^2$, the sum of squares (*SS*).

TABLE 16.3 | Using Family Size and Age of First Marriage among 12 Randomly Selected Cases to Calculate Regression Coefficients, 1971 Census of Canada

Age at first marriage (X)	Family size (Y)	$X - \bar{X}$	$(Y - \bar{Y})$	$(X - \bar{X})(Y - \bar{Y})$	$(X - \bar{X})^2$
21	4	-2.42	0.75	-1.82	5.86
32	2	8.58	-1.25	-10.73	73.62
21	3	-2.42	-0.25	0.61	5.86
25	5	1.58	1.75	2.77	2.50
24	1	0.58	-2.25	-1.31	0.34
17	3	-6.42	-0.25	1.61	41.22
22	5	-1.42	1.75	-2.49	2.02
21	4	-2.42	0.75	-1.82	5.86
22	2	-1.42	-1.25	1.78	2.02
22	5	-1.42	1.75	-2.49	2.02
28	2	4.58	-1.25	-5.73	20.98
26	3	2.58	-0.25	-0.65	6.66
Mean = 23.42	Mean = 3.25			Sum = -20.25	Sum = 168.92

Source: 1971 Census of Canada. STC cat. 12-540.

Now that we have the sum of products, the sum of squares, and the means of the two variables, the equation for the slope, or the line of best fit, is easy to solve:

$$b = \frac{\sum (X - \bar{x})(Y - \bar{y})}{\sum (X - \bar{x})^2}$$

$$= \frac{-20.25}{168.92}$$

$$= -0.12$$

The *y*-intercept is also easy to solve:

$$a = \bar{y} - b\bar{x}$$

$$= 3.25 - (-0.12) * 23.42$$

$$= 6.06$$

To calculate the error, we need to generate **predicted values** for each observation: the predicted family size for each person, given the available information (their age at first marriage). We use our standard regression equation and the values we calculated above to find this:

$$Y = a + bX = 6.06 - 0.12X$$

Next, we calculate predicted values for each observation. By substituting values for X we calculate prediction errors with the following equation: $e = Y - \hat{y}$. (See Table 16.4.) The prediction errors are also called the *residuals*.

TABLE 16.4 | Using Age of First Marriage among 12 Randomly Selected Cases to Calculate Predicted Family Size, 1971 Census of Canada

Age at first marriage (X)	Family size (Y)	($\hat{Y}$)	$e = Y - \hat{Y}$
21	4	3.54	0.46
32	2	2.22	−0.22
21	3	3.54	−0.54
25	5	3.06	1.94
24	1	3.18	−2.18
17	3	4.02	−1.02
22	5	3.42	1.58
21	4	3.54	0.46
22	2	3.42	−1.42
22	5	3.42	1.58
28	2	2.70	−0.70
26	3	2.94	0.06

Source: 1971 Census of Canada. STC cat. 12–540.

Multiple Regression

In the previous example, we used models with only one independent variable and one dependent variable. Although this nicely illustrated the principle of ordinary least squares regression, such small models are rarely used in social science research. Typically, models will have more than one independent variable—in fact, analysts will sometimes include more than 50 independent variables in a single model. This makes for an extremely complex analysis that would be just about impossible without computers and statistical software packages.

The same equations are used for models with several independent variables, but there are more considerations. Any two independent variables will not only be correlated with the dependent variable but probably also with each other. Because of this, the independent effect of one variable cannot be isolated without looking at the impact of the other variables and identifying the relationship between them. For example, the number of years that a Canadian immigrant has been in the country is related to his or her age. Of course, it is impossible to have been in Canada for longer than you've been alive, but older people have had the opportunity to be in Canada for much longer than young people, so we expect there to be a correlation between age and years in Canada. If we use both of these independent variables to predict a dependent variable, such as the number of charities a person participates in, we would not be able to identify the independent impact of either age or years since migration without acknowledging that there is some relatedness, or **correlation**, between them.

BOX 16.2

Regression with One Independent Variable: The Steps

1. Find the mean for each *X* and *Y* variable.
2. Subtract the mean from the value for each observation.
3. Square and sum these terms. Each value is known as the sum of squares.

$$SS_X = \sum (X - \bar{x})^2$$

$$SS_Y = \sum (Y - \bar{y})^2$$

4. Find the sum of products:

$$SP = \sum (X - \bar{x})(Y - \bar{y})$$

5. Calculate *b* (the partial slope):

$$b = \frac{SP}{SS_X}$$

6. Calculate *a* (the *y*-intercept):

$$a = \bar{y} - b\bar{x}$$

7. Calculate *r*:

$$r = \frac{SP}{\sqrt{SS_X SS_Y}}$$

Handling the correlation between the independent variables can become complicated. Since we are dealing with correlations, we can return to bivariate correlations to help us understand regression with more than one independent variable. To remind you, here's the computational equation for a bivariate correlation between variables X and Y:

$$r_{XY} = \frac{N\sum XY - \left(\sum X\right)\left(\sum Y\right)}{\sqrt{\left[N\sum X^2 - \left(\sum X\right)^2\right]\left[N\sum Y^2 - \left(\sum Y\right)^2\right]}}$$

However, we need to change the notation slightly to accommodate the new correlations we're measuring, resulting in the following three equations:

$$r_{X_1Y} = \frac{N\sum X_1Y - \left(\sum X_1\right)\left(\sum Y\right)}{\sqrt{\left[N\sum X_1^2 - \left(\sum X\right)^2\right]\left[N\sum Y^2 - \left(\sum Y\right)^2\right]}}$$

$$r_{X_2Y} = \frac{N\sum X_2Y - \left(\sum X_2\right)\left(\sum Y\right)}{\sqrt{\left[N\sum X_2^2 - \left(\sum X_2\right)^2\right]\left[N\sum Y^2 - \left(\sum Y\right)^2\right]}}$$

$$r_{X_1X_2} = \frac{N\sum X_1X_2 - \left(\sum X_1\right)\left(\sum X_2\right)}{\sqrt{\left[N\sum X_1^2 - \left(\sum X_1\right)^2\right]\left[N\sum X_2^2 - \left(\sum X_2\right)^2\right]}}$$

Technically, it's the same equation, but we denote the dependent variable Y and the independent variables $X_1, X_2, X_3 \ldots X_n$ to help us remember what we're calculating.

These probably look daunting, but they're essentially the same equations that we used to calculate correlation in Chapter 13, except that the notation has changed to indicate that we're looking at different combinations of variables.

Let's see how this works in practice. Recall from Chapter 14 that we used the 1666 Census to illustrate correlation, so let's use that again to elaborate on regression with two independent variables. This time, both the age of the husband and the age of the wife will be independent variables, and we'll use that information to predict the number of children in the household. Let's suppose that we expect the age of the husband to be positively correlated with the age of the wife and for the ages of both husband and wife to be positively associated with the number of children in the household. To get a better sense of what we're trying to do, let's take a look at the data in Table 16.5. (We'll focus on only 12 observations.)

We need to calculate the correlations between all of the variables. (Since we already did that in Chapter 14, let's rely on computers here.)

.corr wifage husage nchildhh (obs=12)

	wifage	husage	nchildhh
wifage	1.0000		
husage	0.7073	1.0000	
nchildhh	0.1485	0.6152	1.0000

TABLE 16.5 | Age of Husband and Wife and Number of Children among 12 Households, 1666 Census of Canada

Obs #	Wifage (*X*1)	Husage (*X*2)	# Child (*Y*)
1	40	41	0
2	47	36	0
3	30	40	5
4	18	27	0
5	37	50	7
6	50	62	4
7	32	30	3
8	40	54	6
9	41	36	1
10	20	27	1
11	36	36	2
12	32	37	5

Note: Data from Ethnic Diversity Survey Public Microdata File, 2002, accessed at http://www5.statcan.gc.ca/bsolc/olc-cel/olc-cel?

As you can see, our expectations are met. Each of the three correlations is positive, suggesting that age of husband is positively correlated with both age of wife and number of children in the household. Similarly, age of wife is positively correlated with age of husband and number of children in the household. Finally, we can infer from the previous two statements that number of children in the household is positively correlated with age of wife and age of husband.

Now that we have the **zero-order correlations** (correlations between variables without assuming any causal order), we need the equations for the partial correlation coefficient:

$$b_1 = \left(\frac{s_Y}{s_{X1}}\right)\left(\frac{r_{X1Y} - r_{X2Y}r_{X1X2}}{1 - r^2_{X1X2}}\right)$$

$$b_2 = \left(\frac{s_Y}{s_{X2}}\right)\left(\frac{r_{X2Y} - r_{X1Y}r_{X1X2}}{1 - r^2_{X1X2}}\right)$$

Where: ***s* refers to the standard deviations of certain variables (which you learned how to calculate in Chapters 6 and 7), and *r* denotes the correlations between them.**

Let's work through these equations to derive the partial slope coefficients:

$$b_1 = \left(\frac{s_Y}{s_{X1}}\right)\left(\frac{r_{X1Y} - r_{X2Y}r_{X1X2}}{1 - r_{X1X2}^2}\right)$$

$$= \left(\frac{2.517}{9.612}\right)\left(\frac{0.1485 - 0.6152 * 0.7073}{1 - 0.7073^2}\right)$$

$$= (0.2618)\left(\frac{0.1485 - 0.4351}{1 - 0.5003}\right)$$

$$b_2 = \left(\frac{s_Y}{s_{X2}}\right)\left(\frac{r_{X2Y} - r_{X1Y}r_{X1X2}}{1 - r_{X1X2}^2}\right)$$

$$= \left(\frac{2.517}{10.765}\right)\left(\frac{0.6512 - 0.1485 * 0.7073}{1 - 0.7073^2}\right)$$

$$= (0.2338)\left(\frac{0.6152 - 0.1050}{1 - 0.5003}\right)$$

When the coefficients are calculated, an interesting difference between these figures and the zero-order correlations emerges. Notice that b_2 remains positive, showing that the age of the husband remains positively correlated with the number of children in the household. At the same time, age of wife is now negatively related to the number of children in the household.

How can this be? The correlation between age of husband and age of wife was so strong that it obscured the negative relationship between the age of wife and the number of children in the household. This nicely illustrates the importance of regression analysis and explains why it is one of the most frequently used techniques in social science data analysis.

Standardized Partial Slopes (Beta Weights)

In our example, the partial slopes (b_1 and b_2) were in the original units of the independent variables. The coefficient refers to the expected increase in the dependent variable when there is a one-unit increase in the value of an independent variable. Since both independent variables were measured in years, it was easy to compare the relative impact of each variable. We know that the age of the husband had a stronger impact than the age of the wife because the coefficient was greater in magnitude.

To compare the relative effects of independent variables that are *not* measured in the same units, a beta weight (b^*) needs to be computed. Beta weights show how much change there is to *standardized* scores of *Y* when there is a one-unit change in the *standardized* scores of each independent variable, and controls for the effects of all other independent variables. With standardized betas, the independent variables do not have to be measured in the same units. Since calculating beta weights is really just a standardization technique that once again relies on standard deviations, the equation is simple:

$$b_{1^*} = b_1\left(\frac{s_1}{s_Y}\right)$$

The Multiple Correlation Coefficient

Recall Chapter 14, where r was used to identify the strength of the relationship between two variables. To get the per cent variation of one variable, we squared r. The same thing is possible using multiple regression, except that we are now able to assess the cumulative effect of several independent variables on one dependent variable, rather than just one-on-one correlations. This is called the **multiple correlation coefficient**.

Although we already calculated the bivariate correlations, adding them together will overestimate the percentage of explained variation because of the correlation that exists between $X1$ and $X2$. Both might be explaining a similar portion of the dependent variable. Consequently, the correlation of independent variables must be accounted for in the calculation of Pearson's r-squared (R^2). The equation to do that is:

$$R^2 = r^2_{YX1} + r^2_{YX1X2}(1 - r^2_{YX1})$$

Or:

$$R^2 = \frac{\sum(\hat{y} - \bar{y})^2}{\sum(Y - \bar{y})^2}$$

We have most of the information we need to solve the equation. The only missing piece is r^2_{YX1X2}, which is calculated as:

$$r^2_{YX1X2} = \frac{r_{YX2} - (r_{YX1})(r_{X1X2})}{\sqrt{1 - r^2_{YX1}}\ \sqrt{1 - r^2_{X1X2}}}$$

Applied to our 1666 example, we get:

$$\begin{aligned} r^2_{YX1X2} &= \frac{r_{YX2} - (r_{YX1})(r_{X1X2})}{\sqrt{1 - r^2_{YX1}}\ \sqrt{1 - r^2_{X1X2}}} \\ &= \frac{0.6152 - 0.1485 * 0.7073}{\sqrt{1 - 0.1485^2}\ \sqrt{1 - 0.7073^2}} \\ &= \frac{0.5102}{0.6991} \\ &= 0.76 \end{aligned}$$

Now, we can solve for R^2:

$$R^2 = r_{YX1}^2 + r_{YX1X2}^2(1 - r_{YX1}^2)$$
$$= 0.1485^2 + 0.6530^2 * (1 - 0.1485^2)$$
$$= 0.0221 + 0.4264 * (0.9780)$$
$$= 0.4881$$

Now we know that age of husband and age of wife together explain about 48 per cent of the variation in the number of children in a household. Note that this result is substantially lower than when the two 0-order correlations are added together [$(0.1485 + 0.6152)^2 = 0.5832$]. This is due to the correlation between age of husband and age of wife.

Requirements/Assumptions of Ordinary Least Squares Regression

There are some fundamental assumptions behind ordinary least squares regression. You'll be familiar with most of them because they've been covered elsewhere (particularly in our discussion of correlation). There are tests that you can perform to ensure that each of these assumptions is met when you are examining an OLS regression equation, but these tests are beyond the scope of this book.

Following is a list of those fundamental assumptions:

1. The dependent variable is interval/ratio, and all independent variables are interval/ratio, dichotomous, or dummy.
2. All error terms are normally distributed.
3. There is linearity between variables. This means that the relationship between an independent variable and a dependent variable is the same across the range of both variables. For example, it is assumed that the relationship between one extra year of schooling and number of hours worked per week is the same for a high school graduate as it is for somebody with a master's degree.
4. There is homoscedasticity. **Homoscedasticity** occurs when the measure of variance in *Y* (likely the standard deviation) is consistent across values of *X*. For example, imagine that you are comparing the weights of people at different ages, and that you included a measure of dispersion (variance, standard deviation, etc.). You would have homoscedasticity if the variances for each age were fairly similar (e.g., the mean weight for a 20-year-old is 60 kg with a variance of 20 kg, and the mean weight for a 30-year-old was 65 kg, also with a variance of 20 kg). **Heteroscedasticity**, on the other hand, refers to a situation where the variance of *Y* differs by *X* value. To illustrate heteroscedasticity, imagine the correlation of university grades with a

person's previous high school grades. Somebody who earned a 65 per cent average in high school could (1) perform even worse in university because the material is more difficult, (2) continue to chug along, earning the same average, or (3) become interested in the curriculum offered in universities, thereby increasing his or her grades. Although the same three options exist for somebody with a 95 per cent average in high school, that person is unable to increase his or her grades dramatically. (It's pretty hard to get more than 100 per cent.) Further, that person is unlikely to fail miserably in university, suggesting that the range of grade for those with a 95 per cent average in high school will be smaller than it is for those with a 65 per cent average. The data in this example are heteroscedastic, but for regression we need them to be homoscedastic.

If these assumptions are not met, it is sometimes possible to adjust your variables so that you can still perform OLS regression. In other cases, where the assumptions are not met, you will have to perform a different kind of analysis (such as logistic regression, which is covered in Chapter 17.)

Creating and Working with Dummy Variables

Many variables of interest in the social sciences are measured at nominal and ordinal levels. For example, a common interest is the differences between men and women. These variables must be treated somewhat differently in regression equations. It is not possible to rank respondents based on their sex, nor is it possible to measure the "distance" between men and women. Compare this to something like age, where we can determine the difference between two respondents based on their response (e.g., we know that a ten-year-old is five years older than a five-year-old). Does this mean that we can't use regression techniques for all variables?

Fortunately, the "distance" between response categories can be measured in a nominal or ordinal variable when there are only two response categories. (Typically, these variables are referred to as **dichotomous variables**, **binary variables**, or **dummy variables**.) If we view the response categories of a dichotomous variable as being on two ends of a continuum (for example, males are on one end, and females are on the other), then we know that a person who identifies as male is 1 and someone who identifies as female is 0. In this indirect way, we are able to quantify the distance between two categories in a dichotomous variable so that it can be used in a regression.

In the case of variables with more than two response categories (e.g., religion, region of residence, visible minority status), a "cheat," or simplifying process, is possible. If nominal and ordinal variables have more than two response categories, they can be broken down into a series of dichotomous variables and then entered into a regression. This process is referred to as *creating dummy variables*.

Let's try an example. Marital status is a variable that sociologists frequently examine. Usually there are four allowable responses to the marital status question in surveys: Married/Common Law (sometimes these are counted separately), Divorced, Single (Never Married),

TABLE 16.6 | Using Marital Status to Illustrate Dummy Variables

	D1	*D2*	*D3*	*D4*
Married/common law	1	0	0	0
Divorced	0	1	0	0
Single/never married	0	0	1	0
Widowed/widower	0	0	0	1

and Widowed/Widower. Without modification, we cannot include this variable in a regression because it is not possible to rank the four response categories or to measure the distance between them, thereby violating OLS assumption one.

To get around the problem, we need to create a series of new variables. Let's call them *D*1 through *D*4. As Table 16.6 illustrates, *D*1 assumes a value of 1 when a respondent identifies as married or common law, and 0 otherwise. *D*2 is set to 1 when a respondent identifies as divorced, and 0 otherwise. In the same way, *D*3 and *D*4 identify single/never married and widowed/widower respondents, respectively.

We've created four new dichotomous variables out of one nominal variable. Since the new variables have only two values, 0 and 1, they can be used in an OLS regression. The new dummy variables use the same logic as the gender example. A person who has a value of 1 on variable *D*1 is "married/common law" while someone who has a value of 0 on variable *D*1 is in one of the three other marital status categories.

Unfortunately, working with dummy variables is complicated by the necessity of having a **reference group**. The four dummy variables don't all need to be included in the regression. Only three are needed. Think about it: if a respondent has a score of 0 on variable *D*1 (indicating that he or she is not married or common law), *D*2 (not divorced), and *D*3 (not single/never married), by default that person must be a widow/widower.

If we assume that everyone responded to the original marital status question, then we can infer a person's marital status using only three variables. Only $k - 1$ dummy variables need to be included in the equation, where k equals the number of response categories in the original variable or the total number of dummy variables.

There is no hard and fast rule for choosing the reference category. Researchers often choose the most common or populous group, although any group that's big enough is acceptable.

Interpreting Dummy Variable Coefficients

Interpreting dummy variable coefficients is more complicated than interpreting coefficients for regular interval/ratio variables. In the case of an interval/ratio variable, each coefficient refers to the effect of a one-unit increase in that particular variable on the dependent variable. In the case of dummy variables, the coefficient refers to the effect of being a "1" on that dummy variable, compared to the reference category. So, for example, the coefficient for somebody who is married/common law represents the increased or decreased value of the dependent variable, *relative to the reference group (whatever group was left out)*. This can be tricky, but an example should help to clarify it.

Returning to the 1666 data, imagine that we believe that families in which the parents are farmers will be larger than families in which the parents work at other occupations. We believe that farmers will have more children because children provide cheap farm labour. To assess this, we create a dichotomous dummy variable "farmer" (1 = yes, 0 = no), and run the regression. This produces a coefficient with a value of 0.405, which can be interpreted this way: "Farmers in 1666 could be expected to have an average of 0.405 more children than non-farmers." Our suspicion is supported.

Inference and Regression

As is the case for nearly all of the material we've covered in this text, regression analysis is typically conducted on samples even though it is the population from which the sample is drawn that is of interest. Given this, we must "infer" our sample results to the population. To do this, we use the material covered in several chapters in this text, most notably Chapter 14.

Up to this point in our discussion of regression, we have not thought about inference or statistical significance, even though it is a central part of what analysts consider when they look at regression results. To illustrate the importance of inference, let's continue with our earlier example of family size in 1971. Below is Table 16.3 again, reproduced here as Table 16.7.

We want to assess the probability that *b*, the partial slope, is significantly different from 0 in the population. In other words, we want to ensure that *b* does not have the same mean

TABLE 16.7 | Using Family Size and Age of First Marriage among 12 Randomly Selected Cases to Calculate Regression Coefficients, 1971 Census of Canada

Age at first marriage (X)	Family size (Y)	$X-\bar{X}$	$(Y-\bar{Y})$	$(X-\bar{X})(Y-\bar{Y})$	$(X-\bar{X})^2$
21	4	−2.42	0.75	−1.82	5.86
32	2	8.58	−1.25	−10.73	73.62
21	3	−2.42	−0.25	0.61	5.86
25	5	1.58	1.75	2.77	2.50
24	1	0.58	−2.25	−1.31	0.34
17	3	−6.42	−0.25	1.61	41.22
22	5	−1.42	1.75	−2.49	2.02
21	4	−2.42	0.75	−1.82	5.86
22	2	−1.42	−1.25	1.78	2.02
22	5	−1.42	1.75	−2.49	2.02
28	2	4.58	−1.25	−5.73	20.98
26	3	2.58	−0.25	−0.65	6.66
Mean = 23.42	Mean = 3.25			Sum = −20.25	Sum = 168.92

value for every value of X. Thinking about it in terms of plotting X and Y values, we do not want our least squares regression line to be perfectly horizontal.

Our calculation of b from the data above shows us that the *sample* value differs from 0, but we want to know how likely it is that the *population value differs from 0.* We once again calculate b as:

$$b = \frac{\sum(X - \bar{x})(Y - \bar{y})}{\sum(X - \bar{x})^2}$$

$$= \frac{-20.25}{168.92}$$

$$= -0.12$$

Next, we need to calculate the standard error of our regression coefficient estimate. To do this, we need to once again recall the information from Table 16.4, reproduced below as Table 16.8, with the addition of the sum of squared error estimates $(Y - \hat{y})^2$.

The sum of squared estimate errors is useful because it gives us a sense of how well our regression estimates approximate actual values. It also tells us how much variation in the dependent variable is *not* explained by the independent variables in the model. Once again, however, this number is sample-size dependent, which means that we must generate a standard error of the estimate by using the following equation:

$$s_{est} = \sqrt{\frac{(Y - \hat{y})^2}{n - 2}}$$

TABLE 16.8 | Using Age of First Marriage among 12 Randomly Selected Cases to Calculate Predicted Family Size, 1971 Census of Canada

Age at first marriage (X)	Family size (Y)	$\hat{Y}$	$e = Y - \hat{Y}$
21	4	3.54	0.46
32	2	2.22	−0.22
21	3	3.54	−0.54
25	5	3.06	1.94
24	1	3.18	−2.18
17	3	4.02	−1.02
22	5	3.42	1.58
21	4	3.54	0.46
22	2	3.42	−1.42
22	5	3.42	1.58
28	2	2.70	−0.70
26	3	2.94	0.06

This equation corrects for the effect of sample size, which is why we call it standardized. Inserting our values above, we get the following:

$$s_{est} = \sqrt{\frac{17.8224}{12 - 2}}$$

$$= 1.335$$

This is a standard error of the estimates, which gets us one step closer to the standard error of the coefficient b. To get this number, we use the following equation:

$$se_b = \frac{\sqrt{\frac{(Y - \hat{y})^2}{n-2}}}{\sqrt{(X - \hat{x})^2}}$$

We have nearly all of the values we need to calculate standard error of b. The numerator in the above equation is the standard error of the estimates, and the denominator is the square root of the sum of squares for X. We can get the value of 168.92 for the sum of squares for X from Table 16.3, giving us the following:

$$se_b = \frac{\sqrt{\frac{(Y - \hat{y})^2}{n-2}}}{\sqrt{(X - \hat{x})^2}}$$

$$= \frac{1.335}{\sqrt{168.92}}$$

$$= \frac{1.335}{12.997}$$

$$= 0.103$$

Finally, b is divided by the standard error of b and is then used to generate a t-statistic, which we usually assess at the 95 per cent confidence level. This can be assessed as a two-tailed test at $df = n - 2$:

$$t = \frac{b}{se_b}$$

$$= \frac{-0.12}{0.103}$$

$$= -1.165$$

Looking at Appendix B, we see a $t_{critical}$ value of 2.228, which is substantially higher than the absolute value of our $t_{observed}$ calculation of –1.165. Thus, based on this sample, we cannot be 95 per cent confident that our results would be seen in the population.

This is a lot of work (see Box 16.3 for a list of the required steps), which is why we rely so heavily on statistical software to produce regression results. Every regression output will have a level of statistical significance and a confidence interval for each coefficient. Unless otherwise specified, the 95 per cent confidence interval will be reported by most programs.

BOX 16.3

Identifying the Statistical Significance of One Regression Slope Coefficient: The Steps

1. Calculate the regression slope coefficient by using the following equation:

$$b = \frac{\sum(X-\bar{x})(Y-\bar{y})}{\sum(X-\bar{x})^2}$$

2. Determine the standard error of the estimate error

$$s_{\text{est}} = \sqrt{\frac{(Y-\hat{y})^2}{n-2}}$$

3. Calculate the standard error for b, using either of the following equations:

$$se_b = \frac{\sqrt{\frac{(Y-\hat{y})^2}{n-2}}}{\sqrt{(X-\hat{x})^2}}$$

or

$$se_b = \frac{s_{\text{est}}}{\sqrt{(X-\hat{x})^2}}$$

4. Generate t_{observed} by dividing b (calculated in step 1) by s_b (calculated in step 3).
5. Compare t_{observed} to t_{critical}, found in Appendix B at $df = n - 2$.

EVERYDAY STATISTICS

Regression and Retirement

In an interesting example of how regression helps sort out complicated processes, Myles, Hou, Picot, and Myers (2007) show that although earnings have risen among Canadian single mothers since 1980, most of the increase has been due to the aging of the baby boomers. This is because the average age of a single mother has been steadily rising since 1980, and since older people are likely to earn more, the change over time is largely due to a change in a key population characteristic of the single mother population. For younger single mothers, there was almost no change in earnings.

Q: What do you think will happen to earnings among Canadian lone mothers once baby boomers start to retire?

Conclusion: A Final Note on OLS Regression

Although we have covered a lot, we have barely scratched the surface of regression. We have also not calculated regression coefficients with more than two independent variables, and we have calculated statistical significance with only one independent variable. Clearly, this chapter is intended to give you only a basic understanding of the most basic type of regression.

You'll encounter more complicated regressions in the more advanced statistics courses that (we hope) you'll take.

Glossary Terms

Binary variables (p. 204)
Correlation (p. 198)
Covariance (p. 194)
Dichotomous variables (p. 204)
Dummy variables (p. 204)
Heteroscedasticity (p. 203)
Homoscedasticity (p. 203)
Least squares regression line (p. 190)
Line of best fit (p. 190)
Multiple correlation coefficient (p. 202)
Partial slope coefficients (p. 191)
Predicted values (p. 197)
Reference group (p. 205)
Zero-order correlations (p. 200)

Practice Questions

1. The following table contains data on the number of rooms and monthly rent for 12 houses. Yan wants to know if there is a relationship between the two variables. Help Yan out by answering the following questions:

Number of rooms (X)	Monthly rent (Y)
3	890
2	568
3	860
1	625
1	775
3	900
3	1095
3	800
2	765
3	629
1	600
3	750

a. Find the mean for each X and Y variable.

b. Subtract the mean from the value for each observation.

c. Find the sum of squares for X and Y, and the sum of cross-products, using the following equations:

$$SS_X = \sum (X - \bar{x})^2$$

$$SS_Y = \sum (Y - \bar{y})^2$$

$$SP = \sum (X - \bar{x})(Y - \bar{y})$$

d. Calculate b (the partial slope):

$$b = \frac{SP}{SS_X}$$

e. Calculate a (the y intercept):

$$a = \bar{y} - b\bar{x}$$

f. Calculate r:

$$r = \frac{SP}{\sqrt{SS_X SS_Y}}$$

g. How much does the accuracy of your prediction of the value of the dependent variable increase if you know a person's score on the independent variable?

h. Generate predicted values for each observation.

i. What is the estimation error for each observation?

j. Is the slope coefficient statistically significant?

2. The city of Abbotsford is launching a study to identify the degree to which the number of people sharing a single dwelling affects the cost of utilities. Use the number of people living in a home to predict the amount charged per month.

Number of people (X)	Monthly charge
3	175
2	130
3	231
3	278
2	40
3	205.83
1	0
1	38.41
1	41.23
3	44.2
3	176
3	315

a. Calculate *r* to measure the association between the two variables above.

b. Calculate the standardized partial slope coefficient by using the following equation to calculate the standard deviations of *X* and *Y* (this is the equation for standard deviation in a sample):

$$s = \sqrt{\frac{\sum (X - \bar{x})^2}{n - 1}}$$

c. Calculate the slope coefficient from the data above.

d. What is the estimated monthly charge for a household that contains four people?

e. What is the standardized value of *b*?

Answers to the practice questions for Chapter 16 can be found in Appendix H.

17 Regression 2—Modelling Discrete/ Dichotomous Outcomes with Logistic Regression

LEARNING OBJECTIVES

Ordinary least squares regression (OLS) is an excellent technique for multivariate analysis with continuous dependent variables. However, there are many situations where outcomes of interest are not continuous. In this chapter, we'll cover:

- why OLS might be inappropriate in these instances; and
- logistic regression, a technique for modelling discrete/dichotomous dependent variables.

Introduction

Suppose you're trying to figure out what factors determine whether or not a person gets a mortgage, and you want to know how their earnings affect their chances. Since getting a mortgage is a dichotomous outcome—you're either approved or you're not—you know that there is an upper limit (getting a mortgage) and a lower limit (not getting a mortgage) to your outcome of interest.

If a person earns $1 an hour, they'll probably be rejected for a mortgage (assuming they have no savings to supplement their earnings). Similarly, if they earn $2 an hour, or even $3, their application is still likely to be rejected. The relationship between earnings and acceptance is relatively consistent at such low wages. Compare this to a scenario where a person earns $101 an hour. That person is *much* more likely to qualify for a mortgage. A person earning $100 an hour is also likely to qualify, suggesting that the relationship between earnings and mortgage may be constant at the upper end of earnings as well. In both cases, we could say that the relationship between earnings and getting a mortgage is "flat."

Somewhere between these two extremes lies a more direct relationship. For example, earning $18 an hour may be below a critical threshold that $19 an hour is above. Earning that extra dollar that didn't matter much in the extremes could make a big difference here.

Using the ordinary least squares regression will oversimplify the relationship between earnings and getting a mortgage, leading to inaccuracies in certain ranges of the dependent variable. This is because ordinary least squares regression assumes that the same relationship exists between both variables at all ranges, even though this is obviously not the case.

In instances where an assumption of linearity cannot be made, there is a family of models that can be used to easily estimate binary outcomes. In this chapter, we'll look briefly at logistic regression, one of the simpler and more commonly used techniques. Logistic regression is a counterpart to OLS for discrete or dichotomous outcomes.

Logistic Regression: The Idea

Although ordinary least squares regression can be used to estimate a binary outcome, and it is easy to do with all software packages, introducing a dichotomous outcome violates the assumptions of ordinary least squares. The most significant assumptions are about the distribution of the dependent variable.[1] For OLS, the dependent variable must be continuous and normally distributed. One of the problems with this assumption for dichotomous outcomes is that OLS coefficients will also assume a normal distribution, with theoretical limits of $\pm\infty$ (plus or minus infinity), rather than a bounded or binomial distribution (which makes more sense for binary outcomes). With logistic regression, the coefficients are calculated with a dichotomous outcome variable that is assumed from the outset. (Thus, the coefficients are bounded.)

Another problem with using OLS regression for dichotomous variables is the assumption of a linear relationship between the outcome and other variables in the model. As the mortgage example demonstrates, this is often not the case with dichotomous outcomes. Results from an OLS regression with a dichotomous outcome variable could contain significant error, depending on the value of the independent variables.

Logistic regression solves these problems, and others, because

1. independent variables do not have to be linearly related to the dependent variable.
2. neither the dependent variables nor the error terms need to be normally distributed. (The dependent variable *does* need to resemble one of the other distributions, but that is beyond the scope of this text.)
3. logistic regression does not assume homoscedasticity in variance across levels of the independent variable.[2]

Logistic Regression: The Formula

Many principles of logistic regression are similar to those of OLS regression, but the language and symbols differ slightly. Logistic coefficients (the log of the probability of an event) replace b coefficients, standardized logistic coefficients correspond with beta weights, and a pseudo-R^2 statistic is available to summarize the strength of the relationship (although the interpretation is not directly comparable to R^2 in OLS).

Practically speaking, logistic regression and least squares regression are almost identical. Both methods produce prediction equations, both have y-intercepts and coefficients (although raw logistic coefficients are much more difficult to interpret and require some translation), and both sets of coefficients measure the predictive capability of independent variables on the outcome of interest.

To recap, OLS **regression equations** take the following form:

$$Y = a + b_1 X_{1i} + b_2 X_{2i} + \cdots + b_n X_{ni} + e_i$$

Where:

Y = the dependent variable
a = the Y intercept
b_1 = the partial slope of X_1 on Y
b_2 = the partial slope of X_2 on Y
b_n = the partial slope of X_n on Y
X_{1i} = the first independent variable for individual i
X_{2i} = the second independent variable for individual i
X_{ni} = the nth independent variable for individual i
e_i = error for individual i

The key differences between logistic regression and OLS are the assumptions about the dependent variable, and the relationship between independent variables and the dependent variable. Instead of predicting the score of *Y*—an observed variable—logistic regression predicts the *probability* of the occurrence of *Y* versus the non-occurrence of *Y* (well, actually, the log odds of *Y* occurring).

Imagine that beneath the surface of every dichotomous outcome variable is a **latent**, or **unobserved, concept**, which we can think of as a propensity score. This score is unobserved, which means that it is not measured and that we cannot see it. A score of 1 on this unobserved propensity variable means that the event will certainly occur (e.g., a person will get a mortgage), while 0 means that it certainly will not (e.g., a person's mortgage application will be rejected). So far, the unobserved dependent variable corresponds with the observed dependent variable. Unlike the observed variable, which has only two values (0 and 1), the unobserved variable has a range of values that are bounded by 0 and 1. Most individuals lie between 0 and 1—there's almost always a slight chance that a person will or will not get a mortgage—and the underlying propensity score is a useful way to understand differences between people.

Logistic regression does not directly predict the probability that our outcome *Y* is equal to 1. Instead, it predicts the *log odds* that an observation will have an indicator equal to 1, where the odds of an event are the ratio of the probability that an event will occur to the probability that it will not, often called an **odds ratio**. Here it is, stated as an equation:

$$\textbf{odds}\ (Y = 1) = \left[\frac{\textbf{Pr}\ (Y = 1)}{\textbf{Pr}\ (Y \neq 1)} \right]$$

This can be read as "the odds of *Y* being equal to 1 are equal to the probability of a positive occurrence over the probability of a negative occurrence." Remember that instead of modelling the odds, we model the natural logarithm of the odds. The following equation articulates that:

$$\textbf{log odds } (Y=1) = \ln\left[\frac{\Pr(Y=1)}{\Pr(Y \neq 1)}\right]$$

You may also see the next equation as representing the log odds, but since exhausted probabilities are always equal to 1, it is actually identical to the last one:

$$\textbf{log odds } (Y=1) = \ln\left[\frac{\Pr(Y=1)}{1-\Pr(Y=1)}\right]$$

The dependent variable is also referred to as the logit. Unlike OLS coefficients, which measure the effect of a one-increment change in the independent variable on the dependent variable, logistic coefficients reflect the effect of a one-increment change in the dependent variable on the log odds of the dependent variable being equal to 1 (or the log odds of the dependent variable occurring).

Why use the odds instead of the probabilities? Sociologist Paul Allison (2000) describes odds as a more sensible scale for multiplicative comparisons (for example, how many *times* bigger is X_1 than X_2) than for additive comparisons (which ask how *much* bigger is X_1 than X_2). Allison illustrates, using voting as an example: if person one has a probability of 0.30 of voting, and person two has a probability of 0.60, it's reasonable to claim that the probability of person two voting is twice as high as it is for person one. But a probability that is twice as high as 0.60 is impossible because of the ceiling of 1 (the event definitely will occur) for the probability of an event.

Allison notes that this is not a problem on the odds scale, since a probability of 0.60 is equivalent to the odds of 1.5:

$$\begin{aligned}\textbf{odds} &= \frac{\Pr(Y=1)}{1-\Pr(Y=1)}\\ &= \frac{0.60}{1-0.60}\\ &= 1.5\end{aligned}$$

So we could say that one person is 1.5 times more likely to vote than the other person.

Converting the probability to the odds removes the upper bound of the dependent variable (because it is no longer constrained by the value of 1). Now, only the lower value is constrained at 0, which can be addressed by using the log odds instead of the odds.

To illustrate, compare the three values in Table 17.1.

Several points on the table are noteworthy. First, notice how a probability of 0.5 (where an event is just as likely to occur as not to occur) corresponds with an odds ratio of 1. Odds

TABLE 17.1 | Probabilities Compared to Odds and Log Odds

Probability	Odds	Log odds
0.01	0.01	–4.60
0.05	0.05	–2.94
0.10	0.11	–2.20
0.15	0.18	–1.73
0.20	0.25	–1.39
0.30	0.43	–0.85
0.40	0.67	–0.41
0.50	1.00	0.00
0.60	1.50	0.41
0.70	2.33	0.85
0.80	4.00	1.39
0.85	5.67	1.73
0.90	9.00	2.20
0.95	19.00	2.94
0.99	99.00	4.60

lie between 0 and $+\infty$ (positive infinity), with one as a neutral value at which both outcomes are equally likely (which is why it corresponds with a probability of 0.5). Positive infinity means that odds can have any positive value with no limit to how large that number may be. Second, when the roles of the two outcomes are switched with odds, each value in the range 0 to 1 is transformed by taking its inverse ($\frac{1}{\text{value}}$) to a value in the range 1 to $+\infty$. For example, if the odds of getting some form of cancer for males are one in nine, the odds of not getting cancer are nine to one.

On the other hand, log odds are completely symmetrical, and lie in the range of $-\infty$ to $+\infty$. The value where both outcomes are equally likely is 0. (This might be considered the "neutral value.") When the roles of the two outcomes are switched, the log odds are multiplied by –1, but the number remains the same. So if the log odds of getting some form of cancer for males are 2.20, the odds of not getting it are –2.20.

To illustrate how the log odds remove the upper and lower bounds, look at probability values of 0.01 and 0.99, the two extreme values. Although they are close to their theoretical limits of 0 and 1, it is possible to get much, much closer. We could have values of, say, 0.000,000,001 and 0.999,999,999, which would yield a log odds ratio of ± 20.72. Although the increases would get smaller, it is possible to approach 0 and 1 even more closely, so we give the log odds a range of $-\infty$ to $+\infty$.

As a further illustration, consider the bar charts of the distribution of a variable with propensity scores ranging between 0 and 1 on 1000 observations, found in Figures 17.1 to 17.3.

Figure 17.1 contains the probability scores. As you can see, variables are scattered almost evenly across the range, and the chart does not resemble a normal distribution. Instead, it is heavily bounded by 0 and 1, and, given that OLS models have an assumption of $\pm\infty$, you can see that there will be estimation error. Let's compare this with a plot of the odds in Figure 17.2.

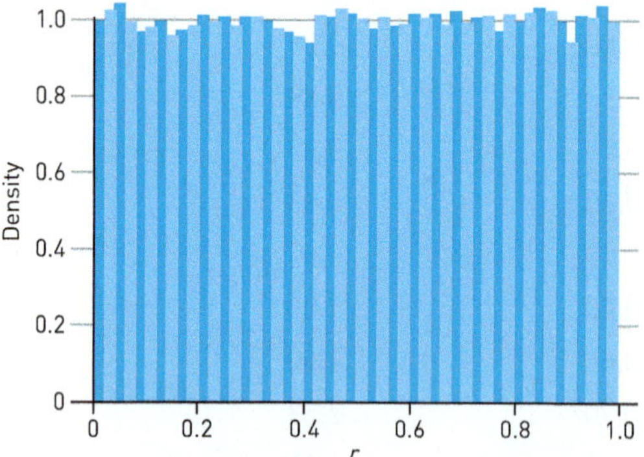

FIGURE 17.1 | Plot of Hypothetical Variable *r*, Stated as a Probability

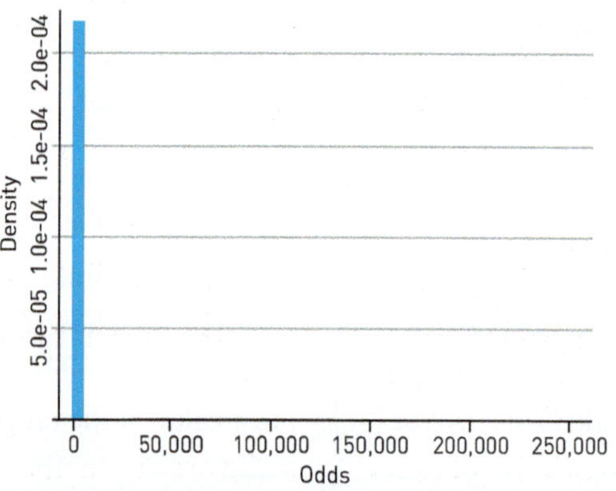

FIGURE 17.2 | Plot of Hypothetical Variable *r*, Stated as Odds Ratios

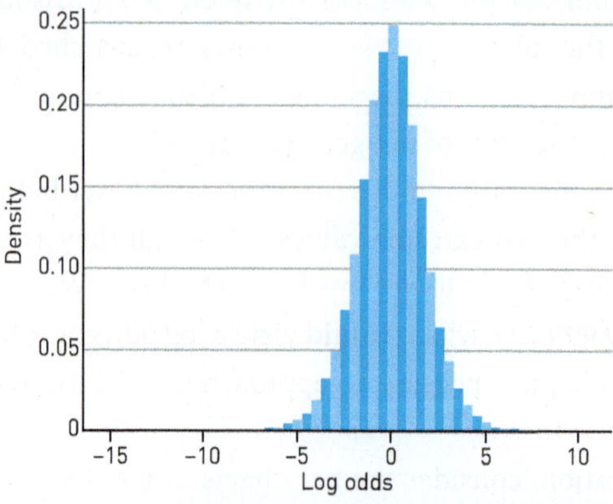

FIGURE 17.3 | Plot of Hypothetical Variable *r*, Stated as Log Odds Ratios

Figure 17.2 is better, in a way, because the constraint of an upper bound has been removed. The problem is that it still does not look like a normal distribution. Compare this to the log odds in Figure 17.3. Figure 17.3 is an almost perfectly normally distributed variable. By now, you should see that the log odds ratio of a dichotomous variable is the best option for a dependent variable.

Modelling Logistic Regression

Thanks to log odds, the theoretical upper and lower bounds on the dependent variable in logistic regression have been removed, and the similarities of logistic regression to OLS are now more obvious. The equation now looks like this:

$$\text{log odds}\ (Y = 1) = a + b_1 X_{1i} + b_2 X_{2i} + \cdots + b_n X_{ni} + e_i$$

Where:

log odds (Y = 1) is the natural logarithm of the odds of the dependent variable being 1
a = the Y intercept
b_1 = the partial slope of X_1 on Y
b_2 = the partial slope of X_2 on Y
b_n = the partial slope of X_n on Y
X_{1i} = the first independent variable for individual i
X_{2i} = the second independent variable for individual i
X_{ni} = the nth independent variable for individual i
e_i = error for individual i

To further illustrate logistic regression, let's look at an example. Suppose that we use the Ethnic Diversity Survey to fit a logistic regression equation that identifies the degree to which a person feels a strong sense of belonging in Canada, where $Y = 1$ if a person experiences a strong sense of belonging and $Y = 0$ if they don't. We use sex of respondent ("female," where 1 = female, 0 = male) and place of birth ("Canada," where 1 = Canada, 0 = other country) as our two independent variables, producing the following predictive equation:

$$\text{log odds}\ (Y = 1) = \alpha + \text{female} + \text{Canada}$$

Or:

$$\text{log [Pr (belong)/Pr (not belong)]} = \alpha + \text{female} + \text{Canada}$$

Where α = constant term.

Estimating the equation in STATA gives us the results found in Table 17.2.

TABLE 17.2 | Gender and Place of Birth as Predictors of Belonging, Stated as Raw Coefficients

• logit belong female Canada

Iteration 0: log likelihood = –28161.371

Iteration 1: log likelihood = –28128.081

Iteration 2: log likelihood = –28128.079

Logistic regression

Number of obs = 41695

LR chi2(2) = 66.58

Prob > chi2 = 0.0000

Pseudo R2 = 0.0012

Log likelihood = –28128.079

belong	Coef.	Std. Err.	z	P . |z|	[95% Conf.	Interval]
female	.1600882	.0199796	8.01	0.000	.120929	.1992474
canada	.0337001	.0225676	1.49	0.135	–.0105316	.0779318
Cons	.2711552	.0219193	12.37	0.000	.2281941	.3141163

Source: Data from Ethnic Diversity Survey Public Microdata File, 2002, accessed at http://www5.statcan.gc.ca/bsolc/olc-cel/olc-cel?catno=89M0019XCB&lang=eng

The interpretation of log coefficients is different from those of OLS regression. We can determine the direction of the relationship by looking at the sign of the coefficients. Since the coefficient for female is significant and positive, the log odds (and, therefore, the probability) of belonging is higher for females. The coefficient for those born in Canada is not statistically significant, so it has no effect on the dependent variable.

Interpreting the Coefficients of a Logistic Regression Equation

To get a better sense of the magnitude of difference, and to make interpreting the coefficients easier, we can exponentiate the results, which is written as exp (b). Remember from Chapter 2 that exponentiation is the inverse function of the logarithm, so the log odds will be changed into the odds. This can be done using *e*, which is the inverse of the natural logarithm. The equation from the last example:

$$\log [\text{Pr (belong)/Pr (not belong)}] = \alpha + \text{female} + \text{Canada}$$

Becomes:

$$[\text{Pr (belong)/Pr (not belong)}] = \exp (\alpha + \text{female} + \text{Canada})$$

Or, once the coefficients from the output are inserted:

$$\textbf{log [Pr (belong)/Pr (not belong)]} = \alpha + \mathbf{0.160} * \textbf{female} + \mathbf{0.034} * \textbf{Canada}$$

Becomes:

$$\textbf{[Pr (belong)/Pr (not belong)]} = \textbf{exp} (\alpha + \mathbf{0.160} * \textbf{female} + \mathbf{0.034} * \textbf{Canada})$$

Alternatively, it could also be listed as:

$$\textbf{Pr}\left(\textbf{belong} = \mathbf{1}\right) = \frac{\textbf{exp} (\alpha + \mathbf{0.160} * \textbf{female} + \mathbf{0.034} * \textbf{Canada})}{\mathbf{1} + \textbf{exp} (\alpha + \mathbf{0.160} * \textbf{female} + \mathbf{0.034} * \textbf{Canada})}$$

This final example, however, is just a more cumbersome form of the same equation.

Exponentiating the coefficients will give the odds ratio, which corresponds with a one-unit change in each independent variable. For example, we could convert the coefficient of 0.160 for female as:

$$\begin{aligned} \textbf{odds ratio}_{\textbf{female}} &= e^{\mathbf{0.160}} \\ &= \mathbf{1.174} \end{aligned}$$

For "Canada," we'd have:

$$\begin{aligned} \textbf{odds ratio}_{\textbf{Canada}} &= e^{\mathbf{0.034}} \\ &= \mathbf{1.035} \end{aligned}$$

Thus, the odds of a female feeling a sense of belonging is 1.174 times that of a male. Stated differently, a female is about 17 per cent more likely to feel a sense of belonging to Canada than a male.

For the variable "Canada," those born in the country are 1.035 times, or 3.5 per cent, more likely than those born elsewhere to feel a sense of belonging. Notice that we cannot be 95 per cent confident that this result exists in the Canadian population, given the low *z*-score of 1.49, and the corresponding significance (denoted in the output by the column $P > |z|$) of 0.135.

With odds ratios, we don't usually convert the intercept, because there isn't really a comparable reference group. For females, the odds ratio of 1.174 is the likelihood of females feeling a sense of belonging relative to males. For Canada, 1.035 compares people born in Canada to those born outside Canada. The intercept simply denotes the point where the regression line crosses the *y*-axis, and there isn't any reason to state that as an odds ratio.

To convince you of the accuracy of our calculation of the odds ratios, let's compare them with the same odds calculated using STATA. (See Table 17.3.)

TABLE 17.3 | Gender and Place of Birth as Predictors of Belonging, Stated as Odds Ratios

logit belong female Canada, or

Iteration 0: log likelihood = −28161.371

Iteration 1: log likelihood = −28128.081

Iteration 2: log likelihood = −28128.079

Logistic regression	Number of obs	=	41695
	LR chi2(2)	=	66.58
	Prob > chi2	=	0.0000
Log likelihood = −28128.079	Pseudo R2	=	0.0012

belong	Odds Ratio	Std. Err.	z	P > |z|	[95% Conf.	Interval]
female	1.173614	.0234483	8.01	0.000	1.128545	1.220484
Canada	1.034274	.0233411	1.49	0.135	.9895236	1.081049

Source: Ethnic Diversity Survey Public Microdata File, 2002

Most software packages include an option to report the odds instead of the log odds (also referred to as the raw coefficients). Our calculations are identical to those produced in STATA. Also notice that the intercept is not reported.

As with OLS, logistic regression is far more intricate and complicated than the cursory overview provided here. The primary purpose of this chapter was to enhance your awareness and to provide some basic information on a frequently used technique.

A Note on Estimating Logistic Regression Equations

In OLS regression, a regression line is calculated with the explicit goal of minimizing the average distance between an observed value and a predicted value. This is done through a series of calculations, which we covered in Chapter 12. Unlike OLS, the coefficient estimates of a logistic regression are obtained through an iterative and complicated process called **maximum likelihood**. For many social scientists, the intricacies of maximum likelihood are treated like a "black box," which means that we don't know or don't question how the results are obtained. Maximum likelihood is a complicated procedure (and there are no truly gentle introductions to the topic that I'm aware of), so we will limit our discussion of it.

The goal of most statistical analysis is to produce a model that predicts our outcome of interest as often and/or as accurately as possible. For our sense of belonging example, above, we want intercept "female" and "Canada" coefficient values that predict whether a person feels belonging as accurately as possible (that is, where there are as few misses as possible).

Maximum likelihood works by fitting a trial equation to the data (often an OLS equation), and comparing the fitted equation to the observed values. Fitting an equation means deriving coefficient estimates. The first equation probably won't maximize the likelihood that we could replicate the results observed in our data set, since the OLS coefficients are likely inaccurate, so the coefficient estimates are tweaked over and over to improve the fit. Iterations stop when the improvement from one step to the next is suitably small, suggesting that the likelihood of replicating the results observed in our data set has been maximized.

Maximum likelihood is widely used in statistical analysis and is not limited to logistic regression alone. In fact, maximum likelihood is so flexible that it is possible to estimate an OLS regression by using maximum likelihood techniques. However, most of the software packages recommend against that since standard OLS techniques allow for greater flexibility.

EVERYDAY STATISTICS

Using Logistic Regression to Determine Mortgage Eligibility

Regression 2—Modelling Discrete/Dichotomous Outcomes with Logistic Regression

As mentioned at the beginning of this chapter, a logistic curve is a more appropriate way to model a dichotomous outcome than a linear regression model. This is because the relationship between an independent variable and a dependent variable is not the same at all values of the independent variable. To illustrate, consider the following figure:

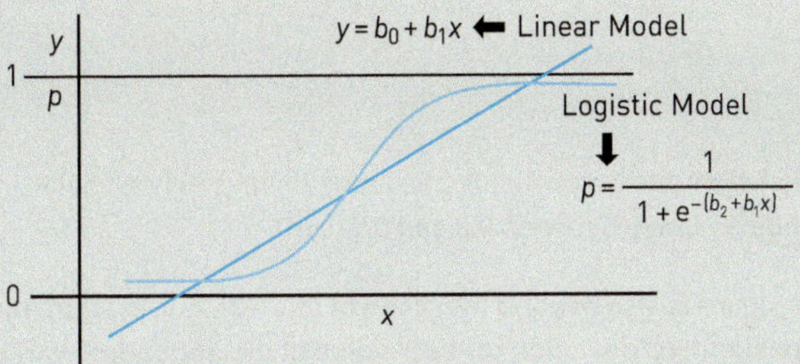

Q: If 1 denotes mortgage acceptance and 0 denotes rejection, do you think people with income (shown on the x-axis) below the mean would be more or less likely to be accepted for a mortgage if ordinary least squares was used to determine eligibility?

Conclusion

This chapter has introduced you to logistic regression, a modelling technique for dichotomous dependent variables. In future statistics classes, you will learn how to expand upon this logic in order to model nominal dependent variables with more than two response categories.

Glossary Terms

Latent concept (p. 215)
Maximum likelihood (p. 222)
Odds ratio (p. 215)
Regression equations (p. 215)
Unobserved concept (p. 215)

Practice Questions

1. Convert the following probabilities to odds ratios:

   a. 0.01
   b. 0.05
   c. 0.1
   d. 0.2
   e. 0.5
   f. 0.667
   g. 0.75
   h. 0.888
   i. 0.999

2. Convert the following logistic regression coefficients to odds ratios.

   a. −1.113
   b. 0.223
   c. 2.78
   d. 0
   e. −2.33
   f. −7.8

Answers to the practice questions for Chapter 17 can be found in Appendix H.

Notes

1. At least in theory. Using OLS for dichotomous outcomes is becoming increasingly popular in certain circumstances, particularly when outcome probabilities range between 0.3 and 0.7.

2. A detailed explanation of what this means is also beyond the scope of this text, but, in short, independent variables do not need to have equal variances or standard deviation for each value of the dependent variable.

PART IV

Advanced Topics

18 Regression Diagnostics

LEARNING OBJECTIVES

Although regression (both OLS and logistic) is a fairly straightforward technique (at least with the help of a computer), numerous hard-to-detect problems can arise in the data. This chapter will cover a few simple diagnostic techniques for regression, focusing in particular on the techniques for ordinary least squares (OLS) regression. These diagnostic techniques help you ensure that your data meet the assumptions required to use regression (as explained in Chapter 16). We'll discuss how to look and correct for:

- influential cases: outliers and leveraged observations;
- heteroscedasticity; and
- collinearity/multicollinearity.

Introduction

Suppose you ran an OLS regression to estimate the factors that determine a person's blood pressure, using the independent variables of body mass index, calories consumed per day, and whether a person drinks and/or smokes (both of these are dummy variables). You might get sensible coefficient estimates for the slope and the intercept, calculations for Pearson's *r*, standard errors, and just about everything else, but there could still be a problem with your model. How can that be?

The problem might be with how your summary measures—essentially, the various components of a regression—represent the data. The model you've estimated may be an incorrect summary of the relationship between a series of independent variables and your dependent variable.

This chapter will cover how this can happen and, more importantly, what can be done about it.

When Ordinary Least Squares Regression Goes Wrong

In a 1973 article in *The American Statistician*, Francis Anscombe demonstrated how it was possible to have similar coefficients, correlations, and standard errors—just about everything between regressions—with very different variables.[1] To illustrate how that's possible, consider his original example in Table 18.1.

Anscombe demonstrated how it is possible to get identical univariate statistics (means, standard deviations) between *X* variables (*X*, *X4*) and *Y* variables (*Y1*, *Y2*, *Y3*, *Y4*), as well as bivariate correlations (r^2), mean squared errors, sums of squares, etc. For our purposes, the most important revelation is that regression intercepts and coefficients can be identical, even though the data, when they are graphed on a scatterplot, look radically different.

To illustrate this better, we'll continue with Anscombe's original example, using four regressions and scatterplots for the data in Table 18.1. (These four plots are sometimes referred to as the "Anscombe Quartet.") In each figure, the regression output appears first, followed by the scatterplot (see Figures 18.1 to 18.4).

TABLE 18.1 | Anscombe's Original 1973 Data for Illustrating Regression Diagnostics

X	Y1	Y2	Y3	X4	Y4
10.00	8.04	9.14	7.46	8.00	6.58
8.00	6.95	8.14	6.77	8.00	5.76
13.00	7.58	8.74	12.74	8.00	7.71
9.00	8.81	8.77	7.11	8.00	8.84
11.00	8.33	9.26	7.81	8.00	8.47
14.00	9.96	8.10	8.84	8.00	7.04
6.00	7.24	6.13	6.08	8.00	5.25
4.00	4.26	3.10	5.39	19.00	12.50
12.00	10.84	9.13	8.15	8.00	5.56
7.00	4.82	7.26	6.42	8.00	7.91
5.00	5.68	4.74	5.73	8.00	6.89

Source: Anscombe, *The American Statistician*, 1973

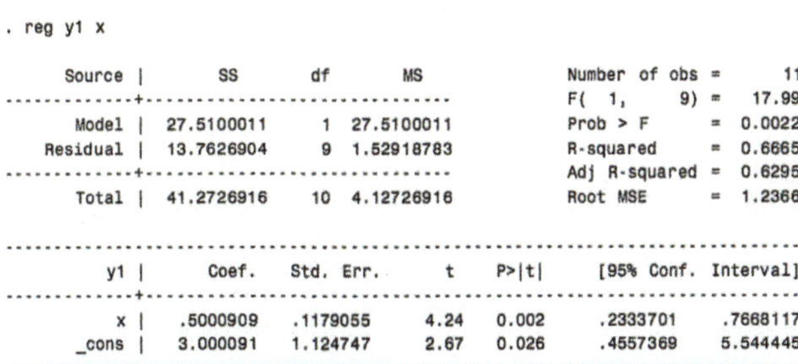

```
. reg y1 x

      Source |       SS       df       MS              Number of obs =      11
-------------+------------------------------           F(  1,     9) =   17.99
       Model |  27.5100011     1  27.5100011           Prob > F      =  0.0022
    Residual |  13.7626904     9  1.52918783           R-squared     =  0.6665
-------------+------------------------------           Adj R-squared =  0.6295
       Total |  41.2726916    10  4.12726916           Root MSE      =  1.2366

------------------------------------------------------------------------------
          y1 |      Coef.   Std. Err.      t    P>|t|     [95% Conf. Interval]
-------------+----------------------------------------------------------------
           x |   .5000909   .1179055     4.24   0.002     .2333701    .7668117
       _cons |   3.000091   1.124747     2.67   0.026     .4557369    5.544445
------------------------------------------------------------------------------
```

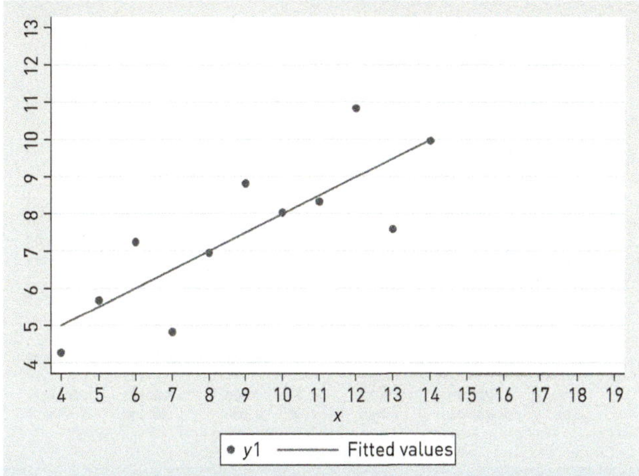

FIGURE 18.1 | Anscombe's Original Demonstration of the Need for Regression Diagnostics, Data—Example #1

Source: Copyright 1973. Figure adapted from "Graphs in Statistical Analysis" by F.J. Anscombe. Reproduced by permission of the American Statistical Association.

In each example, the regression results are identical. This means that the line of best fit is the same in each case, even though the observations have different values. In Figure 18.1, the relationship between the independent variables appears straightforward and relatively linear, with what appear to be randomly distributed errors. However, Figures 18.2 to 18.4 have very different relationships between the *X* and *Y* variables.

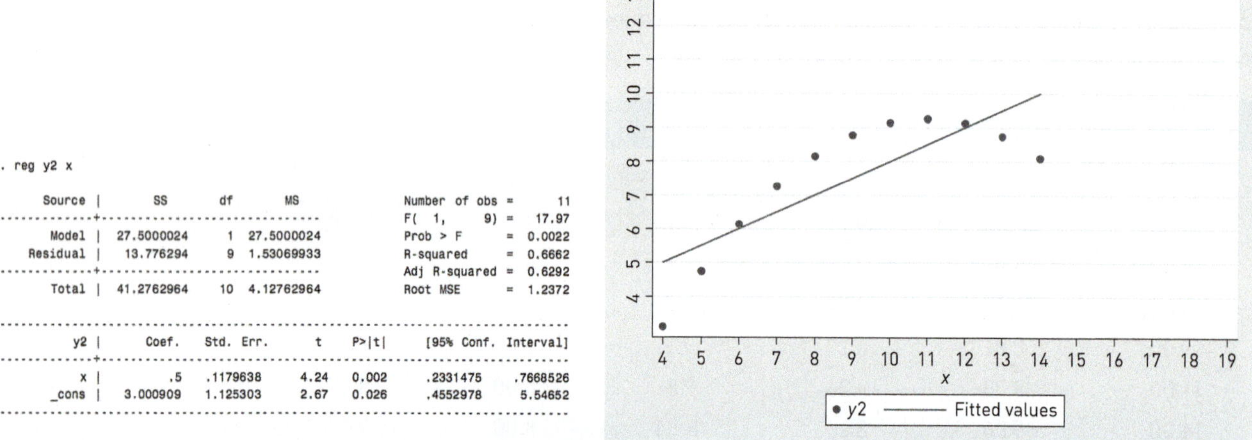

. reg y2 x
Source | SS df MS
Model | 27.5000024 1 27.5000024
Residual | 13.776294 9 1.53069933
Total | 41.2762964 10 4.12762964
Number of obs = 11
F(1, 9) = 17.97
Prob > F = 0.0022
R-squared = 0.6662
Adj R-squared = 0.6292
Root MSE = 1.2372
y2 | Coef. Std. Err. t P>|t| [95% Conf. Interval]
x | .5 .1179638 4.24 0.002 .2331475 .7668526
_cons | 3.000909 1.125303 2.67 0.026 .4552978 5.54652
4 5 6 7 8 9 10 11 12 13
4 5 6 7 8 9 10 11 12 13 14 15 16 17 18 19
x
• y2 Fitted values

FIGURE 18.2 | Anscombe's Original Demonstration of the Need for Regression Diagnostics, Data Example #2
Source: Copyright 1973. Figure adapted from "Graphs in Statistical Analysis" by F.J. Anscombe. Reproduced by permission of the American Statistical Association.

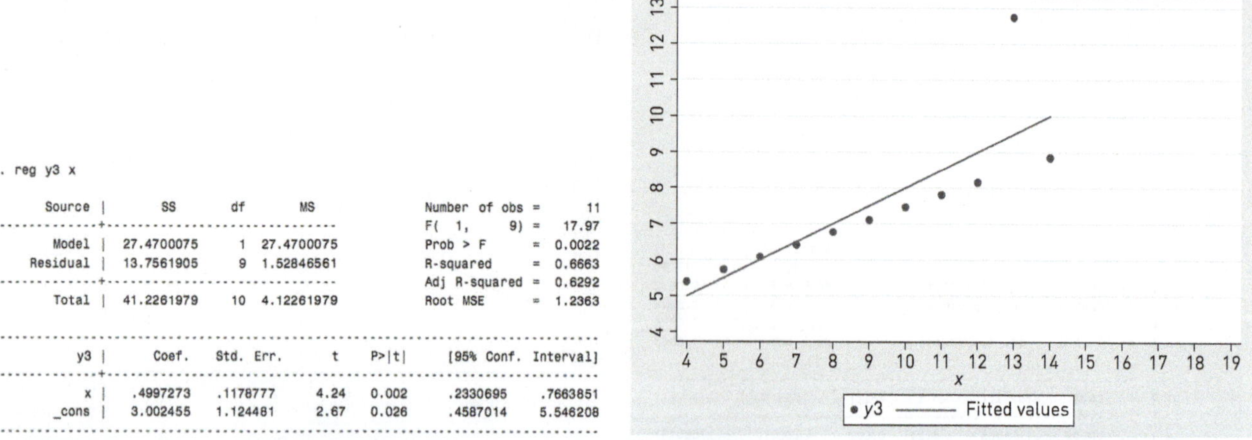

FIGURE 18.3 | Anscombe's Original Demonstration of the Need for Regression Diagnostics, Data Example #3
Source: Copyright 1973. Figure adapted from "Graphs in Statistical Analysis" by F.J. Anscombe. Reproduced by permission of the American Statistical Association.

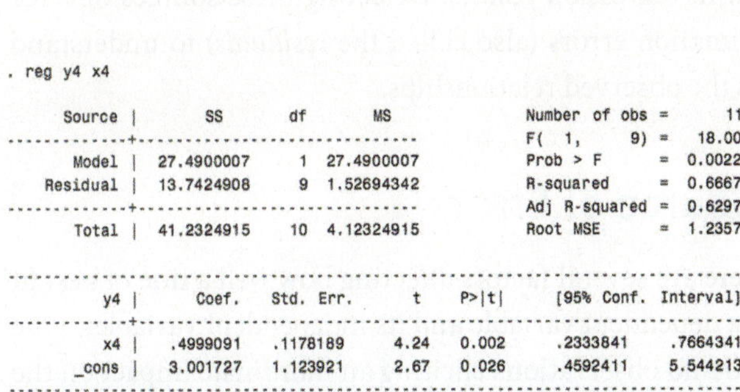

```
. reg y4 x4

      Source |       SS       df       MS              Number of obs =      11
-------------+------------------------------           F(  1,     9) =   18.00
       Model |  27.4900007     1  27.4900007           Prob > F      =  0.0022
    Residual |  13.7424908     9  1.52694342           R-squared     =  0.6667
-------------+------------------------------           Adj R-squared =  0.6297
       Total |  41.2324915    10  4.12324915           Root MSE      =  1.2357

------------------------------------------------------------------------------
          y4 |      Coef.   Std. Err.      t    P>|t|     [95% Conf. Interval]
-------------+----------------------------------------------------------------
          x4 |   .4999091   .1178189     4.24   0.002     .2333841    .7664341
       _cons |   3.001727   1.123921     2.67   0.026     .4592411    5.544213
------------------------------------------------------------------------------
```

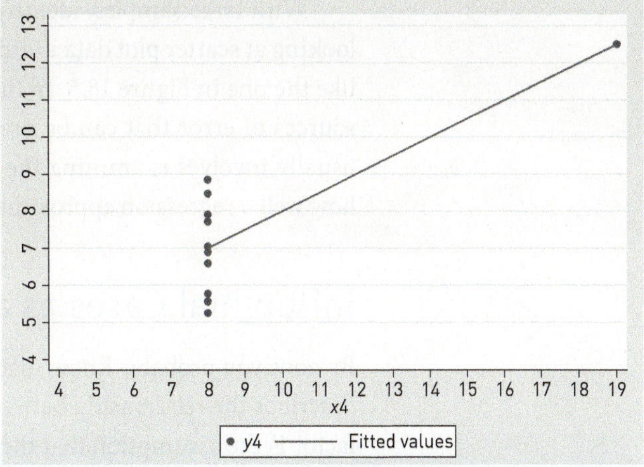

FIGURE 18.4 | Anscombe's Original Demonstration of the Need for Regression Diagnostics, Data Example #4

Source: Copyright 1973. Figure adapted from "Graphs in Statistical Analysis" by F.J. Anscombe. Reproduced by permission of the American Statistical Association.

As a statistical issue, the similarity between the regression equations is problematic because it glosses over the obvious differences in the data. Without looking at the plots, we might conclude that each example does an equally good job of summarizing the nature of the relationship.

The scatterplots reveal something else. Ideally, we'd draw scatterplots whenever possible, but large samples make visualizing relationships very difficult, simply because there are so many data points on the graph. Consider the plot of age of respondent and age of spouse found in the 1666 Census in Figure 18.5.

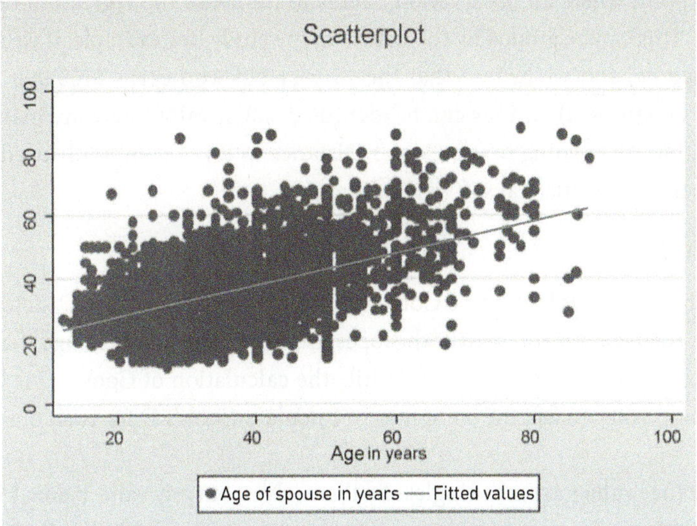

FIGURE 18.5 | Age of Respondent, Age of Spouse, 1666 Census of Canada

Source: Library and Archives Canada. Census of Canada, 1666. MIKAN no. 2318856.

With large samples, identifying all the relationships between different variables just by looking at scatter plot data is difficult. We need to develop a set of tools to identify situations like the one in Figure 18.5. In the following sections, we'll cover some of the more common sources of error that can be present in regression results. Detecting these sources of error usually involves examining the estimation errors (also called the *residuals*) to understand how well a regression approximates the observed relationships.

Influential Cases as a Source of Error

By now, you probably know that there are several factors affecting how well a line of best fit describes the relationship between a dependent variable and its independent variables. One factor is the assumption that there are no observations eliciting an inordinate impact on the calculation of regression coefficients. For example, the observation with an *X*-value of 19 and a *Y*-value of 13.5 in Figure 18.4 appears to have a greater effect on the calculation of the line of best fit than the other observations do. If that observation weren't there, the line would probably have a different slope and intercept.

These observations are called influential cases. Although "influential case" has no firm definition, the term refers to any case that exerts an extraordinary amount of influence on the slope and intercept. There are two types of influential cases: (1) outliers and (2) leveraged observations. An **outlier** refers to an observation that lies far from its estimated *Y*-value (the distance between Y_i and $\hat{y}$ is huge), while **leverage** measures the distance between an *X*-value and the mean for that variable (the gap between X_i and $\bar{x}$). Both can greatly affect the regression coefficients.

To detect an influential case (influential cases have both high leverage and high distance), find the point where an observation begins to influence the regression results to an unusual degree. This can be guided by the nature of the study (for example, if we're studying housing and income, and we believe that the 15-year-old who earns $100,000 a year and owns a home is exceptional), or they can be identified statistically. There are many different ways to do this, but we're going to use Cook's Distance as it is a commonly used technique that is available in most statistical software packages.

Cook's Distance

Introduced in the late 1970s by Dennis Cook, Cook's Distance, or Cook's *D*, determines the extent to which coefficient estimates (both slopes and intercepts) will change if a particular observation is removed from the analysis. While the calculation of Cook's *D* is beyond the scope of this book, you can ask the computer to calculate Cook's *D* for each observation in your sample.

Cook's Distance values are meaningless by themselves, but any value below 1 is regarded as being tolerable for influence. Any observation that has a value exceeding 1 should be examined closely and possibly deleted. The higher than 1 a Cook's *D* value is, the stronger the case for deleting a particular observation. Remember that you should always report on any regression diagnostics you performed when you write up your research results. You should also inform the reader if you decided to remove any influential cases from your analysis (and how many you removed).

Heteroscedasticity as a Source of Error

As discussed in Chapter 16, **homoscedasticity** is examined by looking at the distribution of estimation errors (the difference between the predicted Y's and the actual Y's) across X-values. A regression is considered homoscedastic if the standard deviation of the estimation error for each value of X is roughly similar. If that condition is not met (we want it to be met), the model is **heteroscedastic**. Suppose you had an independent variable with 10 response categories; having a similar error variance across each response category would be important.

A real-life example is the relationship between total income and disposable income. If you wanted to determine whether people with large total incomes have more or less disposable income, you could run an OLS regression, treating disposable income as the dependent variable, and total income as the independent variable. Most programs will estimate this model easily, even though the error variances are likely to be heteroscedastic. To demonstrate, think of how much more disposable income a person earning $100,000 total income could have compared to someone who earns $1000 total income. There is greater potential for us to "miss" with our estimate of disposable income (Y) for the $100,000 earner ($X = 100{,}000$) than there is for the $1000 earner ($X = 1000$), suggesting that the error variance (the difference between our predicted Y and the actual Y) will not be consistent across all values of X.

When the condition of homoscedasticity is not met for a particular variable, the accuracy of that coefficient can be questioned. Heteroscedasticity is complicated and often requires the use of different statistical models (such as weighted least squares or heteroscedastic standard error) or for variables to be transformed by using a power transformation (particularly a logarithmic transformation). These corrections are beyond the scope of this book, so we'll only focus on identifying heteroscedasticity, not on correcting for it.

The easiest way to detect heteroscedasticity is probably to look at a residual versus a fitted value plot. To do that, use your statistical software to predict values for each observation, and plot the disparity across values of the independent variable. Figure 18.6 is an example conducted in STATA.

This plot shows the residuals from a model using the age of an individual to estimate the number of cigarettes that individuals smoke per day. The data includes only respondents aged 12 and above. We might expect there to be wider variation in the number of cigarettes smoked per day for people at younger ages because, although there are likely to be some heavy smokers across all ranges, there are likely to be more casual smokers among young people and fewer casual smokers among older people.

To reduce the number of data points, only women aged 12 and over living in Prince Edward Island are plotted in Figure 18.6. Notice how the dispersion of residual values (the y-axis) is not consistent across the values of the independent variable (the x-axis). This suggests

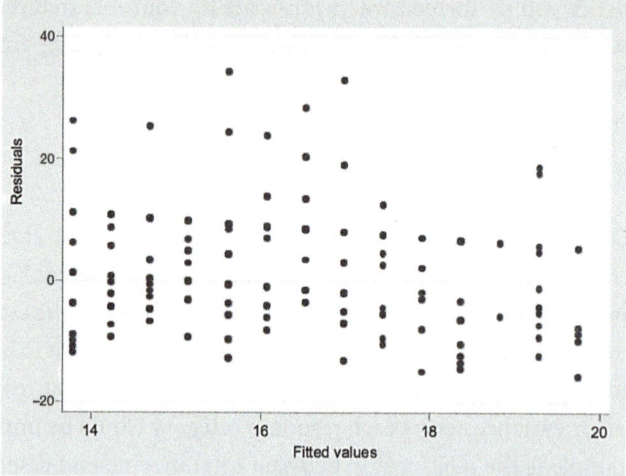

FIGURE 18.6 | Heteroscedastic Data in the Canadian Community Health Survey

Source: Canadian Community Health Survey Public Use Data, Wave 2.1

that there is greater variance at certain values of the independent variable than others. The variance gets smaller as age increases, which suggests that, at least in this sample, there is less variance in the number of cigarettes older people smoke compared to the variance in the number of cigarettes younger people smoke.

There are numerous statistical tests for heteroscedasticity (White's test, Bartlett's test, etc.); most are available in major statistical software packages, but they go beyond what is appropriate for an introductory statistics course.

Multicollinearity as a Source of Error

Collinearity, or **multicollinearity**, is found when independent variables, such as the two from our example, share a common line when graphed; in other words, they are correlated. To illustrate collinearity, let's continue with the example from the introduction: identifying the factors that affect a person's blood pressure. Suppose we suspect that there is a correlation between body mass index (BMI) and calories consumed per day, two of our independent variables. If there was a perfect bivariate correlation between those two variables, how might that affect our regression coefficient estimates? First, including both variables as predictors is unnecessary because the information in one variable (calories consumed per day) can be used to perfectly predict a score on the other variable (BMI), and vice versa. The variables are essentially duplicates of one another.

Remember that for an ordinary least squares regression, coefficients represent the independent impact on the dependent variable of a one-unit increase in the independent variable of interest. Collinearity is important because the coefficients are calculated when all other independent variables are held to zero. When collinearity is strong, identifying the independent impact of BMI will be difficult, because whenever BMI is increased by one increment, so, too, is the number of calories consumed. Finding the independent impact of either variable is virtually impossible because they are so intertwined.

Most of the time there isn't perfect collinearity between two variables. However, even moderate correlation between two independent variables can make it difficult to estimate the independent effect of each independent variable. When any two independent variables are correlated at 0.5 or above, it is important to examine them. Correlation between one variable and several others also poses a problem in the regression because collinearity is calculated across all independent variables.

EVERYDAY STATISTICS

How Do You Like Them Apples? Multicollinearity

A classic example of a collinearity issue in demography is known as the "age-period-cohort" problem. An example of the problem would be as follows: Suppose that you meet three members of a family in a room—a grandfather (age 70), a father (age 40), and a son (age 5)—and each of them is eating an apple. The son tells you that he plans to consume three apples, the father tells you that he will eat two, and the grandfather will stop after one apple. If you were interested in explaining the differences across people, you'd face the following three possibilities:

1. The son is eating three apples because of his age. As he gets older, he will eat fewer apples. This would point to aging as the primary explanation for variation.
2. There is currently a convention in the society in which these three individuals live that allocates a disproportionate number of apples to the young. If the grandfather and father were permitted to eat as much as the son, they would. This suggests a period effect, in other words an effect that is in place because of the historical time period.
3. Since each person grew up in radically different eras, each may have different conventions about the suitable number of apples per consumer. Even when they were younger, the grandfather ate one apple, and the father ate two. This is the cohort effect.

These possibilities are straightforward enough, but it is impossible to distinguish between these explanations in a regular regression because of multicollinearity, since the three variables are linear functions of each other (Age = Period − Cohort, Period = Age + Cohort, etc.). So the regression equation will inevitably suffer from collinearity.

Identifying and Dealing with Multicollinearity

The easiest way to identify multicollinearity is probably to look at the variance inflation factor (VIF) for each variable. The VIF is a standardized version of Pearson's multiple correlation coefficient R^2. The equation for VIF is:

$$\text{VIF}_{Xi} = \frac{1}{1 - R^2_{Xi}}$$

where R^2_{Xi} is equal to the multiple correlation coefficient from a regression of variable Xi on all other independent variables in a model. This suggests that each independent variable in a regression model will have its own VIF value because it will be the "dependent variable" in a model, with all other independent variables used as predictors. The lowest value of VIF will be 1 (this would correspond to an R^2 value of 0), and the highest value is positive infinity ($+\infty$).

Although there is no clear consensus on what value of VIF points to multicollinearity, typically a conservative value of 5 or higher is deemed too high. However, often a lower value is more commonly used. Variables with VIF values that exceed 4 are usually worthy of further investigation.

However you define multicollinearity, once you determine that your variables are too closely correlated, you have several options. The first option, dropping one of the collinear variables, is the easiest and will immediately solve multicollinearity. The second option is more difficult: combining the two offending variables to form one composite measure. However, this method might not make sense in every situation. Whatever you decide to do, it is important to describe how you performed your regression diagnostics and what you decided to do if you detected multicollinearity in your research report.

Conclusion

This chapter covered some of the diagnostic tools for identifying whether regression results are plausible and valid. As noted in the introduction, this is only a preliminary overview. There are entire courses dedicated to regression diagnostics. The aim of this chapter was to enhance your appreciation of the complexity of ordinary least squares regression, and to introduce you to further topics.

Glossary Terms

Collinearity (p. 232)
Heteroscedasticity (p. 231)
Homoscedasticity (p. 231)
Leverage (p. 230)
Multicollinearity (p. 232)
Outlier (p. 230)

Practice Questions

Here are Anscombe's original fabricated data (Anscombe, *The American Statistician*, 1973):

X	Y1	Y2	Y3	X4	Y4
10.00	8.04	9.14	7.46	8.00	6.58
8.00	6.95	8.14	6.77	8.00	5.76
13.00	7.58	8.74	12.74	8.00	7.71
9.00	8.81	8.77	7.11	8.00	8.84

X	Y1	Y2	Y3	X4	Y4
11.00	8.33	9.26	7.81	8.00	8.47
14.00	9.96	8.10	8.84	8.00	7.04
6.00	7.24	6.13	6.08	8.00	5.25
4.00	4.26	3.10	5.39	19.00	12.50
12.00	10.84	9.13	8.15	8.00	5.56
7.00	4.82	7.26	6.42	8.00	7.91
5.00	5.68	4.74	5.73	8.00	6.89

1. Calculate the means and standard deviation for all of the variables. When calculating the standard deviation, treat the observations as sample data (use $n - 1$ instead of N in your calculations).

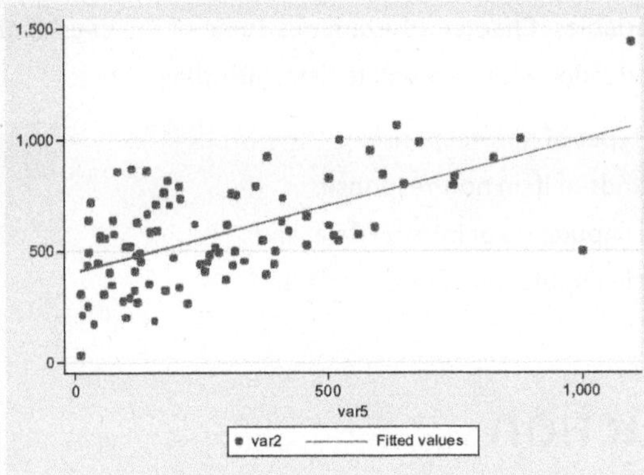

Source: Copyright 1973. Figure adapted From "Graphs in Statistical Analysis" by F.J. Anscombe. Reproduced by permission of the American Statistical Association.

2. Here is a correlation matrix (Pearson's *r* value) for all the variables in the table above.

	X	Y1	Y2	Y3	X4	Y4
X	1.0000					
Y1	0.8164	1.0000				
Y2	0.8162	0.7500	1.0000			
Y3	0.8163	0.4687	0.5879	1.0000		
X4	−0.5000	−0.5291	−0.7184	−0.3447	1.0000	
Y4	−0.3140	−0.4891	−0.4781	−0.1555	0.8165	1.0000

Imagine that you estimated a regression of *X* and *X4* on *Y1*. Calculate VIF for *X* and *X4*. Remember that the table above reports *r*, not r^2.

Answers to the practice questions for Chapter 18 can be found in Appendix H.

Note

1. In 1973, Anscombe was arguing for the importance of graphing raw data alongside regression analysis. As it happens, his example also beautifully illustrates the need for diagnostics.

19 Strategies for Dealing with Missing Data

LEARNING OBJECTIVES

Researchers often have to deal with missing data or with respondents who do not have information on all variables. Chapter 19 examines some of the causes of missing data and some of the strategies for dealing with incomplete data, including:

- three types of non-response;
- four kinds of item non-response;
- single imputation of missing data; and
- multiple imputation of missing data.

Introduction

Paul Allison begins his monograph on missing data by stating, "Sooner or later (usually sooner), anyone who does statistical analysis runs into problems with missing data" (2000: 1). Allison goes on to tell us that when practitioners are faced with these problems they must make several decisions that inevitably affect the conclusions that are drawn from the data.

There are three types of non-response in social surveys that use the household as the unit of analysis:

1. Household non-response
2. Person non-response
3. Item non-response

Household non-response occurs when an entire household fails to complete a questionnaire. This happens for various reasons, such as people being away from home or being unwilling or unable to participate in the survey. Household non-response is difficult to deal with since there is no information available whatsoever.

In the case where a survey is designed to collect information on more than one person in a household, *person non-response* occurs when an interview is obtained from one household member but not from another or others in that household. Like household non-response, person non-response is the result of the unwillingness, inability, or unavailability of a chosen respondent to answer survey questions. Person non-response can be dealt with through editing and imputation of values with reasonable substitutes.

Item non-response occurs when a respondent completes only part of a survey, leaving blanks for some information. Item non-response can occur for many reasons, including the following:

1. A respondent refuses, or is unable, to provide the requested information.
2. A respondent does not identify with any of the response categories and chooses to leave the question blank.
3. An interviewer fails to ask a question or a respondent fails to provide an answer.
4. An interviewer makes an error when recording, or keying in, the response.

Since household and person non-response are usually handled by survey methodologists (which means that by the time you see a data set, the data will have been cleaned up to account for household and person non-response), in this chapter we'll look at item non-response, first by looking at the effects of item non-response, then by looking at the four major reasons for item non-response. Finally, we'll discuss common methods of dealing with these types of missing data.

What Effect Does Item Non-Response Have on Results?

Non-response can have serious statistical consequences. Consider the simple regression equation:

$$Y_i = \alpha + \beta_1 X_i + e_i \qquad (1)$$

Where:

Y_i = **dependent variable**
α = **constant term**
β_1 = **coefficient for** X_i
e_i = **error term**
X_i = **explanatory variable**

Now let's imagine that the explanatory variable X has some error due to non-response. Instead of X_i we observe $X^*_i + u_i$, where u_i = measurement error caused by non-random non-response. So the equation above would need to be modified accordingly:

$$Y_i = \alpha + \beta_1(X_i + u_i) + e_i \qquad (2)$$

Notice that there are now two error terms, e_i, and u_i, which can be simplified to form the compound error term z_i. One of the postulates of OLS regression—that error terms

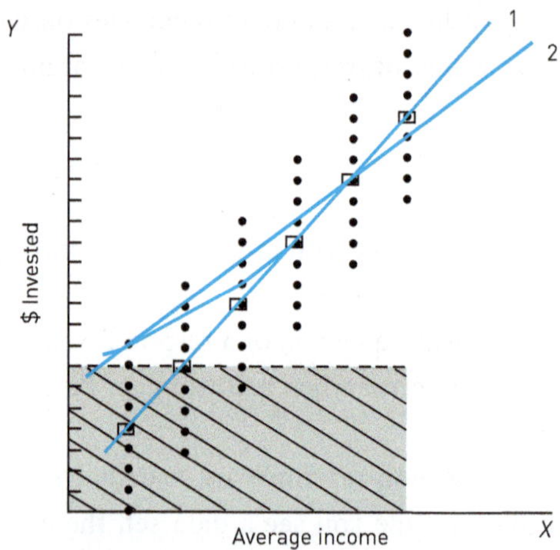

FIGURE 19.1 | How Non-Random Missing Data Affects Regression Estimates
Source: Berk, 1983, p. 387.

are not correlated with explanatory variables—has been violated with the introduction of non-response. The coefficient (β) is now inaccurate.

As a further illustration, consider a simple cross-sectional OLS equation with only one predictor variable, *X*—the average income of an individual—and one outcome, *Y*—the amount of money invested on the stock market per year (see Figure 19.1). Line one demonstrates the linear relationship between those two variables. However, there are problems with missing data: although the sample was randomly drawn, people with lower income levels were less likely to report their income. Therefore, lower-income earners are under-represented in the data, pulling line one upward at lower income amounts (see the curved line in the diagram). If we were to redraw the line of best fit with information on the lower-income people, line two would replace line one as the best fitting line.

Clearly, validity—or the extent to which a variable is measuring what it is supposed to measure—has been compromised when there is non-random missing data. The *X* variable, which measures respondent income, is no longer valid since it doesn't measure what it purports to measure. This can be seen in the new line of best fit.

The Four Kinds of Item Non-Response[1]

To motivate our discussion of the four kinds of item non-response, we will start with a simple regression equation with two independent variables, denoted as:

$$Y_i = \alpha + \beta_1 X1_i + \beta_2 X2_i + e_i \qquad (3)$$

Where:

Y = average grade of a child in the last school year

$X1$ = mother tongue of the person most knowledgeable (PMK) of the respondent (i.e., the person who responds about the child in the survey—usually the mother)

$X2$ = income of PMK for the previous year

***e* is an error term**

$X2$, the income of PMK for the previous year, is a problematic variable. Several respondents are missing values on $X2$, for reasons unknown. We could assume at least four possible reasons for the missing values (Little and Rubin, 1987):

1. $X2$ (income) is "missing completely at random" (MCAR), meaning that the non-response on this question is entirely independent of patterns in either Y (child's grades) or $X1$ (mother tongue). (Here, if the missing information is entirely at random, we can say that the reason for "missingness" is contained in the error term *e*.)
2. $X2$ (income) is "missing at random" (MAR) but dependent on explanatory variable $X1$ (mother tongue), so with certain values of $X1$, $X2$ is more likely to be missing. (For example, perhaps people whose mother tongue is neither French nor English are more likely to be missing on the income variable.)
3. $X2$ (income) is "missing at random" but dependent on the focal outcome (grades), and certain values of Y (grades) increase the probability that $X2$ will be missing. (For example, perhaps those with average grades below 65 per cent are more likely to be missing on income.)
4. $X2$ (income) is "non-ignorable missing" (NIM): that is, $X2$ (income) is often missing when it is a certain value. (For example, income is often missing when it is $< \$10,000$.)

Each of these four types of missing data is a non-response with different underlying causes, and each will have a different effect on coefficient estimates if left unaddressed. MCAR (#1) is not likely to have any effect at all, MAR is likely to result in underestimated coefficients of $X1 + X2$ (for #2) or underestimated coefficients of $X1$ and values of Y (for #3), and NIM (#4) is likely to result in a biased estimated relationship between income and average grades—low-income parents with high aspirations for their children would be under-represented in the analysis.

The next section deals with various missing data-imputation methods. Because of space considerations, we do not discuss weighting in this textbook. You should know, however, that in many nationally representative surveys, **weights** are usually invoked to correct for household or person non-response. Weights are not usually used to correct for item non-response.

What to Do about Missing Data

Missing data can be dealt with in several ways. Some are very simple, and others are more complicated.

1. Do Nothing: List-Wise and Pair-Wise Deletion

List-wise deletion: Delete all observations with missing data. In the previous example, anyone who didn't report their income was removed from the study completely.

Pair-wise deletion (or available case analysis): Use all available data to compute these means. For example, when creating a covariance matrix for two variables, only valid values are used. In a regression equation, observations with missing values will still contribute to coefficient estimates, so the observations with missing $X1$ or $X2$ values will still be included in some calculations. For example, observations with missing $X1$ values would be included when calculating the coefficient for $X2$.

The impact of missing data is slightly lower with pair-wise deletion than with list-wise deletion. However, the two techniques tend to produce similar results in practice. List-wise deletion is probably the most popular method of dealing with missing data. In a survey of recent research articles in political science by King et al., list-wise deletion was found to be the method of choice in 94 per cent of all papers (2001: 49).[2]

List-wise deletion is problematic for several reasons. First, sample size is greatly reduced if all observations with missing data on any variable are deleted. Although sample size is not usually an issue with large, national-level data sets, it still places unnecessary constraints on the types of questions that can be asked.

The second, more serious, problem with list-wise and pair-wise deletion is that both hinge on the assumption that data are "missing completely at random" (MCAR). This is often not the case. "Missingness" often stems from the nature of the question, or survey, itself. To continue with the example, assuming that a random sample of people chose not to report their income is presumptuous. Usually, people with either very high or very low incomes are less likely to report their income on surveys. There is almost always an underlying pattern to missing data, and the MCAR assumption is rarely justified.

2. Do Something: Single Imputation Strategies[3]

Best-guess imputation: The researcher views missing values in a quasi-subjective manner based on knowledge obtained from other variables. An example of a successful best-guess imputation is what Steven Ruggles and Susan Brower did with the historical US Censuses. Ruggles and Brower were able to determine the relationship of all Census respondents to the household head in the 1880 IPUMS data with remarkable accuracy (~99 per cent), through a series of best-guess imputations using age, sex, and order on the Census schedule, among other things. Although Ruggles and Brower were successful, other data sources do not lend themselves so easily to best-guessing techniques. When compared with subsequent Censuses of Canada, the best-guess estimates of the visible minority indicator variable of the 1981 and 1986 censuses are not as accurate (1996 Census Codebook Online). One of the strengths of best guess is that it requires no assumptions about the nature of "missingness" or its distribution. Data can be MCAR, MAR, or NIM, and values can be imputed without affecting sample size.

Zero imputation: A score of 0 (or some other arbitrary numeric indicator) replaces missing values and a dummy variable is added to control for the imputed value. This is not a true imputation method, however, because no plausible value is provided for the missing data, but researchers can use this method to retain problematic cases. Another use of 0 imputation is as a predictor in a regression model, to determine whether or not the missing data are missing at random in relation to the dependent variable. If the effect of the dummy variable is not significant, a persuasive argument can be made for the appropriateness of list-wise deletion, since no significant effect is elicited on the dependent variable. (This addresses only one type of MAR data; "missingness" could still depend on values of other independent variables).

Mean substitution: This replaces missing values with either the arithmetic average (for continuous data) or the mode or most frequent value (for categorical data) of the variable, based on values from valid cases. This approach is simple; the mean or modal value can be quickly calculated and analysis can proceed. The disadvantages include an underestimation of the standard error (through a reduction in the variance) and attenuation of correlations with other variables, producing overly optimistic fit statistics and significance levels (especially when values are not MAR or MCAR).

Hot deck: Here, the researcher uses a value from another case as a "donor" to replace the missing value. There are many methods for selecting suitable replacements: random selection of an observed value or more complicated methods such as the nearest neighbour imputation (NNI) methodology. For NNI (used for the 2000 Brazilian and US Censuses, as well as the 2001 Canadian, Ukrainian, Swiss, and Italian Censuses), the donor is not drawn at random but selected according to data values on variables that the statistician hypothesizes to be the most salient predictors of the missing value. The assumption is that the data are missing at random and there is a risk of using the same donor many times in small samples. This is a common method, but it's generally used before data are delivered to the end user.

Cold deck: The researcher derives a missing value by using anything other than the same variable of that survey. It is the opposite of hot deck, in which non-respondent values are derived from other respondents' values for the same variable. Values from a covariate, or a previous survey, are often used to impute the missing value. Cold deck relies entirely on the MAR assumption to arrive at estimates, since other correlates form the basis for selection of a missing value.

Y regression imputation: The researcher runs a preliminary regression on all observations, with the problematic variable as the focal outcome. A model for predicting the values of the missing data is derived from the regression. Missing values are filled by predicted regression values. The MAR assumption is heavily relied upon, and there is a significant underestimation of the standard error. (Predicted values are perfectly linear when Y is unobserved, but scattered when it is observed [King, Honaker, Joseph, and Scheve, 2001]).

$\hat{Y}$ regression imputation with random error term: This is similar to regression imputation, except that error term is attached to the imputed value, allowing for an element of uncertainty in the estimate.

3. Do Multiple Things: Multiple Imputation

Multiple imputation: This is the most mathematically abstract and complex method of imputation, but also the most accurate and consistent. The basic idea is simple, although in practice it's less straightforward:

1. Determine the model of interest that incorporates random explanatory variables (missing and non-missing).
2. Make random draws for the missing values from the valid cases that have similar scores on the given focal variables, using one of the other imputation techniques.
3. Do this M times (usually between three and five), creating M complete data sets. Observed values remain the same in all data, but missing values are different in each data set.
4. Perform analysis on M data sets, as though data are not missing.
5. Combine estimates by taking the average of coefficients to produce a single estimate.
6. Calculate standard errors by averaging the squared standard error of M estimates. Calculate the variance of M parameter estimates across samples by taking the square root of the sampling variance mean, plus a coefficient variance multiplied by a "correction factor" of $1 + \frac{1}{M}$ (to reward for increases in M).

EVERYDAY STATISTICS

Is There a Solution for Missing Data?

In 2010, the Government of Canada announced that it would no longer issue the long-form version of the 2011 census (this decision has since been reversed under Prime Minister Justin Trudeau). One of the biggest concerns of statisticians, which subsequently led to the resignation of Chief Statistician Munir Sheikh, was that the government maintained that the data from a voluntary survey would not be of poorer quality than a mandatory census. In other words, the claim was that there would be no difference in sampling error. One of the concerns with imputing missing data is that sampling error is replicated through the imputation. If the sample is already not a perfect microcosm of the population, problems in the sample will be reproduced and magnified by imputing values from other observations in the sample.

Q: Do you think that the increase in sampling error because of switching from a mandatory census to a voluntary survey could be addressed by any of the strategies for dealing with missing data mentioned in this chapter?

Multiple Imputation: Advantages and Disadvantages over Single Imputation

All other single imputation methods (except for the $\hat{Y}$ regression imputation with uncertainty element method) are essentially naive edits: those made with little or no knowledge about the case with missing values. Once imputation has occurred, unknown values are indistinguishable from known values, and analysis proceeds as though the values were never missing.

Multiple imputation "builds in" a level of uncertainty (although there is no way of telling whether that level is the appropriate one), preserving, to some degree, the integrity and accuracy of the standard errors and model fit statistics. By running identical analyses on *M* data sets with slightly differing values for missing data, the non-observed values are less precise than observed values, and when coefficient estimates are combined, model uncertainty is retained.

Multiple imputation has gained popularity with good reason; it is "the only general purpose statistical technique that can validly handle missing data problems" (Rubin, 1987). Unlike single imputation methods and deletion methods, estimates retain uncertainty elements to maintain the imprecision of the model resulting from working with data that are incomplete.

Unfortunately, multiple imputation is complicated to use and computer intensive. King, Honaker, Joseph, and Scheve (2001) found some time ago that on a data set with 1000 observations and 100 variables, which is not uncommonly large, multiple imputation takes anywhere from 4 minutes (imputing values for 5 variables with about 5 per cent missing data) to 3.5 days (with 40 variables with 5 per cent missing data). Although computers today are no doubt faster, computation time can still be substantial. The other problem with multiple imputation is that each time it is used, different estimates are produced. Since quasi-random draws are taken from each variable, different values for each of the *M* data sets are selected each time, producing different results when combined.

Despite the advantages, multiple imputation might be too complex for the average user. Since version 11, SPSS has had a module entitled MVA, which allows users to perform many of the single imputation methods with easy-to-use and intuitive diagnostics to compare means and variance structures before and after imputation, and to assess change in model fit when imputed values are included.

Conclusion

In the past, missing data in surveys were not thought to have much of an impact on generalizing results. This is largely because little work had been done to that point comparing respondents to non-respondents. However, since the early 1990s social scientists have begun to realize that missing data can have a significant effect on accuracy. As interest in correcting for missing data increases, software companies might begin to include more functions for multiple imputation. Whether multiple imputation will become the standard method for handling missing data in public-use data sets remains to be seen.

We hope you have enjoyed this journey through introductory statistics with us. We have attempted to provide you with the foundational knowledge you need to become a "literate" quantitative sociologist. You should now be able to read, understand, and critique quantitative journal articles and studies in your field of interest. We also hope that you will be encouraged to conduct your own research on the subjects that interest you, either by analyzing secondary data available to you, or by collecting and analyzing your own data. We wish you luck as you continue to expand on your statistical knowledge and expertise!

Glossary Term

Weights (p. 239)

Notes

1. Usually the four forms are reduced to three, since the two "missing at random" variants (#2 and #3) are usually handled similarly (Schafer, 1997).
2. Since political scientists, for example, often use countries as their unit of analysis, they are constantly dealing with small sample sizes, making list-wise deletion very "expensive" in terms of sample size. However, list-wise deletion is more common among sociologists, who tend to have data sets with larger numbers of observations (individuals rather than countries, for example).
3. Space permits discussion of only a limited number of imputation techniques. A complete list would include best guess imputation; *zero imputation*; *mean substitution*; *hot, warm, and cold deck methods*; and *regression imputation methods*, although many of the criticisms outlined here also pertain to these methods.

APPENDICES

Appendix A
Area under the Normal Curve

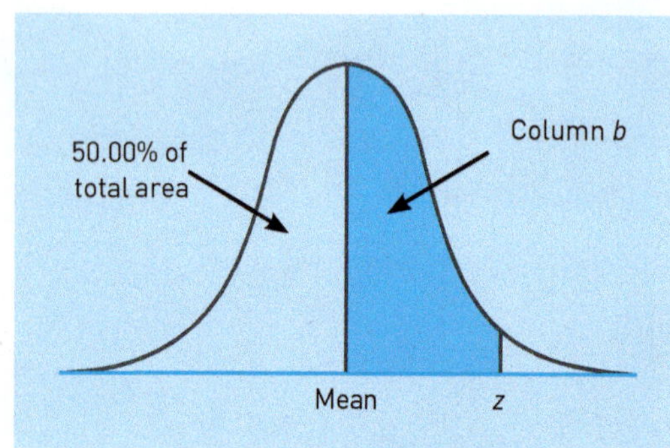

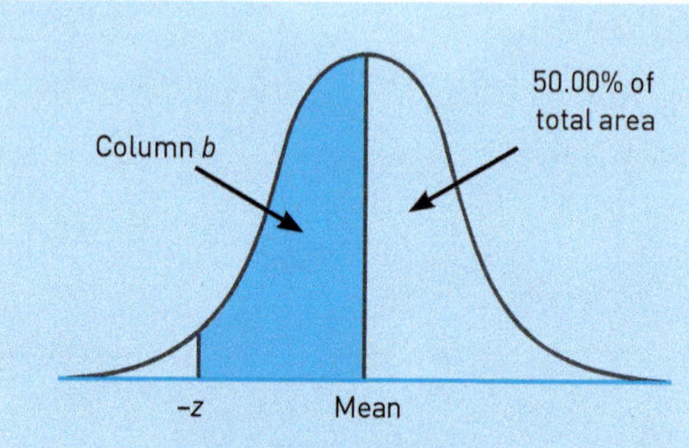

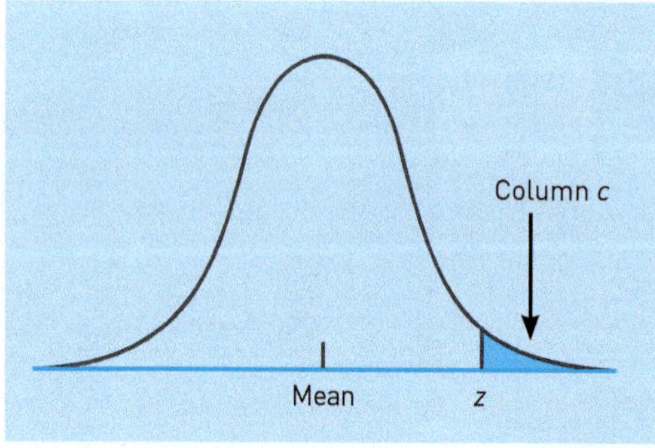

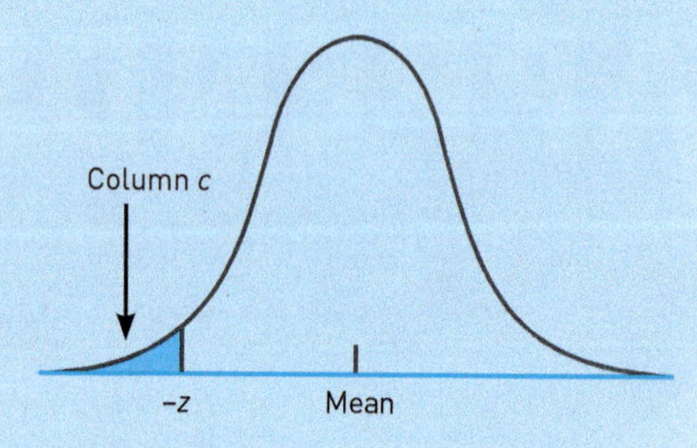

A	B	C	A	B	C
z	Area between mean and z	Area beyond z	z	Area between mean and z	Area beyond z
0.00	0.0000	0.5000			
0.01	0.0040	0.4960	0.41	0.1591	0.3409
0.02	0.0080	0.4920	0.42	0.1628	0.3372
0.03	0.0120	0.4880	0.43	0.1664	0.3336
0.04	0.0160	0.4840	0.44	0.1700	0.3300
0.05	0.0199	0.4801	0.45	0.1736	0.3264
0.06	0.0239	0.4761	0.46	0.1772	0.3228
0.07	0.0279	0.4721	0.47	0.1808	0.3192
0.08	0.0319	0.4681	0.48	0.1844	0.3156
0.09	0.0359	0.4641	0.49	0.1879	0.3121
0.10	0.0398	0.4602	0.50	0.1915	0.3085
0.11	0.0438	0.4562	0.51	0.1950	0.3050
0.12	0.0478	0.4522	0.52	0.1985	0.3015
0.13	0.0517	0.4483	0.53	0.2019	0.2981
0.14	0.0557	0.4443	0.54	0.2054	0.2946
0.15	0.0596	0.4404	0.55	0.2088	0.2912
0.16	0.0636	0.4364	0.56	0.2123	0.2877
0.17	0.0675	0.4325	0.57	0.2157	0.2843
0.18	0.0714	0.4286	0.58	0.2190	0.2810
0.19	0.0753	0.4247	0.59	0.2224	0.2776
0.20	0.0793	0.4207	0.60	0.2257	0.2743
0.21	0.0832	0.4168	0.61	0.2291	0.2709
0.22	0.0871	0.4129	0.62	0.2324	0.2676
0.23	0.0910	0.4090	0.63	0.2357	0.2643
0.24	0.0948	0.4052	0.64	0.2389	0.2611
0.25	0.0987	0.4013	0.65	0.2422	0.2578
0.26	0.1026	0.3974	0.66	0.2454	0.2546
0.27	0.1064	0.3936	0.67	0.2486	0.2514
0.28	0.1103	0.3897	0.68	0.2517	0.2483
0.29	0.1141	0.3859	0.69	0.2549	0.2451
0.30	0.1179	0.3821	0.70	0.2580	0.2420
0.31	0.1217	0.3783	0.71	0.2611	0.2389
0.32	0.1255	0.3745	0.72	0.2642	0.2358
0.33	0.1293	0.3707	0.73	0.2673	0.2327
0.34	0.1331	0.3669	0.74	0.2704	0.2297
0.35	0.1368	0.3632	0.75	0.2734	0.2266
0.36	0.1406	0.3594	0.76	0.2764	0.2236
0.37	0.1443	0.3557	0.77	0.2794	0.2207
0.38	0.1480	0.3520	0.78	0.2823	0.2177
0.39	0.1517	0.3483	0.79	0.2852	0.2148
0.40	0.1554	0.3446	0.80	0.2881	0.2119

[continued]

A *z*	B Area between mean and *z*	C Area beyond *z*	A *z*	B Area between mean and *z*	C Area beyond *z*
0.81	0.2910	0.2090	1.21	0.3869	0.1131
0.82	0.2939	0.2061	1.22	0.3888	0.1112
0.83	0.2967	0.2033	1.23	0.3907	0.1093
0.84	0.2995	0.2005	1.24	0.3925	0.1075
0.85	0.3023	0.1977	1.25	0.3944	0.1056
0.86	0.3051	0.1949	1.26	0.3962	0.1038
0.87	0.3078	0.1922	1.27	0.3980	0.1020
0.88	0.3106	0.1894	1.28	0.3997	0.1003
0.89	0.3133	0.1867	1.29	0.4015	0.0985
0.90	0.3159	0.1841	1.30	0.4032	0.0968
0.91	0.3186	0.1814	1.31	0.4049	0.0951
0.92	0.3212	0.1788	1.32	0.4066	0.0934
0.93	0.3238	0.1762	1.33	0.4082	0.0918
0.94	0.3264	0.1736	1.34	0.4099	0.0901
0.95	0.3289	0.1711	1.35	0.4115	0.0885
0.96	0.3315	0.1685	1.36	0.4131	0.0869
0.97	0.3340	0.1660	1.37	0.4147	0.0853
0.98	0.3365	0.1635	1.38	0.4162	0.0838
0.99	0.3389	0.1611	1.39	0.4177	0.0823
1.00	0.3413	0.1587	1.40	0.4192	0.0808
1.01	0.3438	0.1562	1.41	0.4207	0.0793
1.02	0.3461	0.1539	1.42	0.4222	0.0778
1.03	0.3485	0.1515	1.43	0.4236	0.0764
1.04	0.3508	0.1492	1.44	0.4251	0.0749
1.05	0.3531	0.1469	1.45	0.4265	0.0735
1.06	0.3554	0.1446	1.46	0.4279	0.0721
1.07	0.3577	0.1423	1.47	0.4292	0.0708
1.08	0.3599	0.1401	1.48	0.4306	0.0694
1.09	0.3621	0.1379	1.49	0.4319	0.0681
1.10	0.3643	0.1357	1.50	0.4332	0.0668
1.11	0.3665	0.1335	1.51	0.4345	0.0655
1.12	0.3686	0.1314	1.52	0.4357	0.0643
1.13	0.3708	0.1292	1.53	0.4370	0.0630
1.14	0.3729	0.1271	1.54	0.4382	0.0618
1.15	0.3749	0.1251	1.55	0.4394	0.0606
1.16	0.3770	0.1230	1.56	0.4406	0.0594
1.17	0.3790	0.1210	1.57	0.4418	0.0582
1.18	0.3810	0.1190	1.58	0.4429	0.0571
1.19	0.3830	0.1170	1.59	0.4441	0.0559
1.20	0.3849	0.1151	1.60	0.4452	0.0548

A z	B Area between mean and z	C Area beyond z	A z	B Area between mean and z	C Area beyond z
1.61	0.4463	0.0537	2.01	0.4778	0.0222
1.62	0.4474	0.0526	2.02	0.4783	0.0217
1.63	0.4484	0.0516	2.03	0.4788	0.0212
1.64	0.4495	0.0505	2.04	0.4793	0.0207
1.65	0.4505	0.0495	2.05	0.4798	0.0202
1.66	0.4515	0.0485	2.06	0.4803	0.0197
1.67	0.4525	0.0475	2.07	0.4808	0.0192
1.68	0.4535	0.0465	2.08	0.4812	0.0188
1.69	0.4545	0.0455	2.09	0.4817	0.0183
1.70	0.4554	0.0446	2.10	0.4821	0.0179
1.71	0.4564	0.0436	2.11	0.4826	0.0174
1.72	0.4573	0.0427	2.12	0.4830	0.0170
1.73	0.4582	0.0418	2.13	0.4834	0.0166
1.74	0.4591	0.0409	2.14	0.4838	0.0162
1.75	0.4599	0.0401	2.15	0.4842	0.0158
1.76	0.4608	0.0392	2.16	0.4846	0.0154
1.77	0.4616	0.0384	2.17	0.4850	0.0150
1.78	0.4625	0.0375	2.18	0.4854	0.0146
1.79	0.4633	0.0367	2.19	0.4857	0.0143
1.80	0.4641	0.0359	2.20	0.4861	0.0139
1.81	0.4649	0.0351	2.21	0.4864	0.0136
1.82	0.4656	0.0344	2.22	0.4868	0.0132
1.83	0.4664	0.0336	2.23	0.4871	0.0129
1.84	0.4671	0.0329	2.24	0.4875	0.0125
1.85	0.4678	0.0322	2.25	0.4878	0.0122
1.86	0.4686	0.0314	2.26	0.4881	0.0119
1.87	0.4693	0.0307	2.27	0.4884	0.0116
1.88	0.4699	0.0301	2.28	0.4887	0.0113
1.89	0.4706	0.0294	2.29	0.4890	0.0110
1.90	0.4713	0.0287	2.30	0.4893	0.0107
1.91	0.4719	0.0281	2.31	0.4896	0.0104
1.92	0.4726	0.0274	2.32	0.4898	0.0102
1.93	0.4732	0.0268	2.33	0.4901	0.0099
1.94	0.4738	0.0262	2.34	0.4904	0.0096
1.95	0.4744	0.0256	2.35	0.4906	0.0094
1.96	0.4750	0.0250	2.36	0.4909	0.0091
1.97	0.4756	0.0244	2.37	0.4911	0.0089
1.98	0.4761	0.0239	2.38	0.4913	0.0087
1.99	0.4767	0.0233	2.39	0.4916	0.0084
2.00	0.4772	0.0228	2.40	0.4918	0.0082

[continued]

A	B	C	A	B	C
z	Area between mean and z	Area beyond z	z	Area between mean and z	Area beyond z
2.41	0.4920	0.0080	2.81	0.4975	0.0025
2.42	0.4922	0.0078	2.82	0.4976	0.0024
2.43	0.4925	0.0075	2.83	0.4977	0.0023
2.44	0.4927	0.0073	2.84	0.4977	0.0023
2.45	0.4929	0.0071	2.85	0.4978	0.0022
2.46	0.4931	0.0069	2.86	0.4979	0.0021
2.47	0.4932	0.0068	2.87	0.4979	0.0021
2.48	0.4934	0.0066	2.88	0.4980	0.0020
2.49	0.4936	0.0064	2.89	0.4981	0.0019
2.50	0.4938	0.0062	2.90	0.4981	0.0019
2.51	0.4940	0.0060	2.91	0.4982	0.0018
2.52	0.4941	0.0059	2.92	0.4982	0.0018
2.53	0.4943	0.0057	2.93	0.4983	0.0017
2.54	0.4945	0.0055	2.94	0.4984	0.0016
2.55	0.4946	0.0054	2.95	0.4984	0.0016
2.56	0.4948	0.0052	2.96	0.4985	0.0015
2.57	0.4949	0.0051	2.97	0.4985	0.0015
2.58	0.4951	0.0049	2.98	0.4986	0.0014
2.59	0.4952	0.0048	2.99	0.4986	0.0014
2.60	0.4953	0.0047	3.00	0.4987	0.0013
2.61	0.4955	0.0045	3.01	0.4987	0.0013
2.62	0.4956	0.0044	3.02	0.4987	0.0013
2.63	0.4957	0.0043	3.03	0.4988	0.0012
2.64	0.4959	0.0041	3.04	0.4988	0.0012
2.65	0.4960	0.0040	3.05	0.4989	0.0011
2.66	0.4961	0.0039	3.06	0.4989	0.0011
2.67	0.4962	0.0038	3.07	0.4989	0.0011
2.68	0.4963	0.0037	3.08	0.4990	0.0010
2.69	0.4964	0.0036	3.09	0.4990	0.0010
2.70	0.4965	0.0035	3.10	0.4990	0.0010
2.71	0.4966	0.0034	3.11	0.4991	0.0009
2.72	0.4967	0.0033	3.12	0.4991	0.0009
2.73	0.4968	0.0032	3.13	0.4991	0.0009
2.74	0.4969	0.0031	3.14	0.4992	0.0008
2.75	0.4970	0.0030	3.15	0.4992	0.0008
2.76	0.4971	0.0029	3.16	0.4992	0.0008
2.77	0.4972	0.0028	3.17	0.4992	0.0008
2.78	0.4973	0.0027	3.18	0.4993	0.0007
2.79	0.4974	0.0026	3.19	0.4993	0.0007
2.80	0.4974	0.0026	3.20	0.4993	0.0007

A	B	C	A	B	C
z	Area between mean and *z*	Area beyond *z*	*z*	Area between mean and *z*	Area beyond *z*
3.21	0.4993	0.0007	3.61	0.4998	0.0002
3.22	0.4994	0.0006	3.62	0.4999	0.0001
3.23	0.4994	0.0006	3.63	0.4999	0.0001
3.24	0.4994	0.0006	3.64	0.4999	0.0001
3.25	0.4994	0.0006	3.65	0.4999	0.0001
3.26	0.4994	0.0006	3.66	0.4999	0.0001
3.27	0.4995	0.0005	3.67	0.4999	0.0001
3.28	0.4995	0.0005	3.68	0.4999	0.0001
3.29	0.4995	0.0005	3.69	0.4999	0.0001
3.30	0.4995	0.0005	3.70	0.4999	0.0001
3.31	0.4995	0.0005	3.71	0.4999	0.0001
3.32	0.4995	0.0005	3.72	0.4999	0.0001
3.33	0.4996	0.0004	3.73	0.4999	0.0001
3.34	0.4996	0.0004	3.74	0.4999	0.0001
3.35	0.4996	0.0004	3.75	0.4999	0.0001
3.36	0.4996	0.0004	3.76	0.4999	0.0001
3.37	0.4996	0.0004	3.77	0.4999	0.0001
3.38	0.4996	0.0004	3.78	0.4999	0.0001
3.39	0.4997	0.0003	3.79	0.4999	0.0001
3.40	0.4997	0.0003	3.80	0.4999	0.0001
3.41	0.4997	0.0003	3.81	0.4999	0.0001
3.42	0.4997	0.0003	3.82	0.4999	0.0001
3.43	0.4997	0.0003	3.83	0.4999	0.0001
3.44	0.4997	0.0003	3.84	0.4999	0.0001
3.45	0.4997	0.0003	3.85	0.4999	0.0001
3.46	0.4997	0.0003	3.86	0.4999	0.0001
3.47	0.4997	0.0003	3.87	0.4999	0.0001
3.48	0.4997	0.0003	3.88	0.4999	0.0001
3.49	0.4998	0.0002	3.89	0.4999	0.0001
3.50	0.4998	0.0002	3.90	0.5000	0.0000
3.51	0.4998	0.0002	3.91	0.5000	0.0000
3.52	0.4998	0.0002	3.92	0.5000	0.0000
3.53	0.4998	0.0002	3.93	0.5000	0.0000
3.54	0.4998	0.0002	3.94	0.5000	0.0000
3.55	0.4998	0.0002	3.95	0.5000	0.0000
3.56	0.4998	0.0002	3.96	0.5000	0.0000
3.57	0.4998	0.0002	3.97	0.5000	0.0000
3.58	0.4998	0.0002	3.98	0.5000	0.0000
3.59	0.4998	0.0002	3.99	0.5000	0.0000
3.60	0.4998	0.0002	4.00	0.5000	0.0000

Appendix B

The Student's *t*-Table

For a One-Tailed Test:

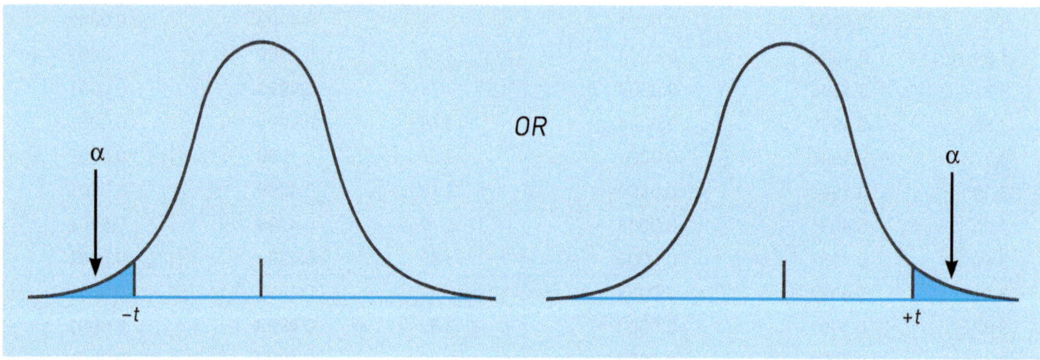

For a Two-Tailed Test:

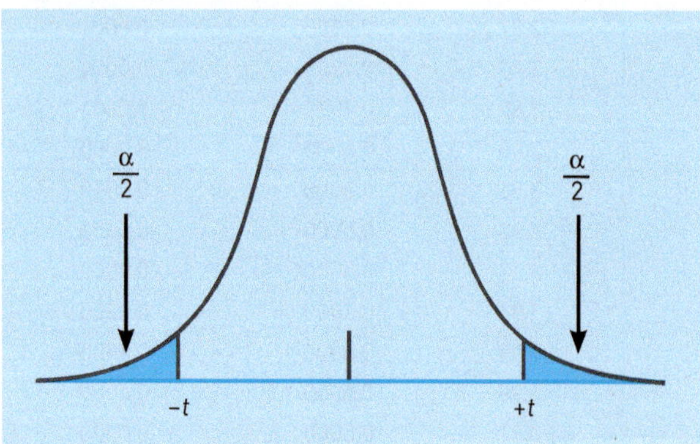

	Level of Significance for One-Tailed Test					
	0.1	0.05	0.025	0.01	0.005	0.001
	Level of Significance for Two-Tailed Test					
df	0.2	0.1	0.05	0.02	0.01	0.002
1	3.078	6.314	12.706	31.821	63.657	318.313
2	1.886	2.920	4.303	6.965	9.925	22.327
3	1.638	2.353	3.182	4.541	5.841	10.215
4	1.533	2.132	2.776	3.747	4.604	7.173
5	1.476	2.015	2.571	3.365	4.032	5.893
6	1.440	1.943	2.447	3.143	3.707	5.208
7	1.415	1.895	2.365	2.998	3.499	4.782
8	1.397	1.860	2.306	2.896	3.355	4.499
9	1.383	1.833	2.262	2.821	3.250	4.296
10	1.372	1.812	2.228	2.764	3.169	4.143
11	1.363	1.796	2.201	2.718	3.106	4.024
12	1.356	1.782	2.179	2.681	3.055	3.929
13	1.350	1.771	2.160	2.650	3.012	3.852
14	1.345	1.761	2.145	2.624	2.977	3.787
15	1.341	1.753	2.131	2.602	2.947	3.733
16	1.337	1.746	2.120	2.583	2.921	3.686
17	1.333	1.740	2.110	2.567	2.898	3.646
18	1.330	1.734	2.101	2.552	2.878	3.610
19	1.328	1.729	2.093	2.539	2.861	3.579
20	1.325	1.725	2.086	2.528	2.845	3.552
21	1.323	1.721	2.080	2.518	2.831	3.527
22	1.321	1.717	2.074	2.508	2.819	3.505
23	1.319	1.714	2.069	2.500	2.807	3.485
24	1.318	1.711	2.064	2.492	2.797	3.467
25	1.316	1.708	2.060	2.485	2.787	3.450
26	1.315	1.706	2.056	2.479	2.779	3.435
27	1.314	1.703	2.052	2.473	2.771	3.421
28	1.313	1.701	2.048	2.467	2.763	3.408
29	1.311	1.699	2.045	2.462	2.756	3.396
30	1.310	1.697	2.042	2.457	2.750	3.385
31	1.309	1.696	2.040	2.453	2.744	3.375
32	1.309	1.694	2.037	2.449	2.738	3.365
33	1.308	1.692	2.035	2.445	2.733	3.356
34	1.307	1.691	2.032	2.441	2.728	3.348
35	1.306	1.690	2.030	2.438	2.724	3.340
36	1.306	1.688	2.028	2.434	2.719	3.333
37	1.305	1.687	2.026	2.431	2.715	3.326
38	1.304	1.686	2.024	2.429	2.712	3.319
39	1.304	1.685	2.023	2.426	2.708	3.313
40	1.303	1.684	2.021	2.423	2.704	3.307

[continued]

	Level of Significance for One-Tailed Test					
	0.1	0.05	0.025	0.01	0.005	0.001
	Level of Significance for Two-Tailed Test					
df	0.2	0.1	0.05	0.02	0.01	0.002
41	1.303	1.683	2.020	2.421	2.701	3.301
42	1.302	1.682	2.018	2.418	2.698	3.296
43	1.302	1.681	2.017	2.416	2.695	3.291
44	1.301	1.680	2.015	2.414	2.692	3.286
45	1.301	1.679	2.014	2.412	2.690	3.281
46	1.300	1.679	2.013	2.410	2.687	3.277
47	1.300	1.678	2.012	2.408	2.685	3.273
48	1.299	1.677	2.011	2.407	2.682	3.269
49	1.299	1.677	2.010	2.405	2.680	3.265
50	1.299	1.676	2.009	2.403	2.678	3.261
51	1.298	1.675	2.008	2.402	2.676	3.258
52	1.298	1.675	2.007	2.400	2.674	3.255
53	1.298	1.674	2.006	2.399	2.672	3.251
54	1.297	1.674	2.005	2.397	2.670	3.248
55	1.297	1.673	2.004	2.396	2.668	3.245
56	1.297	1.673	2.003	2.395	2.667	3.242
57	1.297	1.672	2.002	2.394	2.665	3.239
58	1.296	1.672	2.002	2.392	2.663	3.237
59	1.296	1.671	2.001	2.391	2.662	3.234
60	1.296	1.671	2.000	2.390	2.660	3.232
61	1.296	1.670	2.000	2.389	2.659	3.229
62	1.295	1.670	1.999	2.388	2.657	3.227
63	1.295	1.669	1.998	2.387	2.656	3.225
64	1.295	1.669	1.998	2.386	2.655	3.223
65	1.295	1.669	1.997	2.385	2.654	3.220
66	1.295	1.668	1.997	2.384	2.652	3.218
67	1.294	1.668	1.996	2.383	2.651	3.216
68	1.294	1.668	1.995	2.382	2.650	3.214
69	1.294	1.667	1.995	2.382	2.649	3.213
70	1.294	1.667	1.994	2.381	2.648	3.211
71	1.294	1.667	1.994	2.380	2.647	3.209
72	1.293	1.666	1.993	2.379	2.646	3.207
73	1.293	1.666	1.993	2.379	2.645	3.206
74	1.293	1.666	1.993	2.378	2.644	3.204
75	1.293	1.665	1.992	2.377	2.643	3.202
76	1.293	1.665	1.992	2.376	2.642	3.201
77	1.293	1.665	1.991	2.376	2.641	3.199
78	1.292	1.665	1.991	2.375	2.640	3.198
79	1.292	1.664	1.990	2.374	2.640	3.197
80	1.292	1.664	1.990	2.374	2.639	3.195

	Level of Significance for One-Tailed Test					
	0.1	0.05	0.025	0.01	0.005	0.001
	Level of Significance for Two-Tailed Test					
df	0.2	0.1	0.05	0.02	0.01	0.002
81	1.292	1.664	1.990	2.373	2.638	3.194
82	1.292	1.664	1.989	2.373	2.637	3.193
83	1.292	1.663	1.989	2.372	2.636	3.191
84	1.292	1.663	1.989	2.372	2.636	3.190
85	1.292	1.663	1.988	2.371	2.635	3.189
86	1.291	1.663	1.988	2.370	2.634	3.188
87	1.291	1.663	1.988	2.370	2.634	3.187
88	1.291	1.662	1.987	2.369	2.633	3.185
89	1.291	1.662	1.987	2.369	2.632	3.184
90	1.291	1.662	1.987	2.368	2.632	3.183
91	1.291	1.662	1.986	2.368	2.631	3.182
92	1.291	1.662	1.986	2.368	2.630	3.181
93	1.291	1.661	1.986	2.367	2.630	3.180
94	1.291	1.661	1.986	2.367	2.629	3.179
95	1.291	1.661	1.985	2.366	2.629	3.178
96	1.290	1.661	1.985	2.366	2.628	3.177
97	1.290	1.661	1.985	2.365	2.627	3.176
98	1.290	1.661	1.984	2.365	2.627	3.175
99	1.290	1.660	1.984	2.365	2.626	3.175
100	1.290	1.660	1.984	2.364	2.626	3.174
120	1.289	1.658	1.980	2.358	2.617	3.373
∞	1.282	1.645	1.960	2.326	2.576	3.090

Appendix C

Chi-Square

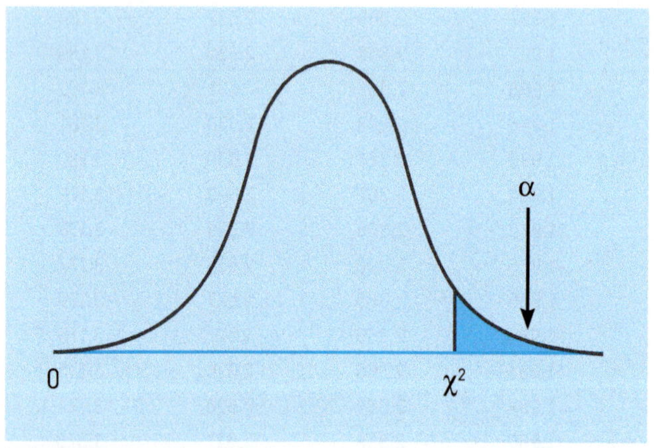

df	Critical Values of Chi-Square Level of Significance for Two-Tailed Test				
	0.1	0.05	0.025	0.01	0.001
1	2.706	3.841	5.024	6.635	10.828
2	4.605	5.991	7.378	9.210	13.816
3	6.251	7.815	9.348	11.345	16.266
4	7.779	9.488	11.143	13.277	18.467
5	9.236	11.070	12.833	15.086	20.515
6	10.645	12.592	14.449	16.812	22.458
7	12.017	14.067	16.013	18.475	24.322
8	13.362	15.507	17.535	20.090	26.125
9	14.684	16.919	19.023	21.666	27.877
10	15.987	18.307	20.483	23.209	29.588
11	17.275	19.675	21.920	24.725	31.264
12	18.549	21.026	23.337	26.217	32.910
13	19.812	22.362	24.736	27.688	34.528
14	21.064	23.685	26.119	29.141	36.123
15	22.307	24.996	27.488	30.578	37.697
16	23.542	26.296	28.845	32.000	39.252
17	24.769	27.587	30.191	33.409	40.790
18	25.989	28.869	31.526	34.805	42.312
19	27.204	30.144	32.852	36.191	43.820
20	28.412	31.410	34.170	37.566	45.315
21	29.615	32.671	35.479	38.932	46.797
22	30.813	33.924	36.781	40.289	48.268
23	32.007	35.172	38.076	41.638	49.728
24	33.196	36.415	39.364	42.980	51.179
25	34.382	37.652	40.646	44.314	52.620
26	35.563	38.885	41.923	45.642	54.052
27	36.741	40.113	43.195	46.963	55.476
28	37.916	41.337	44.461	48.278	56.892
29	39.087	42.557	45.722	49.588	58.301
30	40.256	43.773	46.979	50.892	59.703
31	41.422	44.985	48.232	52.191	61.098
32	42.585	46.194	49.480	53.486	62.487
33	43.745	47.400	50.725	54.776	63.870
34	44.903	48.602	51.966	56.061	65.247
35	46.059	49.802	53.203	57.342	66.619
36	47.212	50.998	54.437	58.619	67.985
37	48.363	52.192	55.668	59.893	69.347
38	49.513	53.384	56.896	61.162	70.703
39	50.660	54.572	58.120	62.428	72.055
40	51.805	55.758	59.342	63.691	73.402

[continued]

df	Critical Values of Chi-Square Level of Significance for Two-Tailed Test				
	0.1	0.05	0.025	0.01	0.001
41	52.949	56.942	60.561	64.950	74.745
42	54.090	58.124	61.777	66.206	76.084
43	55.230	59.304	62.990	67.459	77.419
44	56.369	60.481	64.201	68.710	78.750
45	57.505	61.656	65.410	69.957	80.077
46	58.641	62.830	66.617	71.201	81.400
47	59.774	64.001	67.821	72.443	82.720
48	60.907	65.171	69.023	73.683	84.037
49	62.038	66.339	70.222	74.919	85.351
50	63.167	67.505	71.420	76.154	86.661
51	64.295	68.669	72.616	77.386	87.968
52	65.422	69.832	73.810	78.616	89.272
53	66.548	70.993	75.002	79.843	90.573
54	67.673	72.153	76.192	81.069	91.872
55	68.796	73.311	77.380	82.292	93.168
56	69.919	74.468	78.567	83.513	94.461
57	71.040	75.624	79.752	84.733	95.751
58	72.160	76.778	80.936	85.950	97.039
59	73.279	77.931	82.117	87.166	98.324
60	74.397	79.082	83.298	88.379	99.607
61	75.514	80.232	84.476	89.591	100.888
62	76.630	81.381	85.654	90.802	102.166
63	77.745	82.529	86.830	92.010	103.442
64	78.860	83.675	88.004	93.217	104.716
65	79.973	84.821	89.177	94.422	105.988
66	81.085	85.965	90.349	95.626	107.258
67	82.197	87.108	91.519	96.828	108.526
68	83.308	88.250	92.689	98.028	109.791
69	84.418	89.391	93.856	99.228	111.055
70	85.527	90.531	95.023	100.425	112.317
71	86.635	91.670	96.189	101.621	113.577
72	87.743	92.808	97.353	102.816	114.835
73	88.850	93.945	98.516	104.010	116.092
74	89.956	95.081	99.678	105.202	117.346
75	91.061	96.217	100.839	106.393	118.599
76	92.166	97.351	101.999	107.583	119.850
77	93.270	98.484	103.158	108.771	121.100
78	94.374	99.617	104.316	109.958	122.348
79	95.476	100.749	105.473	111.144	123.594
80	96.578	101.879	106.629	112.329	124.839

df	Critical Values of Chi-Square Level of Significance for Two-Tailed Test				
	0.1	0.05	0.025	0.01	0.001
81	97.680	103.010	107.783	113.512	126.083
82	98.780	104.139	108.937	114.695	127.324
83	99.880	105.267	110.090	115.876	128.565
84	100.980	106.395	111.242	117.057	129.804
85	102.079	107.522	112.393	118.236	131.041
86	103.177	108.648	113.544	119.414	132.277
87	104.275	109.773	114.693	120.591	133.512
88	105.372	110.898	115.841	121.767	134.746
89	106.469	112.022	116.989	122.942	135.978
90	107.565	113.145	118.136	124.116	137.208
91	108.661	114.268	119.282	125.289	138.438
92	109.756	115.390	120.427	126.462	139.666
93	110.850	116.511	121.571	127.633	140.893
94	111.944	117.632	122.715	128.803	142.119
95	113.038	118.752	123.858	129.973	143.344
96	114.131	119.871	125.000	131.141	144.567
97	115.223	120.990	126.141	132.309	145.789
98	116.315	122.108	127.282	133.476	147.010
99	117.407	123.225	128.422	134.642	148.230
100	118.498	124.342	129.561	135.807	149.449

Appendix D

The *F*-Distribution

$df_{between}$	2	3	4	5	6	7	8
df_{within}							
1	199.50	215.71	224.58	230.16	233.99	236.77	238.88
2	19.00	19.16	19.25	19.30	19.33	19.35	19.37
3	9.55	9.28	9.12	9.01	8.94	8.89	8.85
4	6.94	6.59	6.39	6.26	6.16	6.09	6.04
5	5.79	5.41	5.19	5.05	4.95	4.88	4.82
6	5.14	4.76	4.53	4.39	4.28	4.21	4.15
7	4.74	4.35	4.12	3.97	3.87	3.79	3.73
8	4.46	4.07	3.84	3.69	3.58	3.50	3.44
9	4.26	3.86	3.63	3.48	3.37	3.29	3.23
10	4.10	3.71	3.48	3.33	3.22	3.14	3.07
20	3.49	3.10	2.87	2.71	2.60	2.51	2.45
21	3.47	3.07	2.84	2.69	2.57	2.49	2.42
22	3.44	3.05	2.82	2.66	2.55	2.46	2.40
23	3.42	3.03	2.80	2.64	2.53	2.44	2.38
24	3.40	3.01	2.78	2.62	2.51	2.42	2.36
25	3.39	2.99	2.76	2.60	2.49	2.41	2.34
26	3.37	2.98	2.74	2.59	2.47	2.39	2.32
27	3.35	2.96	2.73	2.57	2.46	2.37	2.31
28	3.34	2.95	2.71	2.56	2.45	2.36	2.29
29	3.33	2.93	2.70	2.55	2.43	2.35	2.28
30	3.32	2.92	2.69	2.53	2.42	2.33	2.27
40	3.23	2.84	2.61	2.45	2.34	2.25	2.18
50	3.18	2.79	2.56	2.40	2.29	2.20	2.13
60	3.15	2.76	2.53	2.37	2.25	2.17	2.10
70	3.13	2.74	2.50	2.35	2.23	2.14	2.07
80	3.11	2.72	2.49	2.33	2.21	2.13	2.06
90	3.10	2.71	2.47	2.32	2.20	2.11	2.04
100	3.09	2.70	2.46	2.31	2.19	2.10	2.03
120	3.07	2.68	2.45	2.29	2.18	2.09	2.02
∞	3.00	2.61	2.37	2.22	2.10	2.01	1.94

Appendix E

Area under the Normal Curve: A Condensed Version

A	B	C	A	B	C
z	Area between mean and z	Area beyond z	z	Area between mean and z	Area beyond z
0.0	0.0000	0.5000	2.0	0.4772	0.0228
0.1	0.0398	0.4602	2.1	0.4821	0.0179
0.2	0.0793	0.4207	2.2	0.4861	0.0139
0.3	0.1179	0.3821	2.3	0.4893	0.0107
0.4	0.1554	0.3446	2.4	0.4918	0.0082
0.5	0.1915	0.3085	2.5	0.4938	0.0062
0.6	0.2257	0.2743	2.6	0.4953	0.0047
0.7	0.2580	0.2420	2.7	0.4965	0.0035
0.8	0.2881	0.2119	2.8	0.4974	0.0026
0.9	0.3159	0.1841	2.9	0.4981	0.0019
1.0	0.3413	0.1587	3.0	0.4987	0.0013
1.1	0.3643	0.1357	3.1	0.4990	0.0010
1.2	0.3849	0.1151	3.2	0.4993	0.0007
1.3	0.4032	0.0968	3.3	0.4995	0.0005
1.4	0.4192	0.0808	3.4	0.4997	0.0003
1.5	0.4332	0.0668	3.5	0.4998	0.0002
1.6	0.4452	0.0548			
1.7	0.4554	0.0446			
1.8	0.4641	0.0359			
1.9	0.4713	0.0287			

Appendix F

Random Numbers between 1 and 1000

388	250	87	729	502	962
185	60	160	117	714	496
524	19	360	95	784	800
494	6	951	606	40	530
603	504	919	973	620	320
203	167	195	920	447	756
690	647	821	594	111	918
10	899	275	191	22	988
930	816	335	717	413	470
65	993	620	201	66	949
332	273	128	656	220	326
717	477	37	580	905	421
832	51	422	106	204	42
16	194	916	364	507	405
390	34	243	464	912	453
868	609	556	252	554	487
164	347	163	588	898	889
958	537	166	119	826	880
996	365	168	214	106	644
876	774	271	625	74	821
738	929	550	471	762	994
31	793	556	1	752	300
198	595	960	200	288	362
302	997	409	337	253	310
694	994	8	41	24	456
511	204	686	941	546	70
500	622	64	467	752	448
153	339	447	609	339	938
698	662	134	248	285	532
4	511	701	892	178	943

197	635	960	22	672	440
492	959	399	654	111	174
937	986	741	947	730	734
979	538	226	541	809	445
541	911	105	613	974	961
359	728	913	159	281	951
938	20	224	702	166	998
581	341	124	55	592	881
709	660	216	59	29	162
15	411	172	825	192	174
191	959	230	161	873	915
788	928	56	915	785	881
586	951	839	161	155	239
485	415	627	971	840	824
595	329	648	985	251	18
856	523	860	788	172	525
359	681	246	157	928	997
419	94	519	199	200	593
699	256	903	339	303	140
93	418	735	486	801	21

Appendix G
Summary of Equations and Symbols

Equations

The arithmetic mean of a sample/variable (Chapter 6): $\bar{X} = \frac{\sum_{i=1}^{N} X_i}{N}$

The range (Chapter 6): **Range** $= H - L$

The interquartile range (Chapter 6): **IQR** $= \mathbf{Q_3 - Q_1}$

Mean deviation (Chapter 6): **Mean Deviation** $= \frac{\left|\sum X - \bar{x}\right|}{N}$

Variance

- of a sample (Chapter 6): $s^2 = \frac{\sum (X - \bar{x})^2}{n - 1}$
- of a population: $\sigma^2 = \frac{\sum (X - \mu)^2}{N}$

Standard deviation

- estimated from a sample (Chapter 6): $s = \sqrt{\frac{\sum (X - \bar{x})^2}{n - 1}}$
- of a population: $\sigma = \sqrt{\frac{\sum (X - \mu)^2}{N}}$

- of a difference between observations (Chapter 10): $s_d = \sqrt{\dfrac{\sum (d_i - \bar{x}_d)^2}{n-1}}$

or

$$s_d = \sqrt{\frac{\sum d_i^2}{n-1} - (\bar{x}_1 - \bar{x}_2)^2}$$

- of the difference between sample proportions (Chapter 11):

$$s_{p_1 - p_2} = \sqrt{P_u * (1 - P_u)\frac{n_1 + n_2}{n_1 n_2}}$$

The standard score

- for a sample (Chapter 7): $z = \dfrac{X - \bar{x}}{s}$
- for a population (Chapter 7): $z = \dfrac{X - \mu}{\sigma}$

z with proportions (Chapter 10): $z_{obtained} = \dfrac{p_{sample} - P_{population}}{\sqrt{\dfrac{P_{population}(1 - P_{population})}{n}}}$

z for the difference between sample means (Chapter 11): $z = \dfrac{\bar{x}_1 - \bar{x}_2}{s_{\bar{x}_1 - \bar{x}_2}}$

- where z is obtained (Chapter 11): $z_{obtained} = \dfrac{p_{1_1} - p_{1_2}}{s_{p_1 - p_2}}$

Standard error of the mean

- of a population (Chapter 9): $\sigma_{\bar{x}} = \dfrac{\sigma_x}{\sqrt{n}}$
- of a sample (Chapter 9): $s_{\bar{x}} = \dfrac{s_x}{\sqrt{n-1}}$
- of a sample proportion to a population (Chapter 9): $s_p = \sqrt{\dfrac{p(1-p)}{n-1}}$

Standard error of difference between means (a.k.a. the anticipated level of error between measurements of two-sample means)

- for a large sample (i.e., z) (Chapter 11): $s_{\bar{x}_1 - \bar{x}_2} = \sqrt{\left(\dfrac{s_1^2}{n_1 - 1} + \dfrac{s_2^2}{n_2 - 1}\right)}$

- for a small sample (<120) (i.e., t): $s_{\bar{x}_1-\bar{x}_2} = \sqrt{\left(\frac{n_1 s_1^2 + n_2 s_2^2}{n_1 + n_2 - 2}\right)\left(\frac{n_1 + n_2}{n_1 n_2}\right)}$

Confidence interval

- when z is known (Chapter 9): **Confidence Interval** $= \bar{x} \pm (z_{\text{critical}} * \sigma_{\bar{x}})$

  or

  Confidence Interval $= \bar{x} \pm (z_{\text{critical}} * s_{\bar{x}})$

- for a sample proportion: **Confidence Interval** $= \boldsymbol{P} \pm (\boldsymbol{z}_{\text{critical}} * \boldsymbol{s_p})$

- when t is known: **Confidence Interval** $= \bar{x} \pm (t_{\text{critical}} * s_{\bar{x}})$

  or

  Confidence Interval $= \bar{x} \pm \left(t_{\text{critical}} * \frac{s_x}{\sqrt{n-1}}\right)$

- of a population when μ is unknown (Chapter 10): $CI = \bar{x} \pm t * \frac{\sigma}{\sqrt{N}}$

t-score

- for a sample (Chapter 9): $t = \frac{\bar{x} - \mu}{s_{\bar{x}}}$ or $t = \frac{\bar{x} - \mu}{\frac{s_x}{\sqrt{n}}}$

- for a population (Chapter 9): $t_{\text{observed}} = \frac{\bar{x} - \mu}{\frac{\sigma_x}{\sqrt{n}}}$ or $t_{\text{observed}} = \frac{\bar{x} - \mu}{\sigma_{\bar{x}}}$

- when population mean is known (Chapter 10): $t = \frac{\bar{x} - \mu}{\frac{s_x}{\sqrt{n}}}$

t for the difference between sample means (Chapter 11): $t = \frac{\bar{x}_1 - \bar{x}_2}{s_{\bar{x}_1-\bar{x}_2}}$

t_{observed} (Chapter 14): $t_{\text{observed}} = r\sqrt{\frac{n-2}{1-r^2}}$

Degrees of freedom

- for a t-test (Chapter 9): $df = n - 1$
- for a t-test with a two-sample case (Chapter 11): $df = (n_1 + n_2 - 2)$
- for chi-square test of independence: $df = (\text{rows} - 1)(\text{columns} - 1)$

- within groups (Chapter 15): $df_{\text{within}} = n_{\text{total}} - k$
- between groups (Chapter 15): $df_{\text{between}} = k - 1$

Proportion of a population (Chapter 11): $P_u = \dfrac{n_1 p_{s_1} - n_2 p_{s_2}}{n_1 + n_2}$

Expected frequency (Chapter 12): $f_e = \dfrac{\sum \text{column} * \sum \text{row}}{n}$

Chi-square (Chapter 12): $\chi^2 = \sum \dfrac{(f_o - f_e)^2}{f_e}$

Phi (Chapter 12): $\phi = \sqrt{\dfrac{\chi^2}{n}}$

Cramer's V (Chapter 12): $V = \sqrt{\dfrac{\chi^2}{(n)(min(\text{row} - 1) \mid (\text{column} - 1)}}$

Lambda (Chapter 12): $\lambda = \dfrac{E_1 - E_2}{E_1}$

Kruskal's gamma (Chapter 13): $\gamma = \dfrac{N_{\text{same}} - N_{\text{different}}}{N_{\text{same}} N_{\text{different}}}$

- for testing significance of gamma (Chapter 13): $z_{\text{obtained}} = \gamma \sqrt{\dfrac{N_{\text{same}} + N_{\text{different}}}{N(1 - \gamma^2)}}$

Somers' d (Chapter 13): $d = \dfrac{N_{\text{same}} - N_{\text{different}}}{N_{\text{same}} + N_{\text{different}} + Ties_y}$

- for testing significance of Somers' d (Chapter 13): $z_{\text{obtained}} = d \sqrt{\dfrac{N_{\text{same}} + N_{\text{different}}}{N(1 - d^2)}}$

Kendall's tau-b (Chapter 13):

$$\text{tau-}b = \frac{N_{\text{same}} - N_{\text{different}}}{\sqrt{\left(N_{\text{same}} + N_{\text{different}} + Ties_y\right)\left(N_{\text{same}} + N_{\text{different}} + Ties_x\right)}}$$

- for testing significance of tau-b (Chapter 13): $z_{\text{obtained}} = \text{tau-}b \sqrt{\dfrac{N_{\text{same}} + N_{\text{different}}}{n\left(1 - \left(\text{tau-}b^2\right)\right)}}$

Spearman's *rho* (Chapter 13): $\rho_s = 1 - \dfrac{6 * \sum D^2}{N(N^2 - 1)}$

- for testing significance of Spearman's *rho* (Chapter 13): $t_{\text{obtained}} = r \sqrt{\dfrac{n - 2}{1 - \rho_s^2}}$

Pearson's r illustrational formula (Chapter 14): $r = \dfrac{\sum\left[(X - \bar{x})(Y - \bar{y})\right]}{\sqrt{\left[\sum (X - \bar{x})^2 * \sum (Y - \bar{y})^2\right]}}$

Pearson's r computational formula (Chapter 14): $r_{XY} = \dfrac{N\sum XY - \left(\sum X\right)\left(\sum Y\right)}{\sqrt{\left[N\sum X^2 - \left(\sum X\right)^2\right]\left[N\sum Y^2 - \left(\sum Y\right)^2\right]}}$

or (Chapter 16): $r = \dfrac{SP}{\sqrt{SS_X SS_Y}}$

Sum of squares (Chapter 15):

- total: $SS_{\text{total}} = \sum (X - \bar{x})^2$
- within groups: $SS_{\text{within}} = \sum (X - \bar{x}_{\text{group}})^2$
- between groups: $SS_{\text{between}} = \sum n_{\text{group}} (\bar{x}_{\text{group}} - \bar{x})^2$

or

$$SS_{\text{total}} = \sum X_{\text{total}}^2 - n_{\text{total}} \bar{x}_{\text{total}}^2$$

$$SS_{\text{within}} = \sum X_{\text{total}}^2 - \sum n_{\text{group}} \bar{x}_{\text{group}}^2$$

$$SS_{\text{between}} = \sum n_{\text{group}} \bar{x}_{\text{group}}^2 - n_{\text{total}} \bar{x}_{\text{total}}^2$$

Mean square (Chapter 15):

- within groups: $MS_{\text{within}} = \dfrac{SS_{\text{within}}}{df_{\text{within}}}$
- between groups: $MS_{\text{between}} = \dfrac{SS_{\text{between}}}{df_{\text{between}}}$

F-statistic or F-ratio (Chapter 15): $F = \dfrac{MS_{\text{between}}}{MS_{\text{within}}}$

Regression (Chapter 16): $Y = a + b_1 X_{1i} + b_2 X_{2i} + \cdots + b_n X_{ni} + e_i$

or (Chapter 19): $Y_i = \alpha + \beta_1 X_i + e_i$
or (Chapter 19): $Y_i = \alpha + \beta_1 (X_i^* + u_i) + e_i$

- with two independent variables (Chapter 19): $Y_i = \alpha + \beta_1 X_{1i} + \beta_2 X_{2i} + e_i$

The line of best fit (Chapter 16): $b = \dfrac{\sum (X - \bar{x})(Y - \bar{y})}{\sum (X - \bar{x})^2}$ or $b = \dfrac{SP}{SS_X}$

y-intercept (a) point (Chapter 16): $a = \bar{y} - b\bar{x}$

Error e (Chapter 16): $e = Y - \hat{y}$

Sum of squares for regression with one independent variable (Chapter 16):

$$SS_X = \sum (X - \bar{x})^2$$

$$SS_Y = \sum (Y - \bar{y})^2$$

Sum of products (Chapter 16): $SP = \sum (X - \bar{x})(Y - \bar{y})$

Correlations between X_1 and Y, X_2 and Y, as well as X_1 and X_2 (Chapter 16):

$$r_{X_1Y} = \frac{N\sum X_1Y - \left(\sum X_1\right)\left(\sum Y\right)}{\sqrt{\left[N\sum X_1^2 - \left(\sum X\right)^2\right]\left[N\sum Y^2 - \left(\sum Y\right)^2\right]}}$$

$$r_{X_2Y} = \frac{N\sum X_2Y - \left(\sum X_2\right)\left(\sum Y\right)}{\sqrt{\left[N\sum X_2^2 - \left(\sum X_2\right)^2\right]\left[N\sum Y^2 - \left(\sum Y\right)^2\right]}}$$

$$r_{X_1X_2} = \frac{N\sum X_1X_2 - \left(\sum X_1\right)\left(\sum X_2\right)}{\sqrt{\left[N\sum X_1^2 - \left(\sum X_1\right)^2\right]\left[N\sum X_2^2 - \left(\sum X_2\right)^2\right]}}$$

Correlation between Y, X_1, and X_2 (Chapter 16): $r_{YX1X2}^2 = \dfrac{r_{YX2} - (r_{YX1})(r_{X1X2})}{\sqrt{1 - r_{YX1}^2}\ \sqrt{1 - r_{X1X2}^2}}$

Partial correlation coefficient (Chapter 16): $b_1 = \left(\dfrac{s_Y}{s_{X1}}\right)\left(\dfrac{r_{X1Y} - r_{X2Y}r_{X1X2}}{1 - r_{X1X2}^2}\right)$

$$b_2 = \left(\frac{s_Y}{s_{X2}}\right)\left(\frac{r_{X2Y} - r_{X1Y}r_{X1X2}}{1 - r_{X1X2}^2}\right)$$

Beta weights (Chapter 16): $b_{1*} = b_1\left(\dfrac{s_1}{s_Y}\right)$

Pearson's r^2 (Chapter 16): $R^2 = r_{X1Y}^2 + r_{YX1X2}^2(1 - r_{X1Y}^2)$

or:

$$R^2 = \frac{\sum (\hat{y} - \bar{y})^2}{\sum (Y - \bar{y})^2}$$

Standard error of an estimate (Chapter 16): $s_{\text{est}} = \sqrt{\dfrac{(Y - \hat{y})^2}{n-2}}$

Standard error of coefficient b (Chapter 16): $se_b = \dfrac{\sqrt{\dfrac{(Y-\hat{y})^2}{n-2}}}{\sqrt{(X-\hat{x})^2}}$

Odds ratio (Chapter 17): $\textbf{odds}\ (Y = 1) = \left[\dfrac{\Pr\ (Y = 1)}{\Pr\ (Y \neq 1)}\right]$

Log odds (Chapter 17): $\textbf{log odds}\ (Y = 1) = 1n\left[\dfrac{\Pr\ (Y = 1)}{1 - \Pr\ (Y = 1)}\right]$

Natural logarithm of the odds of the dependent variable being one (Chapter 17):

$$\textbf{log odds}\ (Y = 1) = a + b_1 X_{1i} + b_2 X_{2i} + \cdots + b_n X_{ni} + e_i$$

Variance inflation factor (VIF) for each variable to identify multicollinearity (Chapter 18):

$$\text{VIF}_{Xi} = \frac{1}{1 - R^2_{Xi}}$$

Symbols

a	The Y intercept (Chapter 16, p. 195)
ANOVA	The analysis of variance (Chapter 15, p. 173)
b_n	The partial slope of X_n on Y (Chapter 16, p. 191)
b*	Beta weights (Chapter 16, p. 201)
C	The number of columns in a table (Chapter 12, p. 130)
D	Difference between rankings of variables (Chapter 13, p. 153)
d	Somers' d (Chapter 13, p. 149)
d_i	Difference between scores (Chapter 10, p. 110)
df	Degrees of freedom (Chapter 9, p. 87)
df_{between}	Degrees of freedom between groups (Chapter 15, p. 178)
df_{within}	Degrees of freedom within groups (Chapter 15, p. 178)
e	The transcendental number, 2.718 (Chapter 2, p. 11)
e_i	Error for individual i (Chapter 16, p. 191)
E_1	Classification error of dependent variable (Chapter 12, p. 136)
E_2	Classification error of independent variable (Chapter 12, p. 137)
f_e	Expected frequency (Chapter 12, p. 128)
f_o	Observed frequency (Chapter 12, p. 129)

F	F-statistic or F-ratio (Chapter 15, p. 178)
H	Highest value of a variable (Chapter 6, p. 48)
H_a /H_1	Alternative/research hypothesis (Chapter 10, p. 97)
H_0	Null hypothesis (Chapter 10, p. 97)
k	The number of groups being compared (Chapter 15, p. 178)
L	Lowest value of a variable (Chapter 6, p. 48)
MAR	Missing at random (Chapter 19, p. 239)
MCAR	Missing completely at random (Chapter 19, p. 239)
MS	Mean square (Chapter 15, p. 177)
$MS_{between}$	Mean square between groups (Chapter 15, p. 177)
MS_{within}	Mean square within groups (Chapter 15, p. 177)
n	Number of total observations in a sample (Chapter 7, p. 60)
N	Number of possible total observations (i.e., population size) (Chapter 9, p. 87)
N_{same}	Number of concordant pairs (Chapter 13, p. 144)
$N_{different}$	Number of discordant pairs (Chapter 13, p. 144)
NIM	Non-ignorable missing value (Chapter 19, p. 239)
NNI	Nearest neighbour imputation (Chapter 19, p. 241)
OLS	Ordinary least squares regression (Chapter 16, p. 191)
p	Probability (Chapter 4, p. 26)
P	Proportion (Chapter 9, p. 89)
P_u	The proportion of a population (Chapter 11, p. 121)
R	The number of rows in a table (Chapter 12, p. 130)
r	Pearson's r (Chapter 14, p. 161); correlation between continuous variables (Chapter 16, p. 198)
R^2	Pearson's multiple correlation coefficient r-squared (Chapter 16, p. 202); or the square of the correlation between variables.
s or s_X	The standard deviation of a sample (Chapter 7, p. 60)
s^2	The variance of a sample (Chapter 7, p. 60)
s_{est}	The standard error of the estimate (Chapter 16, p. 208)
s_p	The standard error of a proportion (Chapter 9, p. 89)
$s_{\bar{x}}$	The standard error of a sample (Chapter 9, p. 85)
$s_{\bar{x}_1} - s_{\bar{x}_2}$	The standard deviation of the difference between sample means (Chapter 11, p. 212)
s_{p1-p2}	The standard deviation of the difference between sample proportions (Chapter 11, p. 119)
se_b	The standard error of coefficient b (Chapter 16, p. 208)
$SS_{between}$	The total sum of squares between groups (Chapter 15, p. 176)
SS_{total}	The total sum of squares (Chapter 15, p. 175)
SS_{within}	The total sum of squares within groups (Chapter 15, p. 176)

SP	The sum of products (Chapter 16, p. 196)
t	*t*-distribution/Student's *t*-score (Chapter 9, p. 86)
tau-b	Kendall's tau-*b* (Chapter 13, p. 151)
Ties	The number of ties (Chapter 13, p. 150)
u^{i}	Measurement error due to non-random non-response (Chapter 19, p. 237)
V	Cramer's V (Chapter 12, p. 134)
VIF	Variance inflation factor (Chapter 18, p. 233)
X	An independent variable (Chapter 16, p. 191)
X_{ni}	The *n*th independent variable for individual *i* (Chapter 17, p. 215)
$\bar{x}$	The mean of a variable or sample (Chapter 6, p. 46)
X_i	An explanatory variable with coefficient β (Chapter 19, p. 237)
$\bar{x}_{\text{group}}$	The mean of a group (Chapter 15, p. 176)
Y	A dependent variable (Chapter 16, p. 191)
$\bar{y}$	The mean of variable Y (Chapter 16, p. 194)
$\hat{y}$	A predicted value of Y (Chapter 16, p. 195)
z	The *z*-score or standard score for a sample or a population (Chapter 7, p. 59)
z_i	Compound error ($e_i + u_i$) (Chapter 19, p. 237)
α	Constant term (Chapter 19, p. 237)
β_1	The coefficient for X_i (Chapter 19, p. 237)
γ	Kruskal's gamma (Chapter 13, p. 144)
λ	Lambda (Chapter 12, p. 137)
μ	The mean of a population (Chapter 7, p. 60)
ρ_s	Spearman's *rho* (Chapter 13, p. 152)
σ or σ_x	The standard deviation of a population (Chapter 6, p. 50)
σ^2	The variance of a population (Chapter 6, p. 50)
$\sigma_{\bar{x}}$	The standard error of a population (Chapter 9, p. 83)
Σ	"The sum of" (Chapter 6, p. 46)
ϕ	Phi (Chapter 12, p. 131)
χ^2	Chi-square (Chapter 12, p. 129)

Appendix H

Solution Key

Solution Key for Practice Questions

Chapter 2

1. $10 + 15 = 25$
2. $10 + 15 - 5 = 20$
3. $10 - (-2) = 12$
4. $(10 + 15) - 5 = 20$
5. $(10 - 15) - 2 = -7$
6. $10 * 15 = 150$
7. $10 * 15 - 5 = 145$
8. $10 * (15 - 5) = 100$
9. $10 * 15 - \frac{15}{5} = 147$
10. $\frac{10}{5} * 15 - 5 = 25$
11. $(X * Y)^a + b = X^a * Y^a + b$
12. $(X^a)(X^b) = X^{a+b}$
13. $\sqrt{X} = X^{\frac{1}{2}}$
14. $e^{1.61} = 5$
15. Percentage scores on a math exam: Ratio
16. Letter grades on a math exam: Ordinal
17. Flavours of ice cream: Nominal
18. Fitness training levels on an exercise machine classified as: Easy, Difficult, or Impossible: Ordinal
19. Ethnic origins: Nominal
20. Political parties: Nominal
21. Commuting distances to school in kilometers: Ratio

22. Years between important historical events: Ratio
23. Age (in years): Ratio
24. Amount of money in your savings accounts: Ratio
25. Temperature on the moon, measured in degrees Celsius: Interval

Chapter 3

1. a. Jorge was correct 37 times and incorrect 3 times. This translates into a ratio of 37:3.
   b. His score is calculated as $\frac{37}{40}$, or 92.5 per cent.
   c. His percentile rank is calculated as $\frac{113}{1432}$, which equals 0.0789. Given this, we could say that he placed in the top eighth percentile.
2. a. Ethel's contact rate would be calculated by first determining the percentage of people that she contacts, which is $\frac{432}{541}$, or 79.9 per cent. When stated as a rate (percentages are essentially a rate per 100), we'd need to multiply the numerator and the denominator by 10, which would yield 799 per 1000 people.
   b. Ethel's contact/non-contact rate is 432:109, which is roughly equivalent to 4:1.
   c. The participation rate as a percentage is 112:432, or 25.9 per cent.
3. Using Charles's assumption,
   a. the number of woodchucks who agree is 13.
   b. the approval rate for Charles is calculated as $\frac{13}{30}$, or 43.3 per cent.
   c. the ratio of those who agree to those who disagree is 13:17.

Chapter 4

1. c
2. $\frac{2}{5}$ or 0.4 or a 40 per cent chance
3. $\frac{1}{36}$
4. $\frac{3}{30} = \frac{1}{10}$
5. $(0.52 * 0.52) = 27$ per cent
6. $\left[\left(\frac{4}{52}\right) * \left(\frac{4}{51}\right) * \left(\frac{4}{50}\right)\right] = \frac{64}{132{,}600} = \frac{8}{16{,}575}$
7. There are 52 cards in a deck and $\frac{13}{52}$ chances of getting any one card of a particular suit. As each card is drawn the total number of cards that can be chosen of that particular suit (the numerator) decreases by one, as does the total number of cards in the deck (the denominator).

$$\left(\frac{13}{52}\right) * \left(\frac{12}{51}\right) * \left(\frac{11}{50}\right) * \left(\frac{10}{49}\right) * \left(\frac{9}{48}\right)$$

This results in a probability of 0.000495.

8. The probability that the essay questions Ryan hasn't studied for are on the exam is $\frac{2}{8} = 0.25$.

   Because one essay question has already been chosen for the exam, the probability that the second essay question Ryan has not studied for will appear on the exam is $\frac{1}{7} = 0.143$.

   The probability that both questions Ryan has not studied for will appear on the exam is $0.25 * 0.143 = 0.04$.

9. Since Julie was not a pirate 2 out of the 10 years, the probability that Julie will dress up as something other than a pirate is $\frac{2}{10}$ or $\frac{1}{5}$.

10. a. Your odds have been the same since you started.

    b. Your odds are the same as those around you.

    c. Incorrect. You have a $\frac{1}{36}$ chance of rolling two fives. Do you want to be my friend?

Chapter 5

1. Dr Knifewell might like to know that the data will probably be skewed to the right (positively skewed) and that it will likely be unimodal.
2. The data are unimodal and positively skewed.
3. The outlier is at the value 10,000. It appears to be flattening the normal curve, having a negative effect on the kurtosis value.
4. The distribution will become negatively skewed, or skewed to the left. The unimodal hump will also shift.
5. The impact of weather in Canada becoming more volatile will make the tails thicker because there will be more instances of high and low temperatures.
6. You would tell the cellphone provider that there are only a few customers $\left(\frac{6}{20}\right)$ that exceed 500 air-time minutes a month. The distribution is positively skewed or skewed to the right.
7. You would expect outliers in the lower grade range because when a distribution is skewed to the left, or negatively skewed, the tail on the left is longer and the majority of the scores would actually fall to the right, which when plotted along a normal curve leaves outliers in the lower grade range.
8. Answers will vary.

Chapter 6

1. a. 7

   b. 248,000 (270,000 − 22,000)

2. Mean = 29.4, mean deviation = 30.9

3. Mean = 4.67, variance = 3.39, standard deviation = 1.84
4. The mean would be higher than the median and the histogram would be skewed with a long right tail.
5. The median remains the same, but the mean is increased.
6. Income is almost always positively skewed, which will pull the mean past the median. Very few Canadians earn negative income (expect for some self-employed), suggesting that the distribution of income values hits a wall at $0. On the high side of the distribution there are many high earners, and these factors collectively produce a mean that exceeds the median.
7. All of the values will increase. The mean, median, and range will rise because of a broader set of age values and more people at the older ages, while the standard deviation will increase because of an increase in the average distance from the mean.
8. We would expect that the tails would be longer for a distribution with a larger standard deviation.
9. Remember that the mean deviation, standard deviation, and variance are measures of dispersion. Since Jessica is sending the same number of text messages every day, the values for all three measures of dispersion are therefore zero.
10. Helen invests a lot of time right away and loses interest quickly. Thus, there will be a few large and many small time investments. There will be a larger number of lower values on the curve and a few high values. The curve is positively skewed; therefore, the median value will be lower than the mean.

Chapter 7

1. Answers can be found by looking at column B of Appendix A.
   a. 0.3810
   b. 0.2995
   c. 0.4803
   d. 0.4131
2. Answers can be found in column C of Appendix A and, when z values are positive, subtracting the value from 1.
   a. Percentile rank = 98.78th percentile
   b. Percentile rank = 4.75th percentile
   c. Percentile rank = 92.36th percentile
   d. Percentile rank = 33rd percentile
3. Answers can be found in column C of Appendix A and, when z values are negative, subtracting the value from 1.

a. 40.13 per cent of all cases are above a z-value of 0.25

b. 88.69 per cent of all cases are above a z-value of −1.21 (1 − 0.1131)

c. 11.31 per cent of all cases are above a z-value of 1.21

d. 97.78 per cent of all cases are above a z-value of −2.01 (1 − 0.0222)

4. Answers can be found in column B of Appendix A.

a. 59.64 per cent

b. 12.88 per cent

c. 83.53 per cent

d. 7.69 per cent

5. Sigmund's percentile rank is 8.08th. We get that number by calculating z as:

$$\begin{aligned} z &= \frac{X - \mu}{\sigma} \\ &= \frac{45 - 52}{5} \\ &= -1.40 \end{aligned}$$

The value comes from column C of Appendix A. Clearly, Sigmund didn't do very well.

6. Lesley did a bit better. From column C of Appendix A, z = 0.4 corresponds to 0.3446. Her percentile rank is 0.6554, which comes from subtracting 0.3446 (the area beyond z, or everyone that beat her) from 1. Therefore, 34.46 per cent of the students did better than she did. Here's the calculation for z:

$$\begin{aligned} z &= \frac{X - \mu}{\sigma} \\ &= \frac{54 - 52}{5} \\ &= 0.40 \end{aligned}$$

7. Your child is quite bright, beating 93.7 per cent of all people who wrote the exam. (This value comes from column C of Appendix A.) This corresponds to 0.0630 (1 − 0.0630 = 0.937). Your child scored in the 93.70th percentile.

$$\begin{aligned} z &= \frac{X - \mu}{\sigma} \\ &= \frac{148 - 125}{15} \\ &= 1.53 \end{aligned}$$

8. For Feng:

$$z = \frac{X - \mu}{\sigma}$$
$$= \frac{76 - 80}{8}$$
$$= -0.50$$

This corresponds to 19.15 per cent (use column B).

For Lucy:

$$z = \frac{X - \mu}{\sigma}$$
$$= \frac{94 - 80}{8}$$
$$= 1.75$$

This corresponds to 45.99 per cent (use column B).

Therefore, 65.14 per cent of all people scored between these two scores.

9. Sixteen seconds. Remember that only 5 per cent of all cases resides above or below 2 standard deviations, so knowing that the value for two standard deviations is 32 (120 − 88 = 32, 152 − 120 = 32) makes it easy to determine the value for one standard deviation.

10. a. Mean = 44.7, standard deviation = 13.01
    b. There are six hours that fall above or below ±1 standard deviation. There are no hours that fall above or below 2 or 3 standard deviations.

Chapter 8

1. Since representativeness is the goal, a simple random sample would be preferred.
2. Given the focus on representativeness across facilities, a stratified random sample would be preferred (stratified by faculty).
3. Snowball samples are often used for vulnerable populations. A convenience sample might also be appropriate.
4. Since each instrument is needed in the sample, a stratified random sample would be preferred (stratified by instrument).
5. Given that the sample is random, selected cases will likely differ. Depending on how you chose to stratify (by column, row, etc.), values here will also vary. There should be only six individuals in the sample.

6. a. Assuming that the Canadian population is 34,500,000, approximately 6,900,000 (20 per cent) Canadians would have been administered the long-form census. Since the response was actually higher (94 per cent), approximately 6,486,000 would have completed the survey. This represents about 18.8 per cent of the entire Canadian population.

   b. Assuming that the Canadian population is 34,500,000, approximately 11,385,000 will be administered the National Household Survey. Since the expected response rate is 50 per cent, approximately 5,692,500 will complete the survey. This represents 16.5 per cent of the entire Canadian population.

Chapter 9

1. a.

	Men	Women
Mean	44.50	28.10
Standard deviation	8.68	8.54
Variance	75.39	72.99

   b. Standard error 2.75 2.70

   c. For men (38.29, 50.71)

   For women (21.99, 34.21)

2. First calculate the mean, standard deviation, and the standard error:

   Mean = 10.92

   Standard deviation = 5.60

   Standard error = 1.14

   Then, find the appropriate t-value at $n - 1$ degrees of freedom:

   $t = 2.069$ for a 95 per cent confidence interval at $df = 23$.

   Confidence interval at 95 per cent: 8.56 < population mean < 13.28.

   We can be 95 per cent confident that the actual population mean for the number of hours students worked on their assignment was between 8.56 hours and 13.28 hours.

3. First, calculate the proportion standard of error:

   Standard error = 0.0670

   Confidence interval = 0.170, 0.510

   We can be 99 per cent confident that the real proportion lies between 0.160 per cent and 0.520 per cent.

4. The standard error of a sample decreases as the number of observations increases. The confidence intervals will also decrease as the number of observations increase, so we can be more confident in our results.

5. 71

6. Yes. Once you find the weight of the ninth cat, you have zero remaining degrees of freedom because it is possible to find 10 values by knowing the sum or mean and $n - 1$ individual values.
7. 0.31376, 0.48624
8. a. One-tailed
   b. Two-tailed
   c. One-tailed
   d. Two-tailed
   e. One-tailed
9. Yes. The 95 per cent confidence interval for Rovio is between 38,040 and 41,960, so your observational day is at the low end of a regular day.
10. 42.82, 73.58

Chapter 10

1. a. H_1 = the amount of money raised will be different from the previous year

   H_0 = the amount of money raised will not be significantly different from the previous year

   b. H_1 = the amount of money raised will be greater than the previous year

   H_0 = the amount of money raised will be equal or less than the previous year

2. a. $$z = \frac{\bar{x} - \mu}{\frac{\sigma}{\sqrt{n}}} = \frac{73 - 75}{\frac{8}{\sqrt{5}}} = \frac{-2}{3.5776} = -0.559$$

   b. The absolute value of 0.559 does not exceed the critical z-value of 1.96, suggesting that we cannot be 95 per cent confident in the superiority of Amber's new route.

3. $$z = \frac{\bar{x} - \mu}{\frac{\sigma}{\sqrt{n}}} = \frac{31{,}000 - 29{,}000}{\frac{27{,}000}{\sqrt{16}}} = \frac{2000}{6750} = 0.296$$

Jasper's income was *not* significantly higher than the Canadian average because the $z_{obtained}$ value does not exceed the critical value of 1.96.

4. We can be 95 per cent confident that the population proportion lies between 50.3 per cent (52 − 1.7) and 53.7 per cent (52 + 1.7).

5. There is an average of 4 leaves separating Ted's plants from his neighbour's, and both have a standard deviation of greater than 4, so we can't even be 68 per cent confident of the significance of the difference in means. Unfortunately, Ted cannot make a compelling case for basil superiority.

6. a. H_1 = the girls will do significantly better than their competition

   H_0 = the girls will do the same or worse than the competition

   b. H_a = the girls will do better or worse than their competition

   H_0 = the girls will do the same as the competition

7. First calculate z for samples with proportions:

$$z_{obtained} = \frac{p_{sample} - P_{population}}{\sqrt{\frac{P_{population}\left(1 - P_{population}\right)}{n}}}$$

$$= \frac{0.32 - 0.27}{\sqrt{\frac{0.27(1 - 0.27)}{172}}}$$

$$= \frac{0.05}{\sqrt{0.0011459}}$$

$$= 1.477$$

$z_{obtained}$ = 1.477. Because 1.477 does not exceed 1.96 (the critical value of z at 95 per cent confidence interval), we cannot say that there is a difference between Edmontonians and Albertans regarding their love for their home team at the 0.05 level.

8. Type two error

9. The probability of rejecting the null hypothesis decreases (it becomes increasingly harder to reject the null hypothesis as $z_{critical}$ increases).

10. If you reject the null hypothesis, you are saying that the sample differs from the population on your outcome of interest.

Chapter 11

1. H_0 = there is no difference between men and women on perceived life satisfaction

   H_1 = there is a difference between men and women on perceived life satisfaction

2. Since we have fairly large samples (>120), we can use the z-distribution and, consequently, the simpler calculation for the standard error of the difference between means:

$$
\begin{aligned}
s_{\bar{x}_1-\bar{x}_2} &= \sqrt{\left(\frac{s_1^2}{n_1-1}+\frac{s_2^2}{n_2-1}\right)} \\
&= \sqrt{\frac{7^2}{200-1}+\frac{5^2}{200-1}} \\
&= \sqrt{\frac{49}{199}+\frac{25}{199}} \\
&= 0.610
\end{aligned}
$$

This we use to calculate z_{observed}:

$$
\begin{aligned}
z &= \frac{\bar{x}_1-\bar{x}_2}{s_{\bar{x}_1-\bar{x}_2}} \\
&= \frac{48-45}{0.610} \\
&= 4.918
\end{aligned}
$$

Comparing z_{observed} to z_{critical} value at 95 per cent of 1.96, we can reject the null hypothesis. We can conclude that there are significant differences in the life satisfaction levels of men and women.

3. H_0 = there is no difference between non-homeowners and homeowners concerning the importance of the distance to work

   H_1 = there is a difference between non-homeowners and homeowners concerning the importance of the distance to work

4. First, we need to calculate P_u:

$$
\begin{aligned}
P_u &= \frac{n_1 p_{s_1} - n_2 p_{s_2}}{n_1+n_2} \\
&= \frac{201 * 0.76 + 218 * 0.68}{201+218} \\
&= \frac{301}{419} \\
&= 0.718
\end{aligned}
$$

Next, we used P_u to calculate the standard error of the difference between means:

$$
\begin{aligned}
s_{p_1-p_2} &= \sqrt{P_u * (1-P_u)\frac{n_1+n_2}{n_1 n_2}} \\
&= \sqrt{0.718(1-0.718)\frac{201+218}{201 * 218}}
\end{aligned}
$$

$$= \sqrt{0.718(1-0.718)\frac{419}{43,818}}$$
$$= \sqrt{0.202476 * 0.0095622803414122}$$
$$= \sqrt{0.0019361322744078}$$
$$= 0.044$$

Finally, we calculate z_{obtained}:

$$z_{\text{obtained}} = \frac{p_{s_1} - p_{s_2}}{s_{p_1 - p_2}}$$
$$= \frac{0.76 - 0.68}{0.044}$$
$$= 1.82$$

Comparing z_{obtained} to z_{critical} of 1.96 at 95 per cent, we fail to reject the null hypothesis. We can conclude that the differences are not statistically significant.

5. a. A *t*-test is more appropriate because the sample size is small.

   b. A one-tailed test is more appropriate because we are looking at a difference in one direction ("dogs have greater intelligence than cats").

6. First, we need to calculate the standard error of the difference between means:

$$s_{\bar{x}_1 - \bar{x}_2} = \sqrt{\left(\frac{n_1 s_1^2 + n_2 s_2^2}{n_1 + n_2 - 2}\right)\left(\frac{n_1 + n_2}{n_1 n_2}\right)}$$
$$= \sqrt{\left(\frac{93 * 12^2 + 28 * 12^2}{93 + 28 - 2}\right)\left(\frac{93 + 28}{93 * 28}\right)}$$
$$= \sqrt{\left(\frac{13,392 + 4032}{119}\right)\left(\frac{121}{2604}\right)}$$
$$= \sqrt{146.42 * 0.05}$$
$$= \sqrt{7.321}$$
$$= 2.71$$

We can use this to calculate our *t*:

$$t = \frac{\bar{x}_1 - \bar{x}_2}{s_{\bar{x}_1 - \bar{x}_2}}$$
$$= \frac{79 - 75}{2.71}$$
$$= 1.48$$

To obtain the $t_{critical}$ value for 95 per cent confidence, we need to calculate our degrees of freedom for t as $df = (n_1 + n_2 - 2)$, or 117. Since our value is just below the highest value of 120, we will use the value for 120, which is 1.960. Our $t_{critical}$ is greater than our $t_{observed}$ value of 1.48; therefore we fail to reject the null hypothesis. We do not find a statistically significant difference between the intelligence of dogs and cats.

Chapter 12

1. These variables are nominal because the distance between response categories cannot be identified, nor can the categories be ranked.
2. Chi-square = 1133.1
3. There are 6 degrees of freedom.
4. $\textbf{Phi} = \phi = \sqrt{\frac{\chi^2}{n}} = \mathbf{0.656}$, **Cramer's** $V = \sqrt{\frac{\chi^2}{(n)(min(\text{row} - 1)|(\text{column} - 1))}} = \mathbf{0.464}$.

   It is preferable to use Cramer's V because the table is bigger than two by two.
5. The appropriate statistic to measure the association is phi because the table is two by two. Chi-square = 185.4. Phi = 0.469.
   a. H_0 = People who live in Quebec City are no more likely to identify as Roman Catholic than Church of England.

      H_1 = People who live in Quebec City are more likely to be Roman Catholic than Church of England.
6. H_0 = Male and females believe equally that facial recognition does not constitute an infringement of privacy.

   H_1 = Male and females differ in the belief that facial recognition constitutes an infringement of privacy.
7. First, make two predictions ignoring the independent variable:

   The difference between everyone answering "yes" and the number who actually say "yes" is 2917 − 1075 = 1842. The difference between everyone answering "no" and the number who actually say "no" is 2917 − 1842 = 1075. The lowest of these is 1075, so E_1 = 1075.

   Second, take account of the sex of respondents:

   The difference between all females answering "yes" and those who actually do is 1699 − 699 = 1000. The difference between all females answering "no" and those who actually do is 1699 − 1000 = 699. The difference between all males answering "yes" and those who actually do is 1218 − 376 = 842. The difference between all males answering "no" and those who actually do is 1218 − 842 = 376.

   Add the lowest two values to find E_2 = 699 + 376 = 1075.

   $$\text{Lambda} = \frac{(E_1 - E_2)}{E_2} = \frac{(1075 - 1075)}{1075} = 0$$

8. Since lambda is 0, knowing if someone is male or female will not allow us to predict that person's views on facial recognition. We have no evidence to suggest that sex affects people's views on facial recognition.
9. Chi-square = 2.13. Bo and Jonah's hypothesis is not supported because the calculated value is well below the critical value of 3.841 for a 0.05 level of significance.

Chapter 13

1. H_0 = Exercising every week does not lead to any changes in self-rated health.

   H_1 = People who exercise every week have better scores for self-rated health.
2. N_{same} = 325 and N_{diff} = 187, so gamma = 0.270. This value supports Jerry's research hypothesis.

   The *t*-statistic for this gamma result is 8.367. At degrees of freedom = 64, this is a significant result at 0.01.
3. H_0 = The amount of money Canadian universities received in 2009 is not related to the amount of money they received in 2000.

   H_1 = Canadian universities that received more money in 2000 will receive more money in 2009.
4. Spearman's *rho* = 0.596. This value indicates a strong and positive relationship, suggesting that we should reject the null hypothesis.
5. H_0 = The weight of a squirrel does not impact how much food a squirrel will hide away for the winter.

   H_1 = The more a squirrel weighs, the more food that squirrel will hide away for the winter.
6. Somers' $d = 0.339$

$$\begin{aligned} N_{same} &= a\,(e+f+h+i) + b\,(f+i) + d\,(h+i) + e\,(i) \\ &= 360 + 231 + 84 + 84 \\ &= 759 \end{aligned}$$

$$\begin{aligned} N_{different} &= c\,(d+e+g+h) + b\,(d+g) + f\,(g+h) + e\,(g) \\ &= 90 + 99 + 45 + 21 \\ &= 255 \end{aligned}$$

$$\begin{aligned} Ties_Y &= a\,(b+c) + b\,(c) + d\,(e+f) + e\,(f) + g\,(h+i) + h\,(i) \\ &= 192 + 55 + 96 + 63 + 42 + 24 \\ &= 472 \end{aligned}$$

$$d = \frac{N_{\text{same}} - N_{\text{different}}}{N_{\text{same}} + N_{\text{different}} + Ties_Y}$$

$$= \frac{759 - 255}{759 + 255 + 472}$$

$$= \frac{504}{1486}$$

$$= 0.339$$

The value indicates that there is strong positive relationship between the weight of a squirrel and the amount of food that a squirrel will hide away for the winter. Therefore, we would reject the null hypothesis and conclude that there is a relationship between squirrels' weight and how much food they hide away for the winter.

7. Kendall's tau-*b* = 0.341. To calculate tau-*b*, we have everything we need from #6 above, except for $Ties_X$.

$$Ties_X = a\,(d+g) + d\,(g) + b\,(e+h) + e\,(h) + c\,(f+i) + f\,(i)$$

$$= 108 + 18 + 99 + 14 + 105 + 108$$

$$= 452$$

$$\text{tau-}b = \frac{N_{\text{same}} - N_{\text{different}}}{\sqrt{(N_{\text{same}} + N_{\text{different}} + Ties_y)(N_{\text{same}} + N_{\text{different}} + Ties_x)}}$$

$$= \frac{759 - 255}{\sqrt{(759 + 255 + 472)(759 + 255 + 452)}}$$

$$= \frac{504}{1476}$$

$$= 0.341$$

The result of tau-*b* does not lead us to make a different decision regarding the hypothesis. Because $Ties_X$ and $Ties_Y$ are close, there is little effect on the denominator.

Chapter 14

1. a. H_0 = There is no relationship between family size and family income.

   H_1 = There is a positive relationship between family size and family income.

   H_2 = There is a negative relationship between family size and family income.

   b. $r = -0.2239$

c. We can assess the statistical significance of r by using a t-test:

$$t_{\text{observed}} = r\sqrt{\frac{n-2}{1-r^2}}$$

$$= -0.2239\sqrt{\frac{10-2}{1-(-0.2239)^2}}$$

$$= -0.2239\sqrt{\frac{8}{1-0.05}}$$

$$= -0.2239 * 2.902$$

$$= -0.650$$

Since the t_{observed} value of -0.650 is below the t_{critical} value of 2.306, we cannot be 95 per cent confident that the sample r value of -0.2239 did not occur by chance. We fail to reject the null hypothesis, and we conclude that there is no relationship between family size and family income.

2. The line will slope downwards because of the negative association.

3. H_0 = There is no relationship between how much Chris spends and how much Josie spends when they eat out together

   H_1 = Chris and Josie have a tendency to spend about the same amount as each other when they eat out together.

4. Pearson's $r = 0.281$.

5. We can assess the statistical significance of r by using a t-test:

$$t_{\text{observed}} = r\sqrt{\frac{n-2}{1-r^2}}$$

$$= 0.281\sqrt{\frac{8-2}{1-0.281^2}}$$

$$= 0.281\frac{6}{1-0.079}$$

$$= 0.281 * 2.5524$$

$$= 0.7172$$

   Since the t_{observed} value of 0.7172 is below the t_{critical} value of 2.447, we cannot be 95 per cent confident that the sample r value of 0.281 did not occur by chance.

6. Because of the weak correlation of the results, Chris and Josie's concerns are unwarranted and perhaps there is another explanation for why they are spending too much money at restaurants.

Chapter 15

1. H_0 = Each type of exercise activity burns the same amount of calories.

   H_1 = At least one of the types of exercise burns a different amount of calories than the other types of exercise.

2. a.

	Bicycling		Cleaning		Health club		Yoga		Tennis	
	X	X^2	X	X^2	X	X^2	X	X^2	X	X^2
	236	55,696	207	42,849	325	105,625	236	55,696	413	170,569
	321	103,041	249	62,001	401	160,801	312	97,344	599	358,801
	345	119,025	292	85,264	452	204,304	345	119,025	604	364,816
	292	85,264	302	91,204	474	224,676	281	78,961	434	188,356
	301	90,601	222	49,284	353	124,609	301	90,601	477	227,529
Sum	**1495**	**453,627**	**1272**	**330,602**	**2005**	**820,015**	**1475**	**441,627**	**2527**	**1,310,071**
X	**299**		**254.4**		**401**		**295**		**505.4**	
$n = 25$										
Grand mean = 350.96										

b. $SS_{\text{total}} = \sum X^2_{\text{total}} - n_{\text{total}}\bar{x}^2_{\text{total}}$

$$= (453,627 + 330,602 + 820,015 + 441,627 + 1,310,071) - 25(350.96)^2$$

$$= 3,355,942 - 25(123,172.92)$$

$$= 276,619$$

$$SS_{\text{within}} = \sum X^2_{\text{total}} - \sum n_{\text{group}}\bar{x}^2_{\text{group}}$$

$$= 3,355,942 - \left[5(299)^2 + 5(254.4)^2 + 5(401)^2 + 5(295)^2 + 5(505.4)^2\right]$$

$$= 3,355,942 - [447,005 + 323,596.8 + 804,005 + 435,125 + 1,277,145.8]$$

$$= 3,355,942 - 3,286,877.61$$

$$= 69,064.39$$

$$SS_{\text{between}} = \sum n_{\text{group}}\bar{x}^2_{\text{group}} - n_{\text{total}}\bar{x}^2_{\text{total}}$$

$$= 3,286,877.6 - 3,079,323.04$$

$$= 207,554.56$$

c. $df_{\text{within}} = n_{\text{total}} - k$

$$= 25 - 5$$

$$= 20$$

$$df_{between} = k - 1$$
$$= 5 - 1$$
$$= 4$$

d. $$MS_{within} = \frac{SS_{within}}{df_{within}}$$
$$= \frac{69{,}064.39}{20}$$
$$= 3453.22$$

$$MS_{between} = \frac{SS_{between}}{df_{between}}$$
$$= \frac{207{,}554.56}{4}$$
$$= 51{,}888.64$$

e. $$F = \frac{MS_{between}}{MS_{within}}$$
$$= \frac{51{,}888.64}{3453.22}$$
$$= 15.03$$

f. $F_{calculated} = 15.03$ and $F_{critical} = 2.87$. So $F_{calculated} > F_{critical}$. We can be 95 per cent confident that at least one type of exercise group differs significantly from the other types in terms of the number of calories burned per hour.

3. H_0 = There is no relationship between the brand of jeans and its durability.

H_1 = At least one brand of jeans has a different durability than the other brands.

4. a.

Durability of Jeans								
	Levi's		People's Liberty		Silver		Guess	
	X	X^2	X	X^2	X	X^2	X	X^2
1	182	33,124	209	43,681	1040	1,081,600	260	67,600
2	130	16,900	225	50,625	780	608,400	624	389,376
3	91	8281	156	24,336	520	270,400	416	173,056
4	200	40,000	260	67,600	340	115,600	222	49,284
5	154	23,716	101	10,201	416	173,056	85	7225
Sum	**757**	**122,021**	**951**	**196,443**	**3096**	**2,249,056**	**1607**	**686,541**
$\bar{X}$	**151.4**		**190.2**		**619.2**		**321.4**	
$n = 20$								
Grand mean = 320.55								

$$SS_{\text{total}} = \sum X^2_{\text{total}} - n_{\text{total}}\bar{x}^2_{\text{total}}$$

$$SS_{\text{within}} = \sum X^2_{\text{total}} - \sum n_{\text{group}}\bar{x}^2_{\text{group}}$$

$$SS_{\text{between}} = \sum n_{\text{group}}\bar{x}^2_{\text{group}} - n_{\text{total}}\bar{x}^2_{\text{total}}$$

$$\begin{aligned} SS_{\text{total}} &= \sum X^2_{\text{total}} - n_{\text{total}}\bar{x}^2_{\text{total}} \\ &= (122{,}021 + 196{,}443 + 2{,}249{,}056 + 686{,}541) - 20(320.55)^2 \\ &= 3{,}254{,}061 - 20(102{,}752.30) \\ &= 1{,}199{,}015 \end{aligned}$$

$$\begin{aligned} SS_{\text{within}} &= \sum X^2_{\text{total}} - \sum n_{\text{group}}\bar{x}^2_{\text{group}} \\ &= (122{,}021 + 196{,}443 + 2{,}249{,}056 + 686{,}541) - [5(151.4)^2 \\ &\quad + 5\,(190.2)^2 + 5(619.2)^2 + 5(321.4)^2] \\ &= 3{,}254{,}061 - [114{,}609.8 + 180{,}880.2 + 1{,}917{,}043.2 + 516{,}489.8] \\ &= 3{,}254{,}061 - 2{,}729{,}023 \\ &= 525{,}038 \end{aligned}$$

$$\begin{aligned} SS_{\text{between}} &= \sum n_{\text{group}}\bar{x}^2_{\text{group}} - n_{\text{total}}\bar{x}^2_{\text{total}} \\ &= \left[5(151.4)^2 + 5(190.2)^2 + 5(619.2)^2 + 5(321.4)^2\right] - 20(320.55)^2 \\ &= 2{,}729{,}023 - 2{,}055{,}046.05 \\ &= 673{,}976.95 \end{aligned}$$

$$\begin{aligned} df_{\text{within}} &= n_{\text{total}} - k \\ &= 20 - 4 \\ &= 16 \end{aligned}$$

$$\begin{aligned} df_{\text{between}} &= k - 1 \\ &= 4 - 1 \\ &= 3 \end{aligned}$$

$$\begin{aligned} MS_{\text{within}} &= \frac{SS_{\text{within}}}{df_{\text{within}}} \\ &= \frac{525{,}038}{16} \\ &= 32{,}814.88 \end{aligned}$$

$$MS_{\text{between}} = \frac{SS_{\text{between}}}{df_{\text{between}}} = \frac{673{,}976.95}{3} = 224{,}658.98$$

b.

$$F_{\text{observed}} = \frac{MS_{\text{between}}}{MS_{\text{within}}} = \frac{224{,}658.98}{32{,}814.8} = 6.846$$

c. Since the F_{observed} value of 6.846 exceeds the F_{critical} value of 3.24, we know with 95 per cent certainty that at least one of the brands differs significantly from the others in terms of durability. There is enough evidence to reject the null hypothesis.

5.

	Kingsway		City Centre		Southgate	
	X	X^2	X	X^2	X	X^2
1	45	2025	32	1024	51	2601
2	32	1024	47	2209	55	3025
3	44	1936	55	3025	31	961
4	43	1849	41	1681	30	900
Sum	**164**	**6834**	**175**	**7939**	**167**	**7487**
$\bar{X}$	**41**		**43.75**		**41.75**	
$n = 12$						
Grand mean = 42.17						

$$\begin{aligned} SS_{\text{total}} &= \sum X^2_{\text{total}} - n_{\text{total}}\bar{x}^2_{\text{total}} \\ &= (6834 + 7939 + 7487) - 12(42.17)^2 \\ &= 22{,}260 - 12(1778.31) \\ &= 22{,}260 - 21{,}339.72 \\ &= 920.28 \end{aligned}$$

$$\begin{aligned} SS_{\text{within}} &= \sum X^2_{\text{total}} - \sum n_{\text{group}}\bar{x}^2_{\text{group}} \\ &= 22{,}260 - \left[4(41)^2 + 4(43.75)^2 + 4(41.75)^2\right] \\ &= 22{,}260 - [6724 + 7656.25 + 6972.25] \\ &= 22{,}260 - 21{,}352.5 \\ &= 907.5 \end{aligned}$$

$$SS_{\text{between}} = SS_{\text{total}} - SS_{\text{within}}$$
$$= 920.28 - 907.5$$
$$= 12.78$$

$$df_{\text{within}} = n_{\text{total}} - k$$
$$= 12 - 3$$
$$= 9$$

$$df_{\text{between}} = k - 1$$
$$= 3 - 1$$
$$= 2$$

$$MS_{\text{within}} = \frac{SS_{\text{within}}}{df_{\text{within}}}$$
$$= \frac{907.5}{9}$$
$$= 100.84$$

$$MS_{\text{between}} = \frac{SS_{\text{between}}}{df_{\text{between}}}$$
$$= \frac{12.78}{2}$$
$$= 6.40$$

$$F_{\text{observed}} = \frac{MS_{\text{between}}}{MS_{\text{within}}}$$
$$= \frac{6.40}{100.84}$$
$$= 0.06$$
$$F_{\text{critical}} = 4.26$$

There is no significant difference between the respective malls in terms of visitors because the value of F_{observed} does not exceed the F_{critical} value.

Chapter 16

1. a. $\bar{x} = 2.33$
   $\bar{y} = 771.42$

b and c

X	$X-\bar{X}$	Y	$Y-\bar{Y}$	$(X-\bar{X})(Y-\bar{Y})$	$(X-\bar{X})^2$	$(Y-\bar{Y})^2$
3	0.67	890	118.58	79.45	0.45	14,061.22
2	−0.33	568	−203.42	67.13	0.11	41,379.70
3	0.67	860	88.58	59.35	0.45	7846.42
1	−1.33	625	−146.42	194.74	1.77	21,438.82
1	−1.33	775	3.58	−4.76	1.77	12.82
3	0.67	900	128.58	86.15	0.45	16,532.82
3	0.67	1095	323.58	216.8	0.45	104,704.02
3	0.67	800	28.58	19.15	0.45	816.82
2	−0.33	765	−6.42	2.12	0.11	41.22
3	0.67	629	−142.42	−95.42	0.45	20,283.46
1	−1.33	600	−171.42	227.99	1.77	29,384.82
3	0.67	750	−21.42	−14.35	0.45	458.82
			Σ	838.35	8.68	256,960.96

$$SP = 838.35$$
$$SS_X = 8.68$$
$$SS_Y = 256{,}960.96$$

d. $$\begin{aligned} b &= \frac{SP}{SS_X} \\ &= \frac{838.35}{8.68} \\ &= 96.58 \end{aligned}$$

e. $$\begin{aligned} a &= \bar{y} - b\bar{x} \\ &= 771.42 - (96.58)(2.33) \\ &= 771.42 - 225.03 \\ &= 546.39 \end{aligned}$$

f. $$\begin{aligned} r &= \frac{SP}{\sqrt{SS_X SS_Y}} \\ &= \frac{838.35}{\sqrt{(8.68)(256{,}960.96)}} \\ &= \frac{838.35}{\sqrt{2{,}230{,}421.13}} \\ &= \frac{838.35}{1493.46} \\ &= 0.56 \end{aligned}$$

g. $r_2 = 0.56 * 0.56$
$= 0.31$

You would increase the accuracy of your guess by 31 per cent.

h. and i. Having found values for *a* and *b*, substitute into the following equation to find values for $\hat{Y}$:

$$\hat{Y} = a + bX$$
$$\hat{Y} = 546.39 + 96.58X$$

Number of rooms (X)	Monthly rent (Y)	$\hat{Y}$	$Y - \hat{Y}$	$(Y - \hat{Y})^2$
3	890	836.13	53.87	2901.98
2	568	739.55	−171.55	29,429.40
3	860	836.13	23.87	569.78
1	625	642.97	−17.97	322.92
1	775	642.97	132.03	17,431.92
3	900	836.13	63.87	4079.38
3	1095	836.13	258.87	67,013.68
3	800	836.13	−36.13	1305.38
2	765	739.55	25.45	647.70
3	629	836.13	−207.13	42,902.84
1	600	642.97	−42.97	1846.42
3	750	836.13	−86.13	7418.38
			Σ	175,869.78

j. $$se_b = \frac{\sqrt{\frac{(Y - \hat{Y})^2}{n-2}}}{\sqrt{(X - \hat{X})^2}}$$

$$= \frac{\sqrt{\frac{175{,}869.78}{12-2}}}{\sqrt{8.68}}$$

$$= \frac{132.6159}{2.9462}$$

$$= 45.01$$

$$t_{\text{observed}} = \frac{b}{se_b}$$

$$= \frac{96.58}{45.01}$$

$$= 2.15$$

$$\begin{aligned} df &= n - 2 \\ &= 12 - 2 \\ &= 10 \\ t_{critical} &= 2.228 \end{aligned}$$

With a confidence interval set at 95 per cent, $t_{observed} < t_{critical}$, the coefficient is not statistically significant, so there is insufficient evidence to reject the null that there is no relationship between the two variables.

2. a. $\bar{x} = 2.33$

$\bar{y} = 139.56$

# of people (X)	X − X̄	Bills (Y)	Y − Ȳ	(X − X̄)(Y − Ȳ)	(X − X̄)²	(Y − Ȳ)²
3	0.67	175	35.44	23.74	0.45	1256.00
2	−0.33	130	−9.56	3.15	0.11	91.39
3	0.67	231	91.44	61.26	0.45	8361.27
3	0.67	278	138.44	92.75	0.45	19,165.63
2	−0.33	40	−99.56	32.85	0.11	9912.19
3	0.67	205.83	66.27	44.40	0.45	4391.71
1	−1.33	0	−139.56	185.61	1.77	19,476.99
1	−1.33	38.41	−101.15	134.53	1.77	10,231.32
1	−1.33	41.23	−98.33	130.78	1.77	9668.79
3	0.67	44.2	−95.36	−63.89	0.45	9093.53
3	0.67	176	36.44	24.41	0.45	1327.87
3	0.67	315	175.44	117.54	0.45	30,779.20
			Σ	787.13	8.68	123,755.90

$$\begin{aligned} SP &= 787.13 \\ SS_X &= 8.68 \\ SS_Y &= 123{,}755.90 \end{aligned}$$

$$\begin{aligned} b &= \frac{SP}{SS_X} \\ &= \frac{787.13}{8.68} \\ &= 90.68 \end{aligned}$$

$$\begin{aligned} a &= \bar{y} - b\bar{x} \\ &= 139.56 - (90.68)(2.33) \\ &= 139.56 - 211.28 \\ &= 71.72 \end{aligned}$$

$$r = \frac{SP}{\sqrt{SS_X SS_Y}}$$
$$= \frac{787.13}{\sqrt{(8.68)(123{,}755.90)}}$$
$$= \frac{787.13}{1{,}074{,}201.21}$$
$$= \frac{787.13}{1036.44}$$
$$= 0.76$$

$$r_2 = 0.76 * 0.76$$
$$= 0.31$$

b. Standard deviation for $X = 0.89$.

Standard deviation for $Y = 106.07$.

c. The slope coefficient is 90.83.

d. The estimated monthly charge for a household of four people is $\hat{Y} = -71.72 + 90.83 * 4 = 291$.

e. The standardized value of $b = 0.76$.

$$b_{1*} = b_1\left(\frac{s_1}{s_Y}\right)$$
$$= 90.83\left(\frac{0.89}{106.07}\right)$$
$$= 0.76$$

Chapter 17

1. a. 0.0101
   b. 0.0526
   c. 0.1111
   d. 0.25
   e. 1
   f. 2.003
   g. 3
   h. 7.9286
   i. 999

2. a. 0.3286
   b. 1.2498
   c. 16.119
   d. 1
   e. 0.0973
   f. 0.0004

Chapter 18

1. The mean and standard deviation for the x-variables are 9.00 and 3.32, and for the y-variables they are 7.5 and 2.03, respectively.
2. This question is a little tricky, because there are only two independent variables, X and $X4$. The r value for X on $X4$ is −0.5, and the r value for $X4$ on X is the same, −0.5. So, we will have the same variance inflation factor (VIF) each time.

$$\text{VIF}_X = \frac{1}{1-R_X^2} = \frac{1}{1-0.25} = 1.333 \qquad \text{VIF}_{X4} = \frac{1}{1-R_{X4}^2} = \frac{1}{1-0.25} = 1.333$$

Solution Key for Boxes

Box 7.1: It's Your Turn: Determining the Proportion of Observations at Various Standard Deviation Cut-Points

1. 2.5%
2. 0.5%
3. 50%
4. 16%
5. 97.5%

Box 7.3: It's Your Turn: Converting Standard Scores to Percentile Ranks

1. Range of values ±1 standard deviation from the mean.

   We know that z will equal ±1; therefore, we solve the standard score equation for X.

   For the lower bound:

$$z = \frac{X-\mu}{\sigma}$$

$$-1 = \frac{X-5.7}{5.1}$$

$$5.1 * (-1) = X - 5.7$$

$$-5.1 + 5.7 = X$$

$$0.6 = X$$

For the upper bound:

$$z = \frac{X - \mu}{\sigma}$$

$$1 = \frac{X - 5.7}{5.1}$$

$$5.1 * 1 = X - 5.7$$

$$5.1 + 5.7 = X$$

$$10.8 = X$$

Therefore 68 per cent of Canadians between ages 18 and 29 go out to a restaurant, movie, or theatre approximately 1 (0.6) to 11 (10.8) evenings per month.

2. Value that the lowest 10 per cent of all observations fall below:

   Because we are looking for a value that extends towards the tail (not from the value to the mean), we look in column C (Appendix A) for 0.10 (or column B for 0.40). The z-score value is −1.28. (Remember to change from an absolute value to the end of the distribution you are focused on.)

$$z = \frac{X - \mu}{\sigma}$$

$$-1.28 = \frac{X - 5.7}{5.1}$$

$$5.1 * (-1.28) = X - 5.7$$

$$-6.528 + 5.7 = X$$

$$-0.828 = X$$

   Since a person cannot go out less than zero times a month, we can say that the 10 per cent of people who go out least frequently go out about 0 times per month.

3. Value that the highest 40 per cent of all observations are above:

   We are looking for a value that extends toward the tail (not from the value to the mean). So we will look in column C (Appendix A) for 0.40 (or column B for 0.10). The z-score value is 0.25:

$$z = \frac{X - \bar{x}}{s}$$

$$0.25 = \frac{X - 5.7}{5.1}$$

$$5.1 * (+0.25) = X - 5.7$$

$$1.275 + 5.7 = X$$

$$6.975 = X$$

We can say we expect that the 40 per cent of people who go out the most frequently go out at least seven times per month.

4. The percentage of cases that fall between the values of 4 and 9:

   We will need to calculate the z-score for both the upper and lower limits that we have:

   Lower limit:

$$z = \frac{X - \bar{x}}{s}$$
$$= \frac{4 - 5.7}{5.1}$$
$$= \frac{-1.7}{5.1}$$
$$= -0.333$$

   The z-score of –0.333 is equal to 0.129 or 12.9 per cent. Thus 12.9 per cent of people fall between the mean and going out 4 nights per month.

   Upper limit:

$$z = \frac{X - \bar{x}}{s}$$
$$= \frac{9 - 5.7}{5.1}$$
$$= \frac{3.3}{5.1}$$
$$= 0.647$$

   The z-score of 0.647 is equal to 0.242 or 24.2 per cent. Thus 24.2 per cent of people fall between the mean and going out 9 nights per month.

   Adding the percentage between the mean and 4 and between the mean and 7 will give the total percentage of people in this range. 12.9 + 24.2 = 37.1 per cent of people between the ages of 18 and 29 go out between about 4 and 9 nights per month to a restaurant, movie, or theatre.

5. Value that 75 per cent of all observations fall below:

   This time we are looking at a value above 50 per cent, or more than half of the distribution. Therefore our value must be above the mean. Since we know that 50 per cent covers from the mean to the lowest value, we know that 25 per cent will fall above the mean (75 per cent – 50 per cent). Therefore, we are looking for the z-score at which 25 per cent of cases are between it and the mean.

   We are looking for a value toward the mean (column B) for 0.25. The z-score value is 0.67.

$$z = \frac{X - \bar{x}}{s}$$

$$0.67 = \frac{X - 5.7}{5.1}$$

$$5.1 * (+0.67) = X - 5.7$$

$$3.417 + 5.7 = X$$

$$9.12 = X$$

We can expect that 75 per cent of Canadians aged 18 to 29 go out 9 times or less per month.

Box 10.3: It's Your Turn: *t*-Test for the Same Sample Measured Twice

Individual	# of Partners per year at age 18	# of partners per year at Age 21	$d_i = x_{i1} - x_{i2}$	d^2
1	2	1	1	1
2	0	0	0	0
3	3	2	1	1
4	1	2	−1	1
5	8	1	7	49
6	1	1	0	0
7	2	1	1	1
8	0	2	−2	4
9	0	4	−4	16
10	3	1	2	4
	$\bar{X}_1 = 2.00$	$\bar{X}_2 = 1.50$	$\Sigma d_i = 5$	$\Sigma d^2 = 77$

1. See table for solutions.
2. See table for solutions.
3. $s_d = \sqrt{\frac{\sum d_i^2}{n-1} - (\bar{x}_1 - \bar{x}_2)^2}$

$$= \sqrt{\frac{77}{9} - (2.0 - 1.5)^2}$$

$$= \sqrt{8.56 - 0.25}$$

$$= 2.88$$

4. $$s_{\bar{d}} = \frac{s_d}{\sqrt{n-1}}$$
$$= \frac{2.88}{\sqrt{9}}$$
$$= 0.96$$

5. $$t = \frac{\bar{x}_1 - \bar{x}_2}{s_{\bar{d}}}$$
$$= \frac{2.0 - 1.5}{0.96}$$
$$= 0.52$$

6. There is no significant difference between the number of partners per year at the two ages, because t_{observed} does not exceed the t_{critical} value of 2.262.

Box 11.1: It's Your Turn: The Two-Sample *t*-Test

1. There will be no difference in feelings of attachment between people who are second generation Canadians versus third generation Canadians.
2. Sophia has asked a question that requires a two-sample *t*-test. (Actually, a *z*-test is what's asked for, but given the large sample size, you could do either.)

   First, calculate the standard error of the difference between means:

$$s_{\bar{x}_1-\bar{x}_2} = \sqrt{\left(\frac{n_1 s_1^2 + n_2 s_2^2}{n_1 + n_2 - 2}\right)\left(\frac{n_1 + n_2}{n_1 n_2}\right)}$$
$$= \sqrt{\left(\frac{6799 * 0.776^2 + 23{,}237 * 0.593^2}{6799 + 23{,}237 - 2}\right)\left(\frac{6799 + 23{,}237}{6799 * 23{,}237}\right)}$$
$$= \sqrt{\left(\frac{12{,}265.46}{30{,}034}\right)\left(\frac{30{,}036}{157{,}988{,}363}\right)}$$
$$= \sqrt{0.408 * 0.00019}$$
$$= 0.0088$$

   Then, calculate *t*:

$$t = \frac{\bar{x}_1 - \bar{x}_2}{s_{\bar{x}_1-\bar{x}_2}}$$
$$= \frac{4.59 - 4.78}{0.0088}$$
$$= -21.59$$

The answer may also be –21.56 if no rounding takes place in the previous steps.

3. You would reject the null hypothesis.

Box 11.2: It's Your Turn: The Two-Sample Proportion

1. Calculate P_u the estimate of the proportion of the population in the category of interest (the proportion of somewhat spiritual men and women who use prayer) by using the equation:

$$\begin{aligned} P_u &= \frac{n_1 p_{s_1} - n_2 p_{s_2}}{n_1 + n_2} \\ &= \frac{114{,}734 * 0.381 + 143{,}013 * 0.516}{114{,}734 + 143{,}013} \\ &= \frac{43{,}713.65 + 73{,}794.71}{257{,}747} \\ &= 0.456 \end{aligned}$$

2. Use this value to calculate the standard deviation of the difference between sample proportions:

$$\begin{aligned} s_{p_1 - p_2} &= \sqrt{P_u * (1 - P_u) \frac{n_1 + n_2}{n_1 n_2}} \\ &= \sqrt{0.456(1 - 0.456) \frac{114{,}734 + 143{,}013}{114{,}734 * 143{,}013}} \\ &= \sqrt{0.248 \frac{257{,}747}{16{,}408{,}453{,}542}} \\ &= \sqrt{0.248 * 0.0000157} \\ &= 0.002 \end{aligned}$$

3. Calculate the value of z_{obtained} using:

$$\begin{aligned} z_{\text{obtained}} &= \frac{p_{s_1} - p_{s_2}}{s_{p_1 - p_2}} \\ &= \frac{0.381 - 0.516}{0.002} \\ &= -67.5 \end{aligned}$$

4. As −67.5 is much larger than the one-tailed critical value of −1.65, we can be 95 per cent confident that Aboriginal women who consider themselves very, somewhat, or not very spiritual are more likely to use prayer than Aboriginal men who consider themselves very, somewhat, or not very spiritual to maintain their religion or spirituality.

Box 12.3: It's Your Turn: Phi—Drinking and Daily Exercise

1.

	Regularly Has More than 12 Drinks a Week		
Exercises daily	Yes	No	Total
Yes	4	2	6
No	5	9	14
Total	9	11	20

2. Upper left:

$$f_e = \frac{\sum \text{column} * \sum \text{row}}{n}$$
$$= \frac{9 * 6}{20}$$
$$= \frac{54}{20}$$
$$= 2.7$$

Upper right:

$$f_e = \frac{\sum \text{column} * \sum \text{row}}{n}$$
$$= \frac{11 * 6}{20}$$
$$= \frac{66}{20}$$
$$= 3.3$$

Lower left:

$$f_e = \frac{\sum \text{column} * \sum \text{row}}{n}$$
$$= \frac{9 * 14}{20}$$

$$= \frac{126}{20}$$
$$= 6.3$$

Lower right:

$$f_e = \frac{\sum \text{column} * \sum \text{row}}{n}$$
$$= \frac{11 * 14}{20}$$
$$= \frac{154}{20}$$
$$= 7.7$$

	Regularly Has More than 12 Drinks a Week		
Exercises daily	Yes	No	Total
Yes	4 (2.7)	2 (3.3)	6
No	5 (6.3)	9 (7.7)	14
Total	9	11	20

3.

Group	f_o	f_e	$(f_o - f_e)^2$	$\frac{(f_o - f_e)^2}{f_e}$
YY	4	2.7	1.69	0.63
YN	2	3.3	1.69	0.51
NY	5	6.3	1.69	0.27
NN	9	7.7	1.69	0.22
Total	20	20		$\chi^2 = 1.63$

4. $\phi = \sqrt{\frac{\chi^2}{n}}$

$$= \sqrt{\frac{1.63}{20}}$$
$$= 0.285$$

There is a moderate association between daily exercise and regularly drinking 12 or more drinks a week.

Box 12.5: It's Your Turn: Cramer's V

1. We are given the following information:

Work status in 2000	Marital status			
	Married	Divorced/ separated	Single	Total
Worked mainly full-time weeks	3,485,748 (3,321,615.99)	604,011 (525,916.09)	1,193,175 (1,435,401.92)	5,282,934
Worked mainly part-time weeks	1,291,201 (1,455,333.0)	152,330 (230,424.9)	871,134 (628,907.1)	2,314,665
Total	4,776,949	756,341	2,064,309	7,597,599

In order to calculate chi-square, create a table comparing the variables:

Group	f_o	f_e	$(f_o - f_e)^2$	$\frac{(f_o - f_e)^2}{f_e}$
FM	3,485,748	3,321,615.99	164,133.00	8,110.40
FD	604,011	525,916.09	78,094.91	11,596.58
FS	1,193,175	1,435,401.92	−242,226.92	40,875.299
PM	1,291,201	1,455,333.00	−164,132.00	18,510.76
PD	152,330	230,424.90	−78,094.90	26,467.68
PS	871,134	628,907.10	242,226.90	93,294.97
Total	7,597,599	7,597,599.00		198,856.39

Chi-square = 198,856.

2. To calculate Cramer's V:

$$V = \sqrt{\frac{\chi^2}{(n)(min(\text{row} - 1) \mid (\text{column} - 1)}}$$

$$= \sqrt{\frac{198{,}856.39}{7{,}597{,}559(2-1)}}$$

$$= 0.162$$

Box 12.8: It's Your Turn: Lambda

Wears all protective equipment for in-line skating	Female	Male	Row total
Yes	9	2	11
No	42	47	89
Column total	51	49	100

1. Make the two extreme predictions for the dependent variable.

   Wears protective equipment: 100 – 11 = 89 misclassifications

   Does not wear protective equipment: 100 – 89 = 11 misclassifications

   Based on smaller value $E_1 = 11$.

2. Calculate the extreme predictions by using the independent variable; determine E_2.

   Predicting that a female wears protective equipment: 51 – 9 = 42

   Predicting that a female does not wear protective equipment: 51 – 42 = 9

   Predicting that a male wears protective equipment: 49 – 2 = 47

   Predicting that a male does not wear protective equipment: 49 – 47 = 2

   As lowest classification errors for each sex are for not wearing protective equipment, we sum 9 (females) and 2 (males). Thus $E_2 = 11$.

3. Calculate lambda (or the percentage increase in predictive accuracy):

$$\lambda = \frac{E_1 - E_2}{E_1}$$
$$= \frac{11 - 11}{11}$$
$$= 0$$

Thus, knowing if a person is male or female will not improve our ability to predict whether that person will wear protective gear while in-line skating.

Box 13.2: It's Your Turn: Calculating Gamma

Walk Alone * Quick Justice					
		Courts do good job of quick justice			
		Good	Average	Poor	Total
Walk alone at night	At least once a week	40	77	86	203
	Up to once a month	18	61	51	130
	Never	16	26	38	80
	Total	74	164	175	413

1. Compute N_{same} and $N_{different}$ (use tables):

Cell	# of concordant cells	# of concordant observations	Contribution to N_s
a	4 (e, f, h, i)	61 + 51 + 26 + 38 = 176	40 * 176 = 7040
b	2 (f, i)	51 + 38 = 89	77 * 89 = 6853
c	0		

Cell	# of concordant cells	# of concordant observations	Contribution to N_s
d	2 (h, i)	26 + 38 = 64	18 * 64 = 1152
e	1 (i)	38	61 * 38 = 2318
f	0		
g	0		
h	0		
i	0		N_{same} = 17,363
a	0		
b	2 (d, g)	18 + 16 = 34	77 * 34 = 2618
c	4 (d, e, g, h)	18 + 61 + 16 + 26 = 121	86 * 121 = 10,406
d	0		
e	1 (g)	16	61 * 16 = 976
f	2 (g, h)	16 + 26 = 42	51 * 42 = 2142
g	0		
h	0		
i	0		$N_{different}$= 16,142

2.

$$\gamma = \frac{N_{same} - N_{different}}{N_{same} N_{different}}$$

$$= \frac{17{,}363 - 16{,}142}{17{,}363 + 16{,}142}$$

$$= \frac{1221}{33{,}505}$$

$$= 0.036$$

3. There is a weak relationship between people's views on the efficiency of the courts and walking alone in neighbourhoods after dark.

Box 13.4: It's Your Turn: Calculating Somers' *d*

The Relationship between Educational Status and Hours Spent on Unpaid Labour per Week

Hours on unpaid household labour/week	Educational status			
	Not studying	Part-time student	Full-time student	Total
Fewer than 5	52 (a)	1 (b)	20 (c)	73
5 to 14	59 (d)	2 (e)	5 (f)	66
15 or more	106 (g)	2 (h)	3 (i)	111
Total	217	5	28	250

Source: 2001 Individual Census

1. Compute N_{same}, $N_{\text{different}}$, and Ties_y.

$$
\begin{aligned}
N_{\text{same}} &= a(e + f + h + i) + b(f + i) + d(h + i) + e(i) \\
&= 52(2 + 5 + 2 + 3) + 1(5 + 3) + 59(2 + 3) + 2(3) \\
&= 624 + 8 + 295 + 6 \\
&= 933
\end{aligned}
$$

$$
\begin{aligned}
N_{\text{different}} &= b(d + g) + c(d + e + g + h) + e(g) + f(g + h) \\
&= 1(59 + 106) + 20(59 + 2 + 106 + 2) + 2(106) + 5(106 + 2) \\
&= 165 + 3380 + 212 + 540 \\
&= 4297
\end{aligned}
$$

$$
\begin{aligned}
Ties_y &= a(b + c) + b(c) + d(e + f) + e(f) + g(h + i) + h(i) \\
&= 52(1 + 20) + 1(20) + 59(2 + 5) + 2(5) + 106(2 + 3) + 2(3) \\
&= 1092 + 20 + 413 + 10 + 530 + 6 \\
&= 2071
\end{aligned}
$$

2. Calculate Somers' *d*.

$$
\begin{aligned}
d &= \frac{N_{\text{same}} - N_{\text{different}}}{N_{\text{same}} + N_{\text{different}} + Ties_Y} \\
&= \frac{933 - 4297}{933 + 4297 + 2071} \\
&= \frac{-3364}{7301} \\
&= -0.461
\end{aligned}
$$

3. The value of −0.461 indicates that there is a strong relationship between being a student and how many hours you spend on housework. That it is negative means that as you increase the value on the independent variable (educational status going from not being a student to being a full-time student), you decrease the value on the dependent variable (less time spent on housework).

Box 13.7: It's Your Turn: Calculating Spearman's *rho*

1. Complete table:

Case	Number of drinks a week	Drinks rank	Grade	Grade rank	D	D^2
1	10	2	65	4	−2	4
2	2	5.5	75	2	3.5	12.25
3	2	5.5	52	7	−1.5	2.25
4	0	8	98	1	7	49
5	1	7	45	8	−1	1

Case	Number of drinks a week	Drinks rank	Grade	Grade rank	D	D^2
6	5	3	55	6	−3	9
7	20	1	70	3	−2	4
8	3	4	60	5	−1	1
					0	82.5

2. $$\rho_s = 1 - \frac{6 * \sum D^2}{N(N^2 - 1)}$$
$$= 1 - \frac{6 * 82.5}{8(64 - 1)}$$
$$= 1 - \frac{495}{504}$$
$$= 0.02$$

3. Squared *rho* = 0.0004
4. By knowing the number of drinks a student has, one could reduce Marianne's and Andy's errors of prediction by 0.04 per cent. As this is incredibly low, it doesn't seem that the amount he drinks explains why Andy is doing better than Marianne. Maybe Andy is studying more than he admits?

Box 13.9: It's Your Turn: Kendall's Tau-*b*

1. Compute the value for $Ties_x$.

$$Ties_x = 52(59 + 106) + 59(106) + 1(2 + 2) + 2(2) + 20(5 + 3) + 5(3)$$
$$= 8580 + 6254 + 4 + 4 + 160 + 15$$
$$= 15{,}017$$

2. Calculate Kendall's tau-*b*:

$$\text{tau-}b = \frac{N_{\text{same}} - N_{\text{different}}}{\sqrt{(N_{\text{same}} + N_{\text{different}} + Ties_Y)(N_{\text{same}} + N_{\text{different}} + Ties_X)}}$$
$$= \frac{933 - 4297}{(933 + 4297 + 2071)(933 + 4297 + 15{,}071)}$$
$$= \frac{-3364}{(7301)(20{,}247)}$$
$$= \frac{-3364}{12{,}158.3}$$
$$= -0.277$$

3. By including the ties on the independent variable, $Ties_X$, we can see that the relationship between studying and hours spent on housework is not as strong as it had appeared. There is certainly a moderate (and SPSS tells us significant) relationship, but its strength has decreased once we account for the observations tied on the independent variable.

Box 14.3: It's Your Turn: Calculating Pearson's *r*

1. The first step is to organize the provided information, as well as the sums and products, into a chart:

obs#	X	Y	X^2	Y^2	XY
1	38	2	1444	4	76
2	40	2	1600	4	80
3	60	2	3600	4	120
4	50	2	2500	4	100
5	60	2	3600	4	120
6	50	4	2500	16	200
7	35	1	1225	1	35
8	50	2	2500	4	100
9	36	2	1296	4	72
10	30	3	900	9	90
11	65	1	4225	1	65
12	45	1	2025	1	45
13	48	3	2304	9	144
14	40	2	1600	4	80
15	55	3	3025	9	165
Sum	702	32	34,344	78	1492

2. Next, we have our equation for r:

$$r = \frac{N\sum XY - \left(\sum X\right)\left(\sum Y\right)}{\left[N\sum X^2 - \left(\sum X\right)^2\right]\left[N\sum Y^2 - \left(\sum Y\right)^2\right]}$$

When we combine the two, we get:

$$\begin{aligned} r &= \frac{15 * 1492 - (702)(32)}{\sqrt{\left[15 * 34{,}344 - (702)^2\right]\left[15 * 78 - (32)^2\right]}} \\ &= \frac{-84}{\sqrt{22{,}356 * 146}} \\ &= -0.0465 \end{aligned}$$

3. There is a weak negative correlation between hours of work per week and involvement in organizations.

An Introduction to Statistics for Canadian Social Scientists

IBM SPSS Lab Manual

Contents

Preface

How Does This Manual Work?

This manual is intended to give you the opportunity to apply the concepts and principles you learned in each chapter of the text. Through practical examples and step-by-step instructions, this manual seeks to help you improve your understanding of statistical practices and introduce you to SPSS, a key statistical tool used in the social sciences. Often we learn through practice, and by actively engaging with the examples and assignments covered in this lab, your understanding of the course material and the basics of SPSS will be improved.

For nearly all of the chapters in your textbook, there is a corresponding lab exercise. The labs are approached in a workbook style where you will need to participate in each step in order to successfully complete the final assignment. You may find the labs challenging at times, even frustrating. Take a deep breath! Rome wasn't built in a day and neither is an expertise in statistics. To master a skill, one must practise, be challenged, and then try again. It is the intention of the lab manual to challenge you in this way. The knowledge is cumulative, so if you find a particular section tricky, return to earlier sections and brush up on the basics. You may find you need to do this several times before you are able to make sense of the material and the technical aspects of SPSS. Stay with it! By the end of this lab manual, you will have the ability to efficiently use SPSS and apply the concepts and principles taught in this course in a meaningful way.

The Model

Each lab contains three sections:

1. The first section will briefly reintroduce the topic covered by the corresponding chapter and set the learning objectives for the lab.
2. The second section will provide an example to help focus your thinking and help you to understand the key concepts to be covered. The examples are intended to illustrate real-world applications of the concepts.
3. The third section will include a lab assignment to test your understanding of the concepts and skills.

What's Covered?

This manual is meant to complement the textbook used in this course, not replace it. Through the use of concrete examples, each lab is intended to solidify your understanding of the general concepts and principles taught in each chapter. To help teach you how to analyze data, this lab manual will provide step-by-step instructions on how to use the SPSS analysis software.

The labs cover the following key topics:

- Lab 1: Introduction to SPSS
- Lab 2: Identifying Types of Variables: Levels of Measures
- Lab 3: Univariate Statistics
- Lab 4: Introduction to Probability
- Lab 5: The Normal Curve
- Lab 6: Measures of Central Tendency and Dispersion
- Lab 7: Standard Deviations, Standard Scores, and the Normal Distribution
- Lab 8: Sampling
- Lab 9: Hypothesis Testing: Testing the Significance of the Difference between Two Means
- Lab 10: Hypothesis Testing: One- and Two-Tailed Tests
- Lab 11: Bivariate Statistics for Nominal Data
- Lab 12: Bivariate Statistics for Ordinal Data
- Lab 13: Bivariate Statistics for Interval/Ratio Data
- Lab 14: Analysis of Variance
- Lab 15: OLS Regression: Modelling Continuous Outcomes

The Data Set

The 2013 Alberta Survey (AS) is the twenty-fourth annual provincial survey administered by the Population Research Laboratory (PRL) at the University of Alberta. This annual omnibus survey of households in the province of Alberta enables academic researchers, government departments, and non-profit organizations to explore a wide range of research topics in a structured research framework and environment. Sponsors' research questions are asked together with demographic questions in a telephone interview of Alberta households. In 2013, topics covered by the survey included China's role in Alberta's economy, peoples' interactions with financial institutions/debt management, life experiences and health, opinions about temporary foreign workers, opinions about public health and injuries, and alcohol use. Demographic characteristics that were measured include age, gender, marital status, highest level of education, household income, religion, place of birth, employment status, home ownership, and political party support.

The target population was all persons 18 years of age or older who, at the time of the survey, were living in Alberta and could be contacted by direct dialing. From this population, three samples were drawn to cover Alberta: Edmonton Metropolitan Area, Calgary Metropolitan Area, and the rest of the province. The final data set contains 1207 cases. We have created a version of this data set for use with this textbook, which can be found on the textbook website at www.oupcanada.com/Haan3e. The codebook for the data set can be found in the same place.

This data set is also available online as a public-use data file. You can find it on the Population Research Laboratory's website: www.prl.ualberta.ca/en/AlbertaSurvey.aspx. Click on **Alberta Survey**, then click on the link for Public Release Alberta Survey Data Set 2013. You will also be able to find some previous Alberta Surveys on that website. Note that the public-use data file includes variables and incomplete data that was removed for the purposes of clarity; when completing the labs, it will be easier to use the data file found on the textbook website.

Lab #1: Introduction to SPSS

The focus of this lab is to introduce you to IBM® SPSS® STATISTICS SOFTWARE ("SPSS"). To use SPSS to analyze data, you will need to become familiar with the technical components of this software package. This lab will help familiarize you with the SPSS software, including how to access data files, the various base components (i.e., the syntax, data editor, and output), how to define new variables, and how to enter data.

You should be able to find SPSS on many of the computer terminals at your university, but if you can't, your instructor for the course should be able to help you find SPSS.

LEARNING OBJECTIVES

The following lab is directed at helping you understand how to orient SPSS. Specifically, this lab assignment challenges you to clarify your understanding of:

1. The basic components of SPSS
2. How to define variables
3. How to enter data

Defining SPSS

What Is SPSS?

SPSS stands for Statistical Package for the Social Sciences and it allows you to analyze and describe data. SPSS is a statistical software package, one of the most commonly used among social scientists.

How Does It Work?

SPSS works by taking a series of commands, supplied by you, and applying them to a set of data, also supplied by you. SPSS will produce output displaying the results of the commands. The commands you supply SPSS will determine your output results. Within SPSS, commands are either given in the form of menu selections and by filling in dialogue boxes or by writing your own programming syntax. We'll look at both methods here.

The sequence of commands will occur in the following order:

- Enter your data into SPSS, or open already existing data (which is what we'll do here).
- Tell SPSS to apply commands to the data (menus and dialogue boxes).
- SPSS then produces the output.

How Do I Begin an SPSS Session?

To begin, you need to access a computer that contains the SPSS software. Many universities have the SPSS software on their public use computers within computer labs. Once you have found a computer with a copy of the SPSS software, you will need to click on the **Start Menu** (on the bottom left-hand side of your computer screen), find the SPSS program, and open it.

Once you have opened the SPSS program, a pop-up will appear (like the one in Figure 1.1) asking you what you would like to do. You will have a variety of options, including **Open an existing data source, Open another type of file, Run a tutorial,** or **Type in data.** For the purposes of this lab, you will select **Type in data** (we will look at the Alberta Survey later).

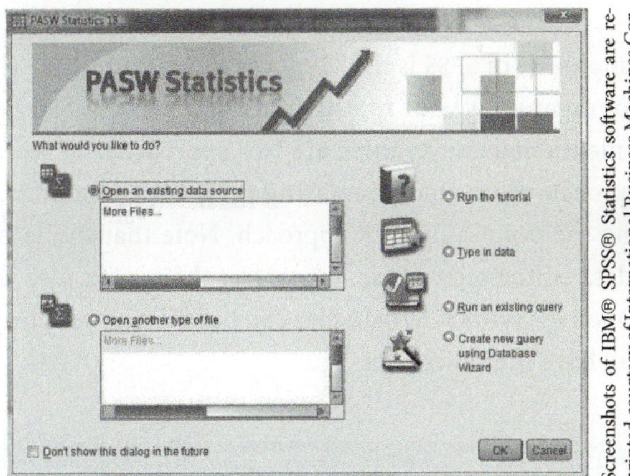

FIGURE 1.1 | SPSS Opening Screen

Once you click **Type in data**, the "Data Editor" screen will appear (as shown in Figure 1.2). The Data Editor window is like a spreadsheet. Within the Data Editor screen, you can both create and edit pre-existing data sets. The data editor window has two tabs on the bottom left-hand side of the window, one for viewing your data, called the "Data View" screen, the second one for viewing information about your variables, called the "Variable View" screen. Figure 1.2 is in Variable View mode. The current data editor screen has the title, Untitled. To save this file, select **File → Save as.** This will allow you to select both a name and location for your file. The saved document will have the suffix ".sav."

Let's take a closer look at the SPSS Data Editor screen. Along the top of the screen you will see the menu bar that is used to access all the commands available. The menu bar has a number of headings that divide the commands into categories of a similar function (**File, Edit, View, Data, Transform, Analyze, Directing Marketing, Graphs, Utilities, Add-ons, Window, Help**). Consider browsing through these folders so you can become familiar with their contents. The more familiar you are with what is contained under each heading, the more comfortable you will be with the SPSS software package.

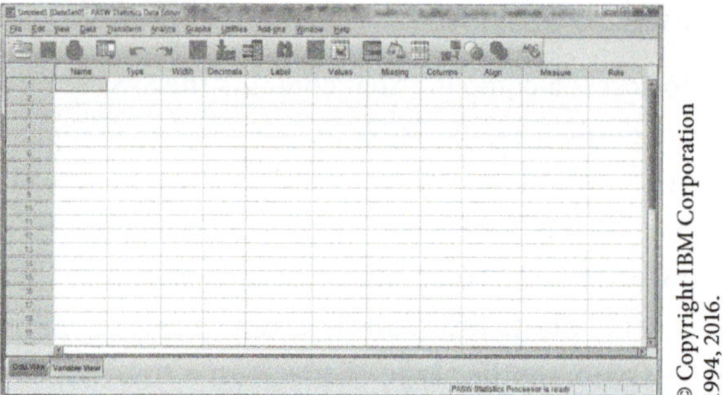

FIGURE 1.2 | Data Editor Window (variable view)

The second component of SPSS is the "Syntax File" (Figure 1.3). The SPSS Syntax File is a very useful tool for organizing your records and analyses. It is a text editor that reads SPSS programming. As mentioned above, there are two approaches to working in SPSS: using a point-and-click approach or manually inputting program commands. In this manual, you will mostly be using the point-and-click approach. Note that the same menu bar that appeared above the data editor screen also appears in the syntax file. To open a new syntax file, select **File → Open → Syntax**. Syntax files can be saved by selecting **File → Save as**. The saved document will have the suffix ".sps."

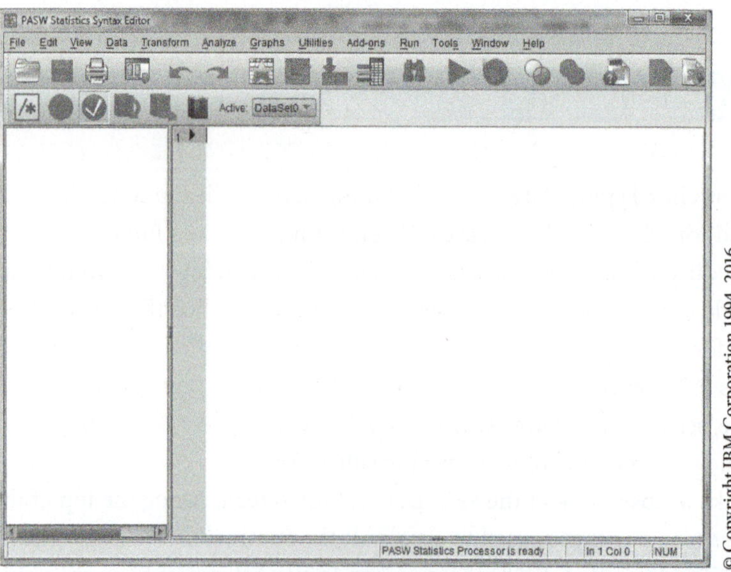

FIGURE 1.3 | The Syntax File

The third component of SPSS that is important to familiarize yourself with is the "Output Window" (Figure 1.4). The Output Window displays the output from the statistical analysis you undertake in the Data Editor window. You will notice a couple of things about this window. First, along the left side of the window is a running log of the commands you have executed during the current session. Second, similar to the Data Editor and Syntax windows, there is a Menu Bar with a series of headings, dividing the commands into categories of a similar function (**File, Edit, View, Data, Transform, Analyze, Graphs, Utilities, Window, Help**). Third, you will notice within the window itself the commands you ran within your syntax file. This is a particularly useful function when you are analyzing large amounts of data with a series of many commands. This can be done by selecting

Edit → Options → Viewer tab → Display commands in log.

To open a new output window, select

File → Open → Output.

The Output Window can be saved by selecting

File → Save as.

The saved document will have the suffix ".spo."

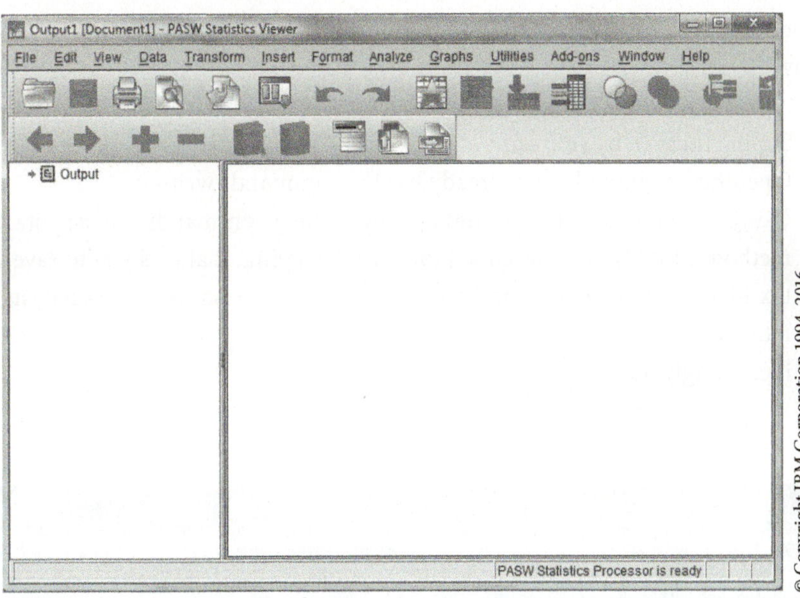

FIGURE 1.4 | Output Window

Entering SPSS Commands

There are two ways of entering commands into SPSS to execute your procedures: The first is known as the point-and-click procedure (Figure 1.5). This procedure can be done via the Data Editor window, the Output window, or the Syntax window by using the headings in the menu bar.

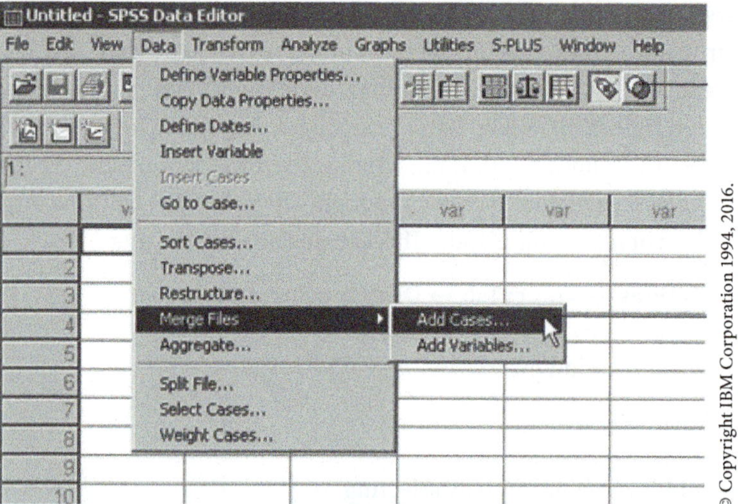

FIGURE 1.5 | The Point-and-Click Technique

The second way is to enter command programming into the syntax file. This can be done in three ways:

1. Typing them in by yourself.
2. Opening a syntax file that already has the commands written.
3. Using the pull-down menus and pasting the text commands into a syntax file. This method is highly recommended for record keeping. It allows you to save your syntax file (*.sps), so that it can be rerun later if you wish. As your analysis becomes more lengthy and complex, the importance of saving your syntax files will become increasingly obvious (Figure 1.6).

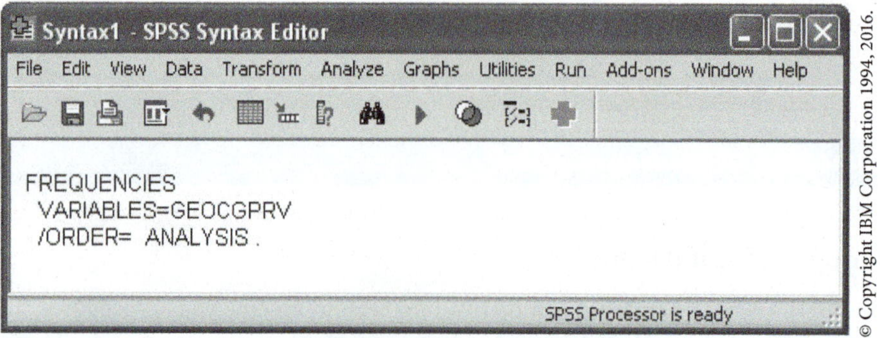

FIGURE 1.6 | Manually Inputting SPSS Commands in a Syntax File

Creating a Data File

Although most of the lab exercises in this manual use data from the Alberta Survey, it is important to understand data structure. By understanding how data files are constructed, you will be in a better position to understand what shape and form the variables within your data set take and how to modify them when you learn how to undertake more advanced analysis.

Imagine that you are interested in whether males or females are more likely to own a cat. You went out into the street and asked five random people two questions:

1. Are you a male or a female? (Male = 1 and Female = 2)
2. How many cats do you own? (0 to 5)

Person 1: Male (1) owns 2 cats
Person 2: Female (2) owns 1 cat
Person 3: Female (2) owns 0 cats
Person 4: Male (1) owns 5 cats
Person 5: Female (2) owns 1 cat

Now that you have your data, you can enter the information into an SPSS data file.

Step 1: Open a new data file.

Step 2: Switch your data file to Variable View from Data View, by clicking on the tab in the bottom left corner of your screen. In this view, notice that several new aspects of your data appear. It is here, for example, where you specify how many decimals your variables have. The rows are going to be your variables (sex and cats), the columns are the characteristics of the variables (name of the variable, type of variable, and so on). For this course, we are primarily concerned with the columns entitled Name, Label, Values, and Missing.

Step 3: Name your variables. You must first decide on the names of your variables; these names must not have spaces, and they must start with a letter. As good practice, it is useful to keep the number of characters to a minimum where possible. Often, a variable name will correspond with the survey question number from which it was derived, such as Q1 or Var01. However, this is not necessary. For our purposes, we will call our variables sex and cats.

Step 4: Define your variable labels. Although we have given our variables names that clearly indicate what the variable is measuring, your variable label can help clarify what the variable is measuring, or, if you have named the variable after survey questions (i.e., Q1), it will allow other users to know what information Q1 is measuring. For sex, we will put "Sex of respondent" and for cats, we will put "Number of cats owned."

Step 5: Define your value labels. This is done by clicking on the cell for the variable you want under the column "value labels." Often, value labels are attached with ordinal and nominal level variables (concepts you will learn about as you proceed through this course). Without labels, it is difficult to know what the values mean. So, for example, we have attached a value of "1" to our male respondents and a value of "2" to our female respondents. You must specify this under the value label column. To do this, click on the values cell for the variable you are creating. Click on the grey button in the cell. A dialogue box like the one found in Figure 1.7 will open. Following our example, assign labels male and female to numeric values "1" and "2," respectively, by entering "1" into the value box and its label, male, into the label box. Click **Add** to make the changes for each value of your variable, then click **OK** when you are finished.

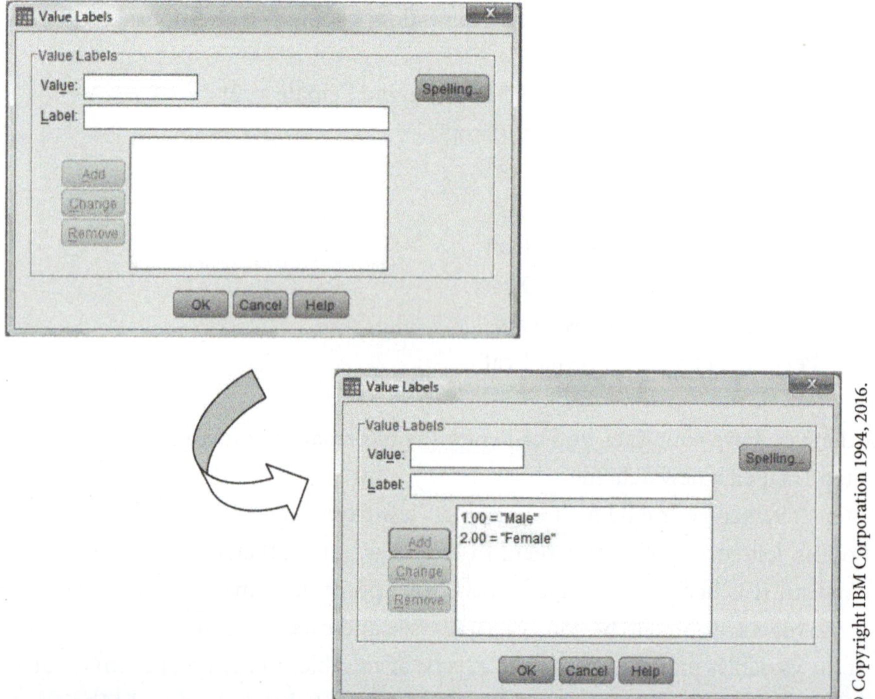

FIGURE 1.7 | Defining Value Labels

Step 6: Define your missing values. Sometimes when you are collecting data, respondents may not want to provide an answer for a variety of reasons. In such cases, you need to have a way of dealing with *missing data*. In our example, we don't have any such cases but if someone had refused to tell us what their sex was or how many cats they owned, we would have to assign a value to their non-response. While the decision how to code this data is up to the individual researcher creating the database, in the Alberta Survey, in most cases, a value of "0" is associated with "No Response," a 9 for "Not Applicable," and a value of "8" is associated with "Don't Know." To declare missing values, while still in "Variable View," click on the missing cell for the variable you are creating. Click on the grey button in the cell. A dialogue box like the one found in Figure 1.8 will open. You can assign up to three discrete missing values, such as 0, 8, and 9, as we find in the Alberta Survey 2013, or you can declare a range of values, such as 99 to 199. Click **OK** when you are finished.

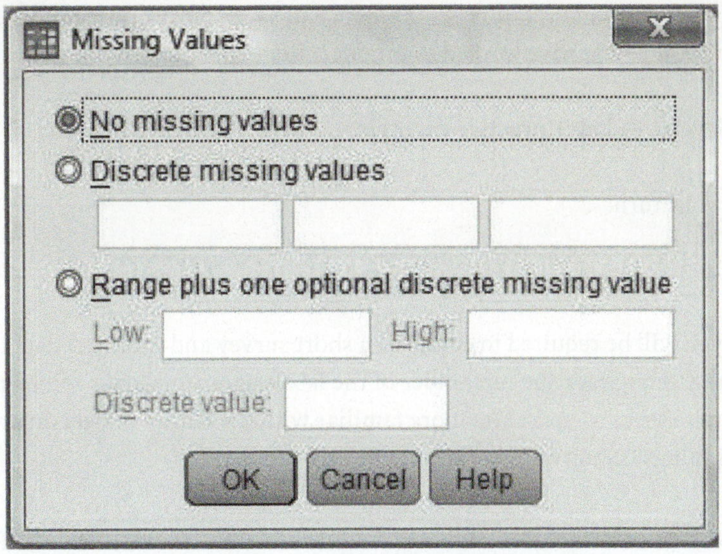

Missing Values

No missing values

Discrete missing values

Range plus one optional discrete missing value

Low: High:

Discrete value:

OK Cancel Help

FIGURE 1.8 | Declaring Missing Values

Now that you have finished creating your variables, you are ready to enter your data. Switch your data editor from Variable View to Data View. Select the first cell under the variable sex and enter the value for your first respondent, who, if you recall, is male and has two cats. Repeat this process for each of your respondents until your data editor screen looks like the one in Figure 1.9.

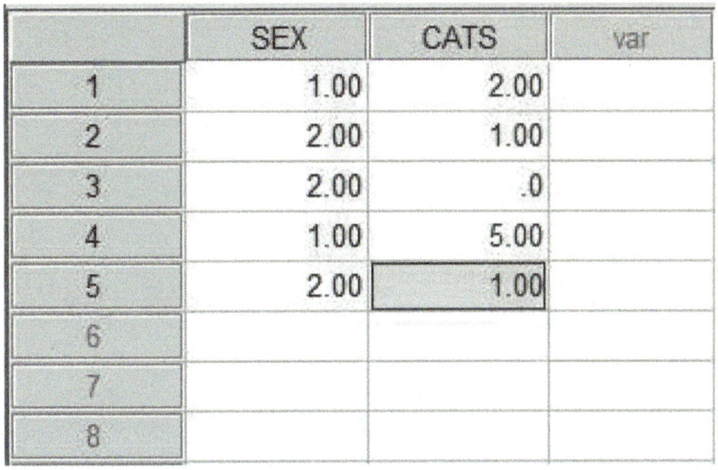

	SEX	CATS	var
1	1.00	2.00	
2	2.00	1.00	
3	2.00	.0	
4	1.00	5.00	
5	2.00	1.00	
6			
7			
8			

FIGURE 1.9 | Entering Data into the Data Editor Screen

Finally, save your new data set. It is good practice to save your work regularly so you don't lose information that you have worked to produce. Remember to save files:

File → Save as → Lab 1 practice example.

Now it's your turn!

Putting the Information into Practice

For this lab, you will be required to complete a short survey and input the data into the data editor screen. Although for the remainder of the lab assignments, data will be provided for you, this exercise seeks to make you more familiar with the nature of data organization and storage within the SPSS software package.

The following questions have been asked in some of the Alberta Surveys in the past.

Part 1: Take the time to answer each question as it relates to your life. Then, ask two friends to answer these questions. Pretend a fourth respondent refused to give you any answers to your survey. You should have a total of four respondents for your short survey.

1. What is your gender?
    1. Male
    2. Female

2. Do you presently have a paid job or are you self-employed?
    1. Yes, paid job
    2. Yes, self-employed
    3. Yes, paid job and self-employed
    4. No, neither

3. What is your *current* marital status?
    1. Never Married (Single)
    2. Married
    3. Common-Law Relationship/Live-In Partner
    4. Divorced
    5. Separated
    6. Widowed

4. Do you presently live in . . .
    1. A City
    2. A Town
    3. A Village
    4. A Rural Area

5. How safe do you feel from crime walking alone in your area after dark? Do you feel . . .
    1. Very safe
    2. Reasonably safe
    3. Somewhat unsafe
    4. Very unsafe

Part 2: Create a data set by using the skills you just learned. Remember to (1) name your variables; (2) define your variable labels; (3) define your value labels; and (4) declare any missing values (hint, this would be the person who refused to answer any of your survey questions). When you have finished creating your data set, which will include four respondents and five variables, save your file. Congratulations, you have just created your first data set. Now let's learn what to do next!

Lab #2: Identifying Types of Variables: Levels of Measures

The focus of this lab is to introduce you to the four different levels of variable measurement, to show you how to identify different types of variables within SPSS, and to help you understand how different levels of measurement are coded and organized within SPSS. This material corresponds with the material presented in Chapter 2.

LEARNING OBJECTIVES

The following lab is directed at helping you understand levels of measurement and how to identify different types of variables within a data set. Specifically, this lab assignment helps you clarify your understanding of:

1. Variable measurement
2. How variables are organized within SPSS

Understanding Levels of Measurement

Variables are measured at four different levels: nominal, ordinal, interval, and ratio. Each of these levels has unique characteristics that define them. For *nominal data*, numeric values are typically used for identification purposes. It is not possible to rank the response categories, and there is no quantifiable difference between categories. For *ordinal data*, the numeric values can signify an inherent ordering, because you can rank the response categories but you cannot measure the distance between those categories. For *interval data*, the data can be organized into an order that can be added or subtracted but not multiplied or divided, because there is no true zero value. *Ratio data* are similar to interval data except they have a true zero value.

There are two ways to collect information about the way in which a variable is measured in SPSS. One is to examine the variable values in the data editor screen and the other is to produce a frequency distribution. Using data from the Alberta Survey, this lab will demonstrate the first way to identify variable values. It will then give you a chance to practise identifying variables within SPSS. You will learn how to produce frequency distributions in the next chapter of this lab manual.

To open the Alberta Survey data, simply select **File → Open → Data**, then select the Alberta Survey 2013 (AS2013) (your instructor will need to tell you where he/she has placed the file). Or, if you are just opening SPSS, choose the "open existing data" option on the startup screen.

When you have opened the file, take a moment to scroll through the list of variables (be sure to confirm that your data editor screen is on Variable View). You will notice that there are 134 variables contained within this data set. If you change your screen to Data View and scroll to the bottom of the data file, you will see there are 1207 respondents. OK, now return to the Variable View screen. You might notice that the names of the variables don't make a lot of sense without a thorough understanding of the background of this survey. (If you want more information about the Alberta Survey 2013, it can be found on the Population Research Laboratory website. We suggest that you download the questionnaire so that you have it handy and can refer to it when you encounter new variables in the data set.) However, in most cases, the variable labels provide key insight into what each of those variables is measuring. For the most part, you will be able to make an informed guess as to what the level of measurement would be for most variables. However, this may not always be accurate.

For example, variable k12a has the label "Total Household income for the past year before taxes and deductions." Given that this is an income variable and that a household can feasibly have no income, you might be inclined to guess that this is a ratio level variable. However, when you click on the cell in the "values" column, you can see that the variable has been collapsed into 33 discrete categories (Figure 2.1). Therefore, this variable is actually an *ordinal* level variable because while you can rank the categories, the values associated with the discrete categories (i.e., 1 refers to under $6000) cannot be meaningfully measured.

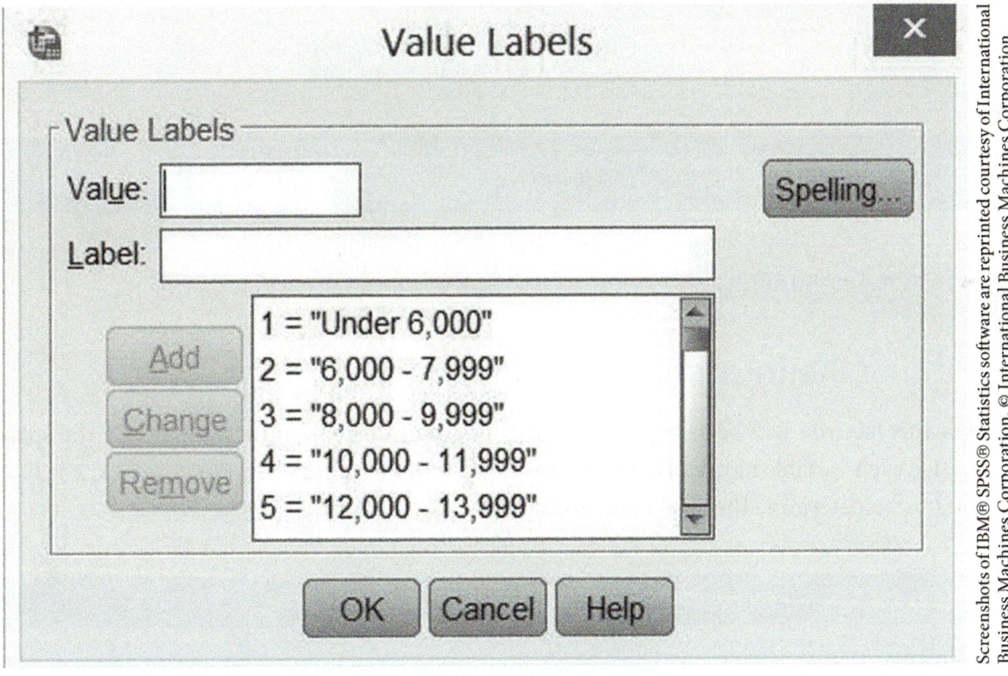

FIGURE 2.1 | Using Value Labels to Identify a Variable's Level of Measurement, Variable k12a

Let's look at one more example. Find the variable named strata with the label "Area of the province." Without going to the codebook found on the Population Research Laboratory website, we do not know how this question was to be answered by respondents. In SPSS, we can check to see how the variable is coded. To do this, again we want to check the values by clicking on the cell corresponding to this variable under the "Values" column.

As we can see in Figure 2.2, this is a nominal level variable and respondents were classified as living either in Edmonton, Calgary, or elsewhere in Alberta.

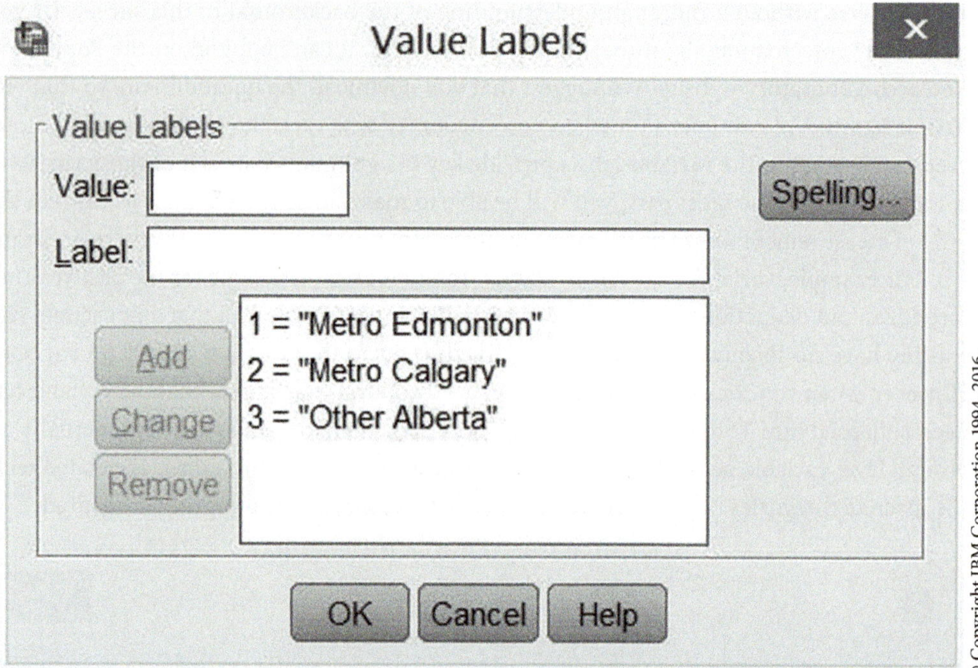

FIGURE 2.2 | Using Value Labels to Identify a Variable's Level of Measurement, Variable Strata

Now it's your turn!

Putting the Information into Practice

For this lab, you will be asked to fill in the missing information in Table 2.1. In the space beside each variable name, provide the variable label, the value labels (i.e., 1 = Yes, 2 = No), and try to determine the level of measurement.

TABLE 2.1

Variable	Variable label	Value label	Level of measurement
sex			
d1			
d9			
e11			
ft1			
ft8			
g6e			
k3a			
age			
agex			

Lab #3: Univariate Statistics

The focus of this lab is to begin to introduce you to analysis with one variable. Generating frequencies is a basic procedure used to obtain a summary of a variable by looking at the number of cases associated with each value of the variable. This material corresponds with the material presented in Chapter 3.

LEARNING OBJECTIVES

The following lab is directed at helping you understand ways of studying the characteristics of data. Specifically, this lab assignment challenges you to clarify your understanding of:

1. How to generate and interpret frequency distributions
2. Data presentation
3. The connection between data presentation and levels of measurement

Part 1: Producing Frequency Distributions

We will first learn to produce a frequency table. Throughout this lab manual, we will use the point-and-click technique through a syntax file. To obtain a frequency distribution, use the following steps.

Step 1: Open a new syntax file, by clicking on

File → New → Syntax file

Step 2: Click on:

Analyze → Descriptive statistics → Frequencies

Step 3: Within the dialogue box that opens, highlight the "Gender" variable (sex) in the left-hand screen; next click the arrow (this will move the variable into the right-hand screen); then click **Paste** (Figure 3.1).

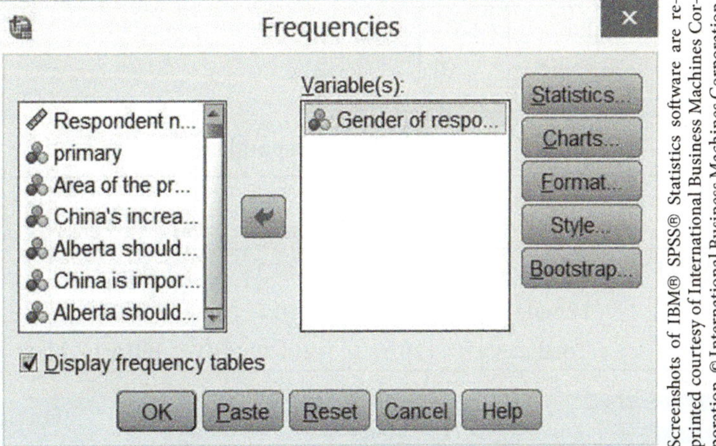

FIGURE 3.1 | Frequencies Dialogue Box

This will paste the SPSS program coding into your syntax file (Figure 3.2).

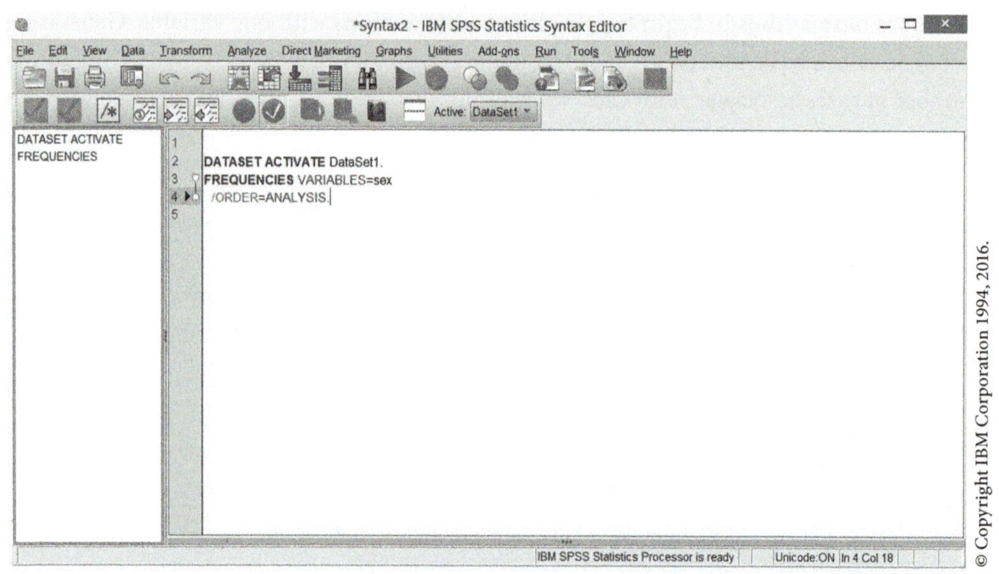

FIGURE 3.2 | Pasting Commands into a Syntax File

Step 4: In the Menu Bar, click on the green arrow to execute the commands. Or, you can put your cursor on the command you want to execute and press **Ctrl** and **R** at the same time. This will run the current analysis and bring it up in an SPSS output window.

This may be a good time to save your syntax file. It is good practice to save your work regularly.

The output from the frequencies procedure will contain these two tables:

Statistics

Gender of respondent

N	Valid	1207
	Missing	0

Gender of respondent

		Frequency	Percent	Valid Percent	Cumulative Percent
Valid	Male	595	49.3	49.3	49.3
	Female	612	50.7	50.7	100.0
	Total	1207	100.0	100.0	

FIGURE 3.3 | Frequency Distribution Output

What should we note about this output?

1. Under statistics, we can see how many cases are *valid* and how many are *missing*. We have 1207 valid cases and no missing cases.
2. In the table, we can see the distribution of "Males" versus "Females" by raw "Frequencies" and by "Percent," "Valid Percent," and "Cumulative Percent." Most often, you will refer to your Valid Percent. In situations where you have missing cases, there will be differences between your Percent and Valid Percent. The Valid Percent uses only those cases that are valid answers to the variable you are examining; the Percent column incorporates missing values when calculating the percentage breakdown. You can see that there are 595 males and 612 females. Is this the same as what you have on the screen before you?

Part 2: Types of Charts

A *pie chart* is a way of summarizing a set of categorical data. Data are categorical when the values or observations belonging to them can be sorted according to groups but not by values. For example, sex is a categorical variable with two categories, "Male" and "Female," and people cannot belong to both categories. We can then refer to sex as being mutually exclusive. A pie chart is a circle that is divided into segments, each of which represents a particular category. The area of each segment is proportional to the number of cases in that category.

Figure 3.4 is an example of a pie chart for a variable called strata, which measures the area of Alberta the respondent lives in. Note that in this example we appear to have an equal distribution of respondents from Edmonton, Calgary, and Other Alberta.

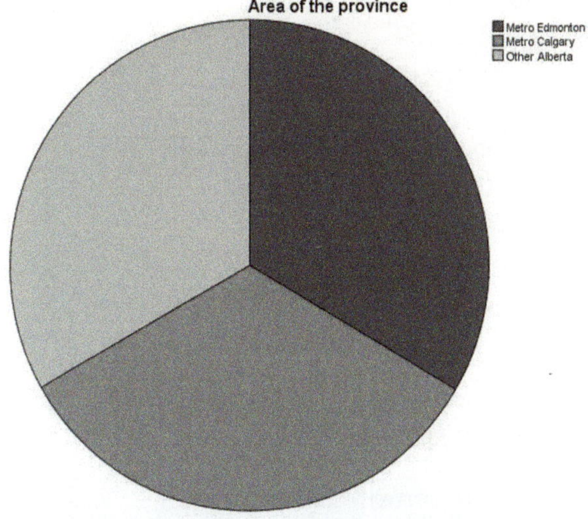

FIGURE 3.4 | A Pie Chart
Source: © Copyright IBM Corporation 1994, 2016.

How can you create a pie chart by using SPSS?

Step 1: At the top of your screen, click on:

Analyze → Descriptive statistics → Frequencies

Step 2: Within the dialogue box that opens, highlight the area of the province variable (strata) in the left-hand screen. Next, click the arrow (this will move the variable into the right-hand screen; see Figure 3.5).

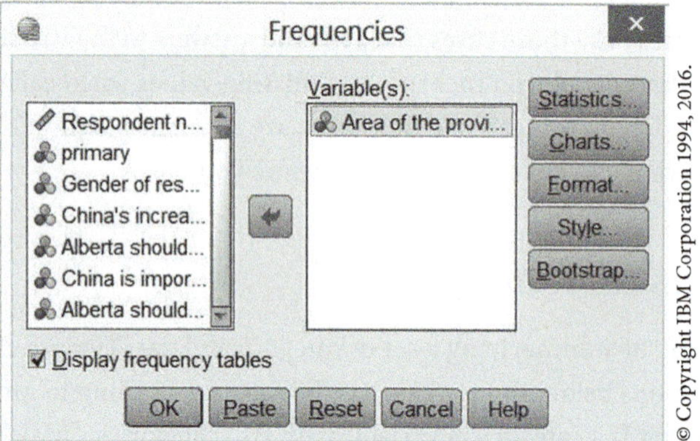

FIGURE 3.5 | Frequencies Dialogue Box

Step 3: Click on **Charts**. This will open a dialogue box that asks you which chart type you would like to present (Figure 3.6). Click on **Pie charts → Continue** (this will close the menu screen) **→ Paste**. The Paste command is optional but is highly recommended for record keeping purposes and for tracking your progress. However, if you decide not to keep a syntax record, you can simply click on **OK**. For the purposes of this lab manual, we will be pasting all commands into a syntax file.

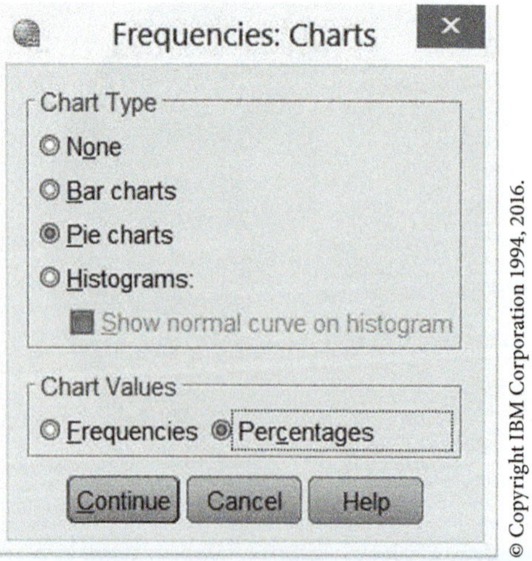

FIGURE 3.6 | Frequencies: Charts Dialogue Box

Now, after running the commands in your syntax file, you should get an output file containing a pie chart that resembles the one in Figure 3.4.

Bar Graphs

A *bar graph* is a way of summarizing a set of categorical data. It displays the data by using a number of rectangles of the same width, each of which represents a particular category. The length of each rectangle is proportional to the number of cases in the category it represents. Figure 3.7 is an example of a bar graph for the question "To the best of your knowledge, what is the leading cause of death for Albertans under the age of 45?" (g1).

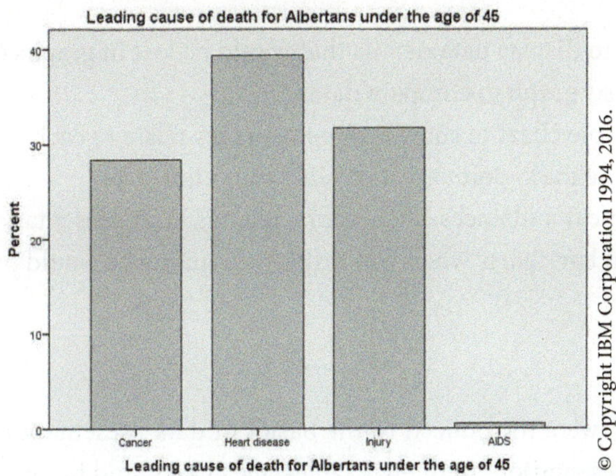

FIGURE 3.7 | A Bar Chart

Bar graphs are obtained in the same way as pie charts, except that where you selected "Pie charts" within the frequency procedure, you will now select "Bar charts" (Figure 3.8).

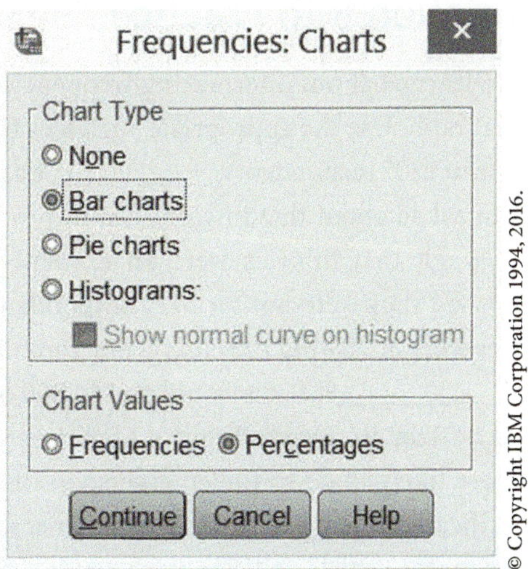

FIGURE 3.8 | Frequencies: Charts Dialogue Box

You will also notice that there's an option to produce a "Histogram." Since we'll be dealing with histograms later in the course, we'll skip over them for now. They are, however, a very important mode of presentation in statistics.

Which Mode of Presentation Is Best?

Deciding in which format to present your data depends on the level of measurement of your selected variable and the clarity of the presentation. For example, if you have a categorical variable, but a large number of categories, a pie chart may make the presentation crowded and confusing, so selecting a bar graph may be more appropriate.

There are some general suggestions you may want to consider:

- Use tables to display data details that would be lost in graphs or charts.
- Opt for a bar graph to compare data.
- Consider a pie chart to show how percentages relate to each other within a whole.
- Focus on the main point and consider your audience.
- Non-technical audiences often appreciate visual representations of data, so try to use pie and bar charts when you think your audience would prefer them.

Summary

In this section, you were introduced to the basics of data presentation. You explored three methods of data presentation: frequency tables, pie charts, and bar graphs. Specifically, you learned how to generate and interpret frequency distributions, how to create three types of modes of presentation, and the connection between data presentation and levels of measurement.

Now it's your turn!

Putting the Information into Practice

1. Using what you've learned about interpreting frequency tables, fill in the blanks of the following paragraph. Use the appropriate variables from the lesson above.
2. In AS2013's sample of 1207 respondents, __________% are males and __________% are females. When asked about the different health problems facing Albertans today, __________% felt that injuries were an extremely serious health problem and __________% felt they were not serious. Respondents were varied in guessing what their chances were of visiting the emergency room in the next year because of an injury. In fact, __________% felt it would be 1 in 500 compared to __________% who felt it would be 1 in 10. Interestingly, a higher percentage of people thought they would be more likely to go to the emergency room because of a motor vehicle collision. Specifically, __________% felt the chances of going to the emergency room because of a motor vehicle collision were one in 500.

3. Produce the specified graphs for the following variables:
   a. el1: Bar chart
   b. d1: Pie chart
   c. g3: Bar chart
   d. k5a: Frequency table

4. Choose one of the variables from Question 2 and provide an interpretation. Specifically, what is the percentage distribution of the categories? What is the total number of respondents who answered the question? How many, if any, missing values are there?

Lab #4: Introduction to Probability

The focus of this lab is to review what you have already learned and to introduce you to the concept of recoding variables. Recoding variables is an important component in conducting analysis because the categories of a variable, as they were asked in the questionnaire, may not work for your specific needs. For example, if you are interested in individuals who have a high school education or less compared with those who have a post-secondary education, you may not require a level of detail that looks at all the specific types of post-secondary education available. This lab will show you how to manipulate variables.

We will also look at how to calculate probabilities from SPSS output. This material corresponds with the material presented in Chapter 4.

LEARNING OBJECTIVES

The following lab is directed at helping you understand how to manipulate variables. Specifically, this lab assignment challenges you to clarify your understanding of:

1. Why you might want to recode variables
2. How to use SPSS to recode variables
3. How to use SPSS output to calculate probability

Recoding Variables

Let's assume you are interested in the opinions Albertans have regarding whether temporary foreign workers are needed to fill jobs in the Alberta labour market. Within the current data set, we have a variable that asks, "Indicate how much you agree or disagree with the following statement: Temporary foreign workers are needed to fill jobs in the Alberta labour market" (ft4). After obtaining a frequency distribution (Figure 4.1) of the variable, you realize you do not require this level of detail. You are only interested in whether people disagree, neither agree nor disagree, or agree. Therefore, you realize that you will need to collapse the first two categories together ("Strongly disagree" and "Somewhat disagree") and the last two categories together ("Somewhat agree" and "Strongly agree"). You plan to leave the middle category ("Neither disagree nor agree") as is.

Step 1: To change the coding of this variable to suit your needs, click on:

Transform → Recode Into Different Variables

It is important to select this option because you are interested in creating a new variable from your original variable rather than altering the existing variable. Although it is possible to recode into the same variable, the original information on the AS2013 will be overwritten, *so always choose to recode into different variables!*

Temporary Foreign Workers are needed to fill jobs in the Alberta labour market

		Frequency	Percent	Valid Percent	Cumulative Percent
Valid	Strongly disagree	95	7.9	8.1	8.1
	Somewhat disagree	172	14.3	14.6	22.7
	Neither disagree nor agree	288	23.9	24.5	47.2
	Somewhat agree	471	39.0	40.1	87.2
	Strongly agree	150	12.4	12.8	100.0
	Total	1176	97.4	100.0	
Missing	No response	2	.2		
	Don't know	29	2.4		
	Total	31	2.6		
Total		1207	100.0		

FIGURE 4.1 | Frequency Distribution for Variable ft4

Step 2: The "Recode into different variables" dialogue box will appear. First, you need to specify the variable you are interested in working with, in this example, Recode: Temporary foreign workers are needed to fill jobs in the Alberta labour market (ft4). Bring that variable into the "Numeric Variable" > "Output Variable box." This can be done either by dragging and dropping the variable, or by using the little arrow between the two boxes.

Now, decide on a name and variable label for your new variable and type them into the Output Variable box. As you can see from Figure 4.2, I chose ft4_recode. You can call your variable whatever you'd like, but make sure that the name is intuitive. Once you've done this, click on **Change**.

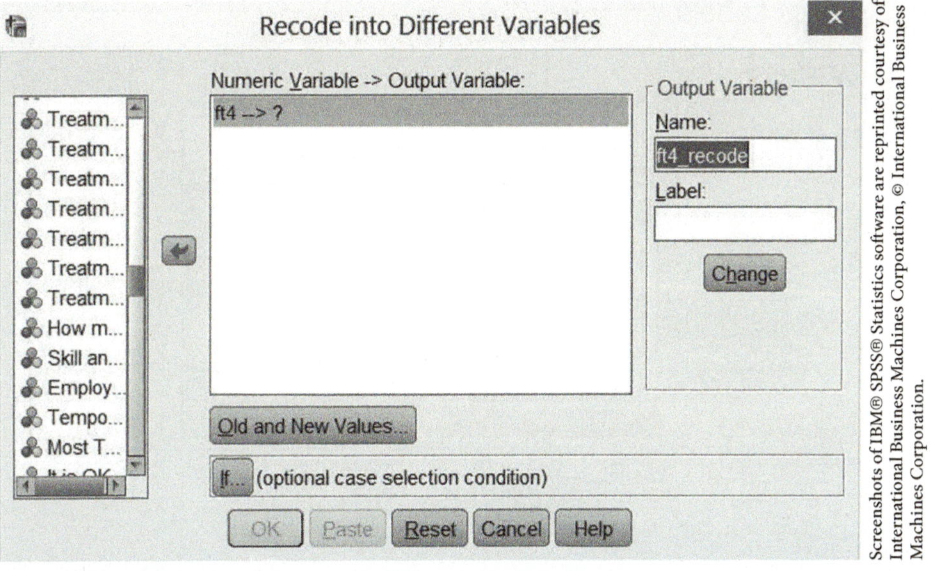

FIGURE 4.2 | Recoding into Different Variables Dialogue Box

Step 3: You now want to define the categories within your new variable, using the old variable values as a basis. So, you need to click on **Old and new values.** Since you are interested in creating a variable where Strongly Disagree and Disagree are in one category and Agree and Strongly Agree are in another, you will need to tell SPSS this is what you want.

Recall from the first lab that you can use your Data Editor screen to determine which values were assigned to particular categories within a variable. In Variable View, go to the row with your variable of interest and click on the cell corresponding with the "Values" column. You can also use the codebook located on the Population Research Lab's website for the Alberta Survey.

Step 4: By clicking on the **Values column** for variable ft4, we see that 1 and 2 correspond with Strongly Disagree and Somewhat Disagree and 4 and 5 correspond with Somewhat Agree and Agree (Figure 4.3). Therefore, we will need to recode our new variable in the following way:

Old value → New value

1, 2 → 1 (Disagree)

3 → 2 (Neither disagree nor agree)

4, 5 → 3 (Agree)

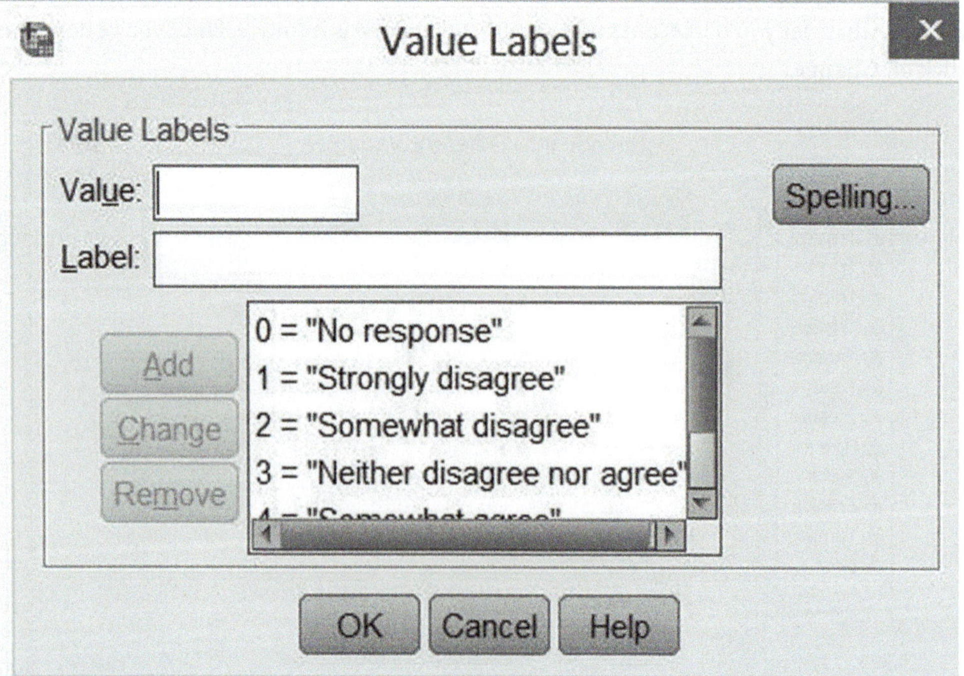

FIGURE 4.3 | Value Labels Dialogue Box

Step 5: Within the dialogue box you opened when you clicked on **Old and new values**, you will transform the values in the original ft4 variable into the new ft4_recode. In Figure 4.4, you can see that 1 and 2 have been coded into 1; 3 has been coded into 2; and 4 and 5 have been coded into 3. Once you have finished defining your new variable, click on **Continue**.

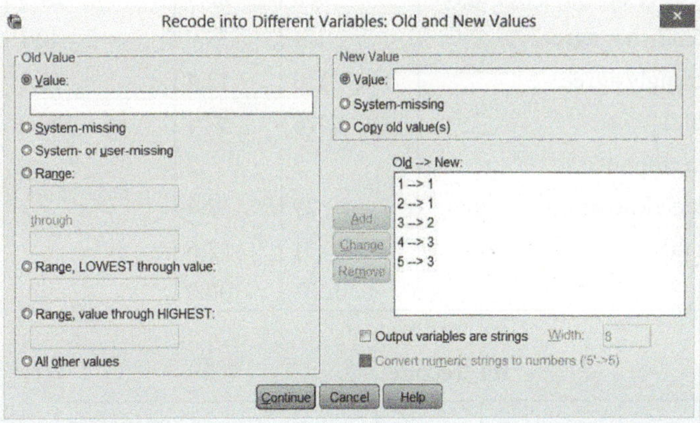

FIGURE 4.4 | Recoding into Different Variables: Old and New Values

Step 6: Click on **Paste** to display the commands in your syntax file. Then click on the green arrow to execute the commands. Or, you can hit **Continue** at the bottom of the screen, and **OK** on the next screen. Either way, your syntax file should look like Figure 4.5.

Now, as a final check, run a frequency distribution on the new variable, ft4_recode, and the original variable, ft4, to confirm that your new values add up to the values in the original variable (Figure 4.6). For example, the values of Strongly Disagree + Somewhat Disagree (95 + 172) should sum the total of the first category of the new variable (267).

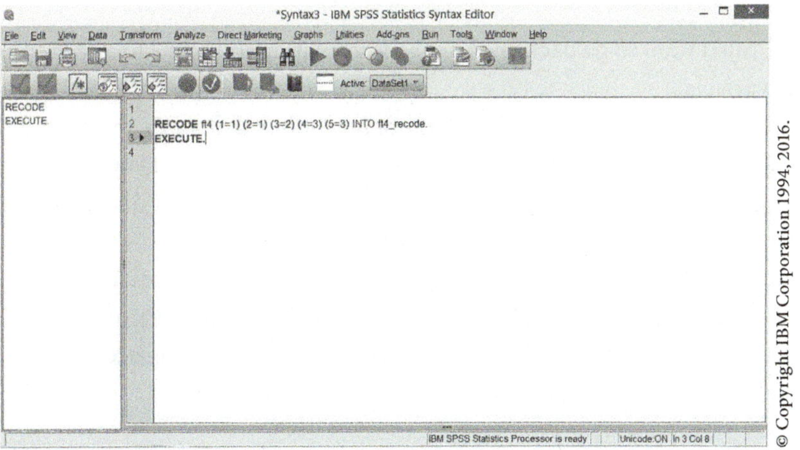

FIGURE 4.5 | Syntax File with Recode Commands

Temporary Foreign Workers are needed to fill jobs in the Alberta labour market

		Frequency	Percent	Valid Percent	Cumulative Percent
Valid	Strongly disagree	95	7.9	8.1	8.1
	Somewhat disagree	172	14.3	14.6	22.7
	Neither disagree nor agree	288	23.9	24.5	47.2
	Somewhat agree	471	39.0	40.1	87.2
	Strongly agree	150	12.4	12.8	100.0
	Total	1176	97.4	100.0	
Missing	No response	2	.2		
	Don't know	29	2.4		
	Total	31	2.6		
Total		1207	100.0		

ft4_recode

		Frequency	Percent	Valid Percent	Cumulative Percent
Valid	1.00	267	22.1	22.7	22.7
	2.00	288	23.9	24.5	47.2
	3.00	621	51.4	52.8	100.0
	Total	1176	97.4	100.0	
Missing	System	31	2.6		
Total		1207	100.0		

FIGURE 4.6 | Frequency Distribution of Variables ft4 and ft4_recode

Naturally, you'd want to go back and use the information in Lab #2 to attach value labels to your new variable.

Chapter 4 discusses probabilities and simply requires that you know the number of observations in each category to be able to calculate probabilities. So, if you wanted to know the probability that someone chosen at random in your sample would disagree with the statement that temporary foreign workers are needed to fill jobs in the Alberta labour market, you would only need to know the frequency of people in the disagree category (267) over the total number of people in the sample (1207). This yields a probability of roughly 22 per cent $\left(\frac{267}{1207}\right)$. Calculating the probability of the other outcomes would proceed in a similar way, except that you'd have a different number in the numerator.

Putting the Information into Practice

1. Using the techniques that you have just been taught, recode variable e13, "I usually expect things to go my way." Create a three-category variable where the categories are "1 Disagree," "2 Neither agree nor disagree," and "3 Agree."
2. Interpret your recoded variable. Specifically, what is the percentage distribution of the categories? What is the total number of respondents who answered the question?
3. Based on the frequency distribution of your recoded variable, what is the probability that individuals will agree that they usually expect things to go their way?
4. Based on the frequency distribution of the original variable, what is the probability that individuals will disagree that they usually expect things to go their way?
5. Based on the frequency distribution of your recoded variable, what is the probability individuals will neither agree nor disagree that they usually expect things to go their way?

Lab #5: The Normal Curve

The focus of this lab is to introduce you to the concept of the normal curve. The distribution of data can take on different shapes. Understanding how data are distributed is important for more complex analysis that you will learn as this course progresses. This lab corresponds with the material presented in Chapter 5.

LEARNING OBJECTIVES

The following lab is directed at helping you understand how to manipulate variables. Specifically, this lab assignment challenges you to clarify your understanding of:

1. How to produce histograms in SPSS, and how to use them to see how data are distributed
2. How to describe and identify distributions

Creating a Histogram in SPSS

Step 1: Within your syntax file, click on:

Analyze → Descriptive statistics → Frequencies

Step 2: Within the dialogue box that opens (Figure 5.1), highlight the "Total years of schooling" variable (k7) in the left-hand screen. Next, click the arrow (this will move the variable into the right-hand screen).

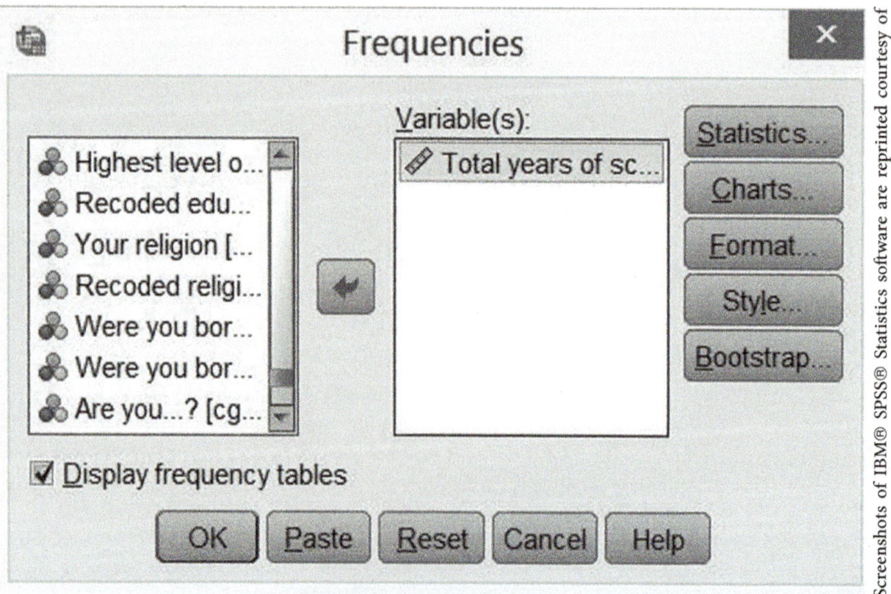

FIGURE 5.1 | Frequencies Dialogue Box

Step 3: Click on **Charts**. This will open a dialogue box that asks you which chart type you would like to present (Figure 5.2). Click on **Histograms → Show normal curve → Continue** (this will close the menu screen) **→ Paste.** The paste command is optional but is highly recommended for record-keeping purposes and for tracking your progress. However, if you decide not to keep a syntax record, you can simply click on **OK**. For the purposes of this lab manual, we will be pasting all commands into a syntax file.

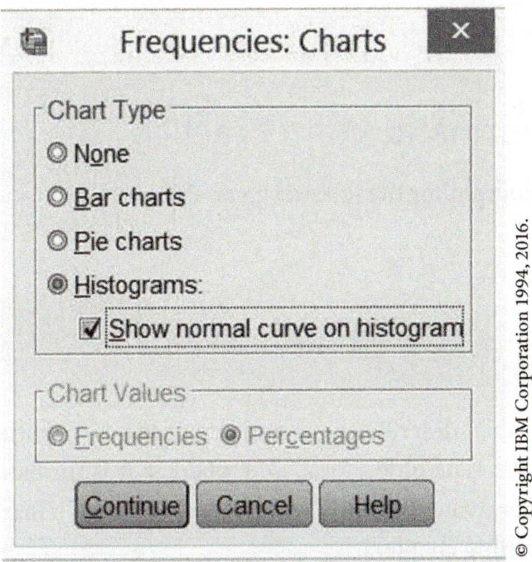

FIGURE 5.2 | Frequencies: Charts Dialogue Box

Now, after running the commands in your syntax file, you should get an output file containing a histogram that resembles the one in Figure 5.3.

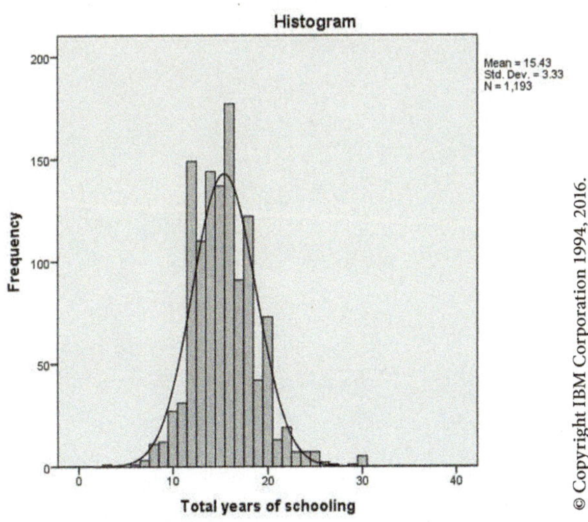

FIGURE 5.3 | Histogram with Normal Curve

Now, what can we say about this distribution? Well, it is *bimodal* because there are two spikes, one at the 12 years of schooling mark and the other at the 16 years of schooling mark. The curve has a slight *positive* or *right skew* because there are more people with higher levels of education than lower levels. The curve is a little more "peaked" than normal. This is hard to know from seeing a distribution, but having two peaks that are higher, and none that are dramatically lower than the rest of the distribution, provides a clue that the distribution has positive kurtosis.

OK, now it's your turn.

Putting the Information into Practice

1. Obtain a histogram for the following variables:
   a. age
   b. k3a
   c. k3b
   d. k3c

2. For each variable, describe the symmetry of the distribution, the skewness of the distribution, its type of kurtosis, and whether it is unimodal, bimodal, or multimodal. For now, you can only do this visually, but we'll learn how to do more than this in the coming chapters.

Lab #6: Measures of Central Tendency and Dispersion

The focus of this lab is to introduce you to various measures of central tendency. Measures of central tendency allow you to further understand the distribution of a variable. In this lab, you will learn how to generate various measures of central tendency within SPSS. This lab corresponds with the material presented in Chapter 6.

LEARNING OBJECTIVES

The following lab is directed at helping you understand how to generate measures of central tendency. Specifically, this lab assignment challenges you to clarify your understanding of:

1. The differences between the mean, median, and mode
2. The relationship between levels of measurement and the mean, median, and mode
3. How to generate measures of central tendency within SPSS

Generating Measures of Central Tendency in SPSS

Step 1: Within your syntax file, click on:

Analyze → Descriptive statistics → Frequencies

Step 2: Within the dialogue box that opens (Figure 6.1), highlight the "Total years of schooling" variable (k7) in the left-hand screen; next, click the arrow (this will move the variable into the right-hand screen).

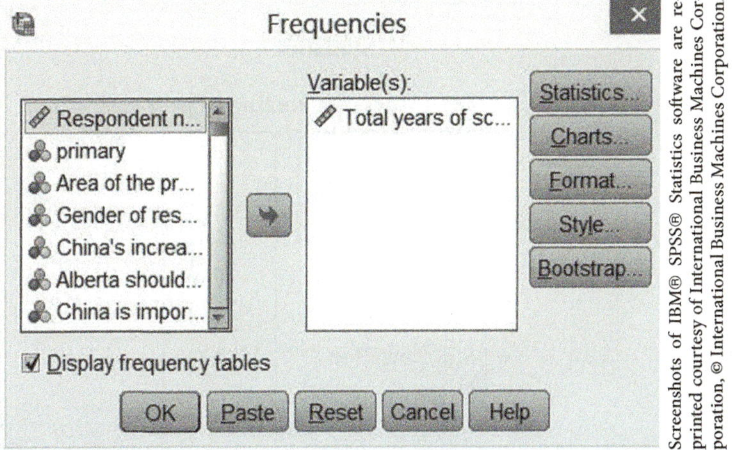

FIGURE 6.1 | Frequencies Dialogue Box

Step 3: Click on **Statistics**. This will open a dialogue box that asks you which statistics you are interested in obtaining (Figure 6.2). Under "Central Tendency," click on **Mean**, **Median**, and **Mode**. Then, under "Dispersion," select **Std. deviation**, **Variance**, and **Range**. Notice too that it's possible to obtain skewness and kurtosis statistics here. Let's leave these unchecked for now, however, and click on **Continue** (this will close the menu screen) and then click **Paste**.

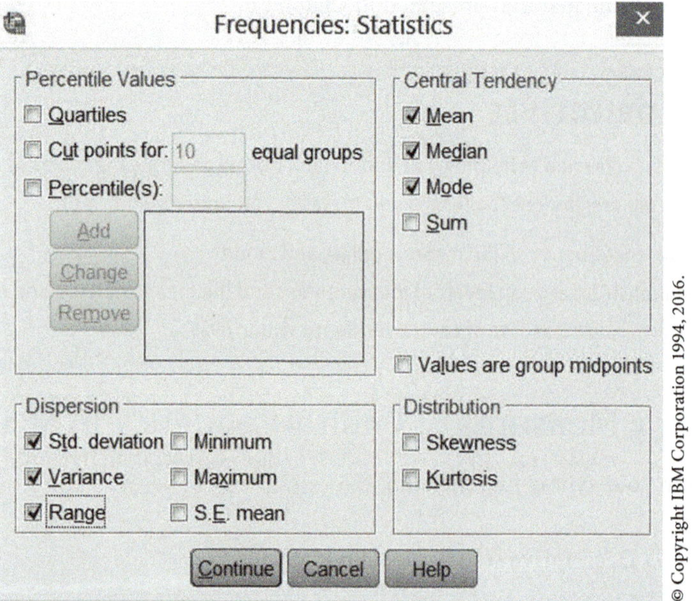

FIGURE 6.2 | Frequencies: Statistics Dialogue Box

Now, after running the commands in your syntax file, you should get an output file containing the table shown in Figure 6.3:

Statistics

Total years of schooling

N	Valid	1193
	Missing	14
Mean		15.43
Median		15.00
Mode		16
Std. Deviation		3.330
Variance		11.088
Range		27

FIGURE 6.3 | Statistics Box

Now, before we continue, we need to ask ourselves if these numbers make sense. In other words, do these measures of central tendency make sense in relation to the level of measurement of our variable? Years of schooling is a ratio level variable because we have a true zero value: individuals can obtain no years of schooling.

This table tells us that, on average, our respondents have 15.43 years of schooling. The Median tells us that if we were to line up all our respondents, from the lowest level of schooling to the highest, the middle score would be 15 years of schooling. The Mode, or most frequently occurring score, is 16 years of school.

We have a standard deviation of 3.330 and a variance of 11.088. What does this mean? As you progress through this course, the meaning of these statistics will become clearer. However, for now, knowing how to generate these statistics in SPSS is sufficient. We also see that we have a Range of 27. This means that the distance between the highest level of education and the lowest level of education is 27 years. Since this number is plausible, we should have faith that the data have no errors, or that we didn't do anything wrong (not trivial occurrences in statistics!).

Let's look at one more example to illustrate the difference between mean, median, and mode. Figure 6.4 shows the frequency distribution for the variable "If an Election Were Held Today, How Would You Vote Federally?" (k16a).

If an election were held today, how would you vote federally?

		Frequency	Percent	Valid Percent	Cumulative Percent
Valid	PC/Tory	502	41.6	60.7	60.7
	Green Party	28	2.3	3.4	64.1
	Liberals	191	15.8	23.1	87.2
	NDP	91	7.5	11.0	98.2
	Other specified	15	1.2	1.8	100.0
	Total	827	68.5	100.0	
Missing	No response/Refused	53	4.4		
	Would not vote	58	4.8		
	Not eligible	21	1.7		
	Don't Know	248	20.5		
	Total	380	31.5		
Total		1207	100.0		

FIGURE 6.4 | Frequency Distribution of Variable k16a

This variable is a nominal level variable because the categories cannot be ranked. Referring back to the material presented in Chapter 6, we know that the mode is most commonly used for nominal or ordinal level data. Though this is a good rule to memorize, here is the reason. When we look at the mean (Figure 6.5), we have a value of 1.9. If we were to translate a value of 1.9 into words, it would mean that on average, people would vote somewhere in between the PC/Tory and the Green Party (closer to the Green Party). If you have ever voted

If an election were held today, how would you vote federally?

N	Valid	827
	Missing	380
Mean		1.90
Median		1.00
Mode		1
Std. Deviation		1.197
Variance		1.433
Range		4

FIGURE 6.5 | Statistics Box

in a federal election, you know you have to pick one, and only one, party or your ballot will be deemed invalid. Therefore, the mathematical mean does not make sense for nominal level data. Instead, we should pick the mode, which tells us that the most frequently selected category is "1," or PC/Tory. Can you see why this is the case?

Now it's your turn!

Putting the Information into Practice

1. For each variable, identify the level of measurement of the variable.
   a. strata
   b. age
   c. sex
   d. b5
   e. b11
   f. e10
   g. ft9
   h. h5
   i. mrelig
   j. k12a

2. For each of the 10 variables, run both a frequency distribution and *appropriate* measures of central tendency and dispersion (i.e., mean, median, mode, standard deviation, variance, and/or range).

3. Based on your analysis of your output, complete the following table by writing in the values of the *appropriate* measures of central tendency and dispersion. Leave the boxes blank where the statistic is inappropriate.

Variable	Mean	Median	Mode	Standard Deviation	Variance	Range
strata						
age						
sex						
b5						
b11						
e10						
ft9						
h5						
mrelig						
k12a						

Lab #7: Standard Deviations, Standard Scores, and the Normal Distribution

The focus of this lab is to introduce you to *z*-scores and help you further understand how the standard deviation relates to the normal curve. The most commonly used standard score, the *z*-score, is a measure of the relative location in a distribution. Specifically, in standard deviation units, *z*-scores give the distance that a particular score is from the mean. In this lab, you will learn how to generate various *z*-scores within SPSS. This lab corresponds with material presented in Chapter 7.

LEARNING OBJECTIVES

The following lab is directed at helping you understand how *z*-scores relate to the normal curve and the standard deviation. Specifically, this lab assignment challenges you to clarify your understanding of:

1. The shape of distributions and the normal curve
2. The relationship between *z*-scores and the normal curve
3. How to use *z*-scores to mathematically calculate the percentage of cases that fall between two values

Part 1: Reviewing the Shape and Characteristics of Distributions

Before learning to calculate *z*-scores, let's first refresh our memories on the shape and characteristics of distributions.

In Figure 7.1, we see a histogram for the variable "In Total, How Many Years of Schooling Do You Have?"

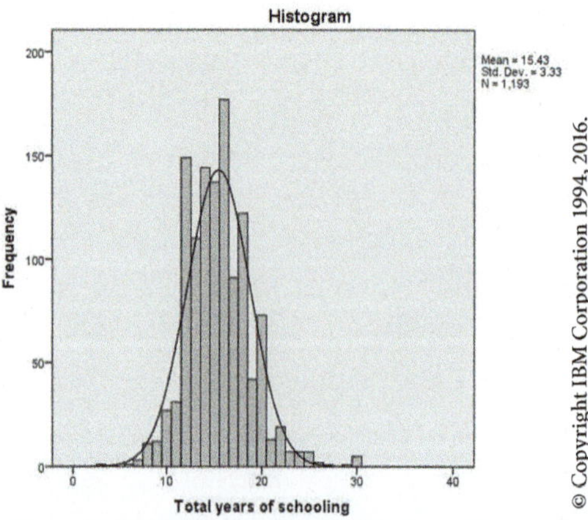

FIGURE 7.1 | Histogram with Normal Curve

Now, what can we say about this distribution? It is *bimodal* because there are two spikes, one at the 12 years of schooling mark and the other at the 16 years of schooling mark. The curve has a slight *positive skew* because there are more people with higher levels of education than lower levels. The two peaks of the distribution are higher than the rest of the distribution, suggesting that the distribution has positive kurtosis.

When we look at the statistics box in Figure 7.2, we see that on average, respondents have 15.43 years of school with a standard deviation of 3.330.

Statistics

Total years of schooling

N	Valid	1193
	Missing	14
Mean		15.43
Median		15.00
Mode		16
Std. Deviation		3.330
Variance		11.088
Range		27

FIGURE 7.2 | Statistics for "In Total, How Many Years of Schooling Do You Have?"

All right, so now we have refreshed our memory regarding the shape and characteristics of distributions. Keeping these elements in mind will help us to further understand *z*-scores and what it means to standardize a distribution.

Part 2: Calculating *z*-Scores

Calculating a *z*-score is similar to other commands you have learned up to this point.

Step 1: Within your syntax file, click on:

Analyze → Descriptive statistics → Descriptives

Step 2: Within the dialogue box that opens (Figure 7.3), highlight the "Total years of schooling" variable (k7) in the left-hand screen. Next, click the arrow (this will move the variable into the right-hand screen). The key to this step is to click the box that says, **Save standardized values as variables**. This will create a new variable at the bottom of your data editor screen called zk7 where all the values of the original k7 variable are now standardized scores, or *z*-scores. Then, click **Paste** and run your Syntax commands.

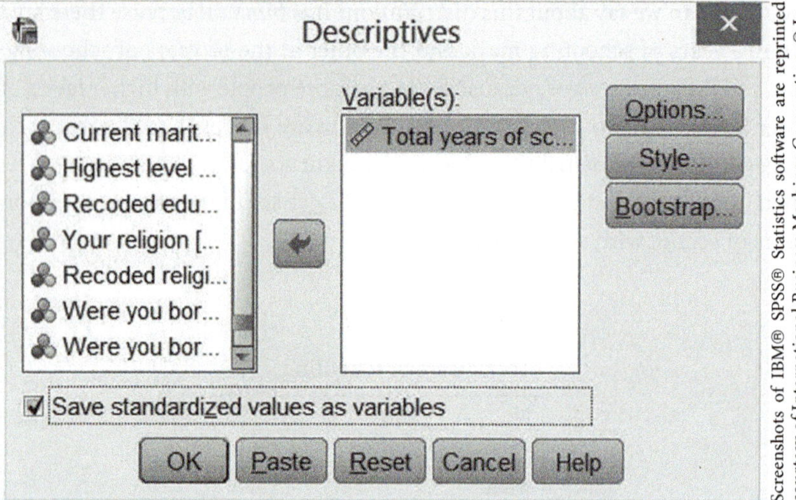

FIGURE 7.3 | Descriptives Dialogue Box

Step 3: Run a frequency distribution on your new variable zk7. Be sure to include the mean, standard deviation, and variance.

In our statistics box in Figure 7.4, you will first notice that we have a mean of 0, a standard deviation of 1, and a variance of 1.

To convince yourself of the accuracy of the calculations, feel free to manually calculate the *z*-score for an observation, using the equation presented below (and taken from Chapter 7). Remember that you need to use the standard deviation from the actual variable, not the *z*-score variable:

$$z = \frac{X - \mu}{\sigma}$$

Statistics

Zscore: Total years of schooling

N	Valid	1193
	Missing	14
Mean		.0000000
Median		-.1301435
Mode		17017
Std. Deviation		1.00000000
Variance		1.000
Range		8.10842

FIGURE 7.4 | Statistics Box for *z*-Scores of Variable k7

In Figure 7.5, you will notice that some of the scores are negative numbers. These scores now measure the distance of each original value (years of schooling) from the mean.

Zscore: Total years of schooling

		Frequency	Percent	Valid Percent	Cumulative Percent
Valid	-3.73389	1	.1	.1	.1
	-2.83295	1	.1	.1	.2
	-2.53264	3	.2	.3	.4
	-2.23233	11	.9	.9	1.3
	-1.93201	12	1.0	1.0	2.3
	-1.63170	27	2.2	2.3	4.6
	-1.33139	31	2.6	2.6	7.2
	-1.03108	149	12.3	12.5	19.7
	-.73077	110	9.1	9.2	28.9
	-.43046	144	11.9	12.1	41.0
	-.13014	137	11.4	11.5	52.5
	.17017	177	14.7	14.8	67.3
	.47048	91	7.5	7.6	74.9
	.77079	122	10.1	10.2	85.2
	1.07110	42	3.5	3.5	88.7
	1.37142	73	6.0	6.1	94.8
	1.67173	13	1.1	1.1	95.9
	1.97204	19	1.6	1.6	97.5
	2.27235	7	.6	.6	98.1
	2.57266	7	.6	.6	98.7
	2.87297	7	.6	.6	99.2
	3.17329	2	.2	.2	99.4
	3.47360	1	.1	.1	99.5
	4.07422	1	.1	.1	99.6
	4.37453	5	.4	.4	100.0
	Total	1193	98.8	100.0	
Missing	System	14	1.2		
Total		1207	100.0		

FIGURE 7.5 | Frequency Table for z-Scores of Variable k7

Now it's your turn!

Putting the Information into Practice

1. Run a frequency distribution for "Age." Be sure to include a histogram, the mean, standard deviation, and variance.
   a. Are your respondents' ages normally distributed? What evidence supports your answer (consider the shape and characteristics of your distribution)?

2. Create a new variable called zage.
3. In your Data Editor window, on Data View, find respondent #61. (Hint: Use the variable called respnum. This variable assigns a number that is unique for each respondent who participated in the survey.) What are respondent #61's values for age and zage?
4. In your Data Editor window, on Data View, find respondent #33. What are respondent #33's values for age and zage?
5. What percentage of respondents are older than respondent #61?
6. What percentage of respondents are younger than respondent #33?
7. What percentage of respondents are older than respondent #61 but younger than respondent #33?

Lab #8: Sampling

The focus of this lab is to introduce you to the select cases function within SPSS and to help you to further understand how larger sampling distributions improve data accuracy. A sample that is accurately and carefully selected without a lot of sampling error allows for a more precise analysis without including the full population. Because we are often unable to survey every individual, we make decisions about how much of the population to include based on our knowledge of the population parameter. In this lab, we are going to pretend that the total number of respondents who participated in this survey represent the entire population of Alberta (as though the survey was a census). This lab corresponds with the material presented in Chapters 8 and 9.

LEARNING OBJECTIVES

The following lab is directed at helping you understand what effect sample size has on the accuracy of sample values. Specifically, this lab assignment challenges you to clarify your understanding of:

1. The relationship between samples and populations
2. How increasing a sample size will reduce sampling error

How to Select Cases in SPSS

Before we begin, run a frequency distribution on the sex variable.

Gender of respondent

		Frequency	Percent	Valid Percent	Cumulative Percent
Valid	Male	595	49.3	49.3	49.3
	Female	612	50.7	50.7	100.0
	Total	1207	100.0	100.0	

FIGURE 8.1 | Frequency Distribution on the Sex Variable

Note that we have 595 males and 612 females. This will be the basis of our population parameters. However, remember that these numbers, in reality, do not represent the true population of Alberta. We are only using this as a population for illustrative purposes.

Step 1: Within your syntax file, click on:

Data → Select cases

Under this window, you have a variety of options (Figure 8.2). For our purposes, you are going to select the option "Random sample of cases."

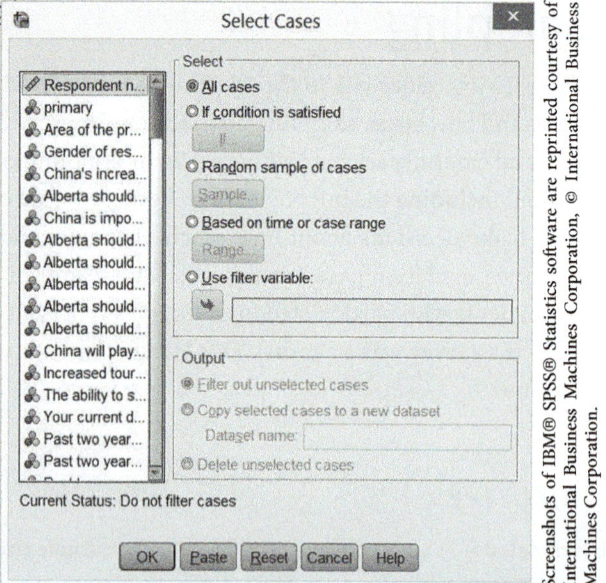

FIGURE 8.2 | Select Cases Dialogue Window

Step 2: Click on:

Random sample of cases → Sample

A dialogue box like the one in Figure 8.3 will appear. Here you will decide what percentage of cases you are interested in selecting. For this example, we will select 5 per cent of the cases. To do so, type **5** into the box next to the words "% of all cases." Notice that an option at the bottom-right corner of the box now illuminates, asking what you'd like to do with the cases you don't select. Typically, the best thing to do is to filter out the cases because otherwise you won't be able to draw the unselected cases back in if you delete.

Once you have taken a look at the options and are satisfied with what you are doing, click **Continue**, then **Paste**. Next, run these commands in your Syntax file.

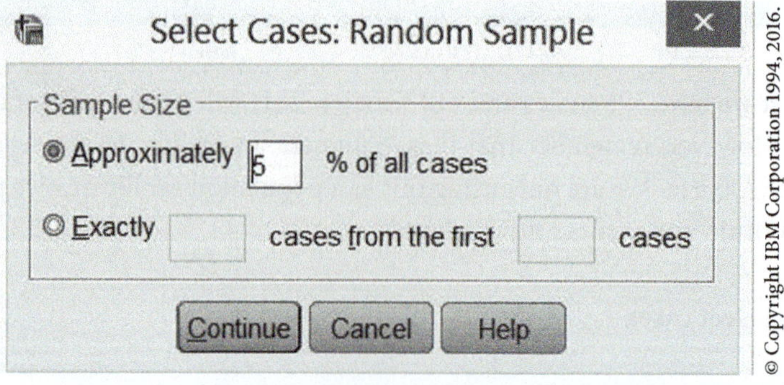

FIGURE 8.3 | Select Cases: Random Sample Dialogue Box

Now, run a frequency distribution on the sex variable.

Sex of respondent

		Frequency	Percent	Valid Percent	Cumulative Percent
Valid	Male	38	57.6	57.6	57.6
	Female	28	42.4	42.4	100.0
	Total	66	100.0	100.0	

FIGURE 8.4 | Frequency Distribution, Sex Variable

Can you see how your sample size is much smaller than what is in the full data set? What's more, given that the sample is randomly drawn each and every time, it is possible that your frequencies differ from those above. Can you see why this might be the case? If you'd like to generate another sample, you can go to **Data → Select cases → Reset**, then generate another sample. Is this sample different yet again? It might be, because once again you've created another random sample.

Standard Error of a Sample Mean

The next thing we're going to do is use SPSS to help us calculate the standard error of a sample mean. Recall from Chapter 9 that the equation is

$$s_{\bar{x}} = \frac{s_X}{\sqrt{n-1}}$$

Using this, it is possible to estimate the distance that your sample is likely to be from a population mean. You can do this even though you don't know what the population mean actually is, using statistical theory and what we know about the normal distribution (which is how we're assuming the data from the Alberta Survey are distributed).

Suppose that you wanted to know the average age of your 5 per cent sample. Remember that you would do this by selecting **Analyze → Descriptives**, then selecting your variable of interest (age). The resulting output will give you the mean, the standard deviation, and the number of observations necessary to calculate the standard error.

Now it's your turn!

Putting the Information into Practice

1. Run a frequency distribution for the years of schooling variable (k7) for the population.
2. Use the select cases procedure to take a random sample of 2 per cent of the cases from the "Population."

3. Run summary statistics on k7 and record the mean, *N*, and the standard deviation. Calculate the standard error of the sample mean, using the standard deviation estimates generated by SPSS.
4. Repeat this process to complete the following table.

%	Mean years of schooling	Sampling error
5%		
10%		
25%		
50%		
75%		

5. Which percentage of cases would you choose if you were under budget and time constraints to complete your survey? In other words, which percentage begins to most closely resemble your population, and at what point does your distribution start to level off?

Lab #9: Hypothesis Testing: Testing the Significance of the Difference between Two Means

The focus of this lab is to introduce you to the "one sample *t*-tests" function within SPSS and help you to further understand how we use confidence intervals to determine generalizability of our samples to populations. This lab corresponds with material presented in Chapter 10.

LEARNING OBJECTIVES

The following lab is directed at helping you understand what effect sample size has on the accuracy of sample values. Specifically, this lab assignment challenges you to clarify your understanding of:

1. How to use a *t*-test to approximate the mean for a population from your sample
2. The relationship between confidence intervals and statistical significance

Calculating a One-Sample *t*-Test in SPSS

Let's pretend we are interested in knowing whether the average years of schooling of our sample differs significantly from the population mean of 16 years (high school plus an undergraduate degree) at the 95 per cent confidence level. Suppose that we got this number from the Canadian census and that it accurately represents the entire population.

Step 1: Within your syntax file, click on:

Analyze → Compare means → One-sample *t*-test

Step 2: Within the dialogue box that opens (Figure 9.1), highlight the "Total years of schooling" variable (k7) in the left-hand screen; next, click the arrow (this will move the variable into the right-hand screen).

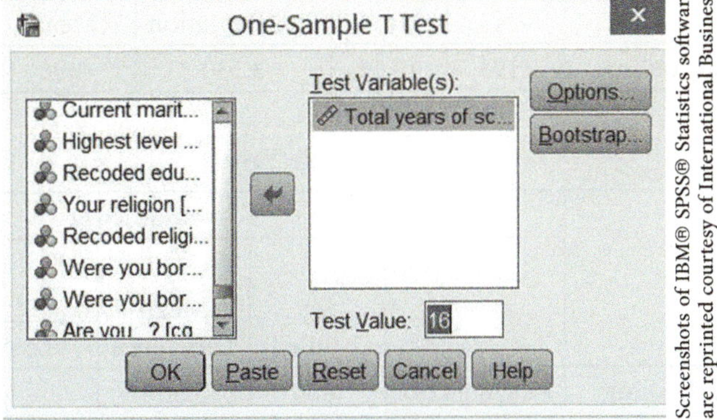

FIGURE 9.1 | One-Sample *t*-Test Dialogue Box

Step 3: Insert the population mean in the test value box. This option is useful because it allows you to construct confidence intervals from *any* number, not just a population mean.

Step 4: Click on **Options.** In this dialogue box (Figure 9.2), you want your "Confidence Interval Percentage" to equal 95 per cent. This way, you can be 95 per cent confident your sample mean is generalizable to the population mean. Then click **Continue**, **Paste**, and run your Syntax commands.

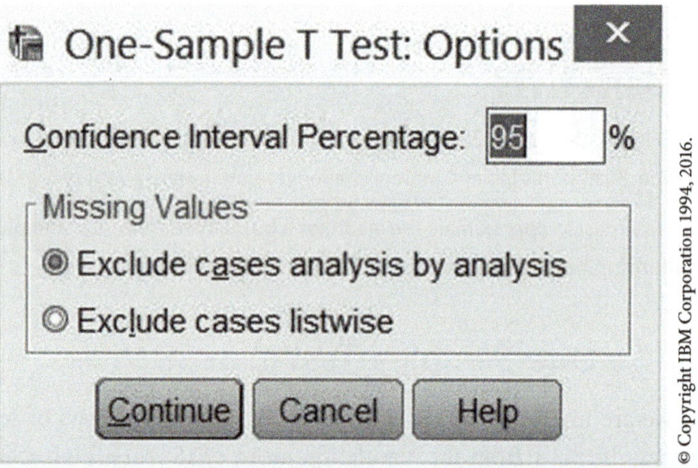

FIGURE 9.2 | One-Sample *t*-Test: Options Dialogue Box

The following tables should come up in your output file:

One-Sample Statistics

	N	Mean	Std. Deviation	Std. Error Mean
Total years of schooling	1193	15.43	3.330	.096

One-Sample Test

	Test Value = 16					
					95% Confidence Interval of the Difference	
	t	df	Sig. (2-tailed)	Mean Difference	Lower	Upper
Total years of schooling	-5.878	1192	.000	-.567	-.76	-.38

FIGURE 9.3 | Output File for One-Sample *t*-Test

First, we can note that we have 1193 cases and a mean of 15.43 years of schooling. The standard deviation is 3.330. In the next piece of output, we see we have a t value of −5.878 with 1192 degrees of freedom. This is the equivalent of our t-obtained value. By looking at the Sig. (two-tailed), we can see that we have a significance level of 0.000. This means that we can be more than 99.9 per cent confident that the average of our sample is significantly different from a sample with a mean of 16 years of schooling. Can you use these numbers to do the t-test manually, by looking at the student's t-table in Appendix B of your textbook?

Don't despair if you can't because SPSS does the work for you. If you go to the far right-hand side of your output, you can see at the 95 per cent confidence interval that our lower boundary is −0.76 and our upper boundary is −0.38. This means that 95 times out of 100, our sample mean difference will be between −0.76 and −0.38.

If you wanted to (and who wouldn't want to?), you could use the information in the SPSS output to calculate the confidence intervals yourself and compare them to the numbers that SPSS generated.

OK, now it's your turn!

Putting the Information into Practice

1. Let's imagine that the average age of someone living in Alberta is 27 years old. We want to know if the average age of our sample differs significantly from the average age of someone living in Alberta at the 95 per cent confidence level. Conduct a one-sample *t*-test.
2. Can you be 95 per cent confident that the average age of our respondents differs significantly from the average age of someone living in Alberta?
3. What is your confidence interval? What does this mean?

Lab #10: Hypothesis Testing: One- and Two-Tailed Tests

The focus of this lab is to introduce you to the "*t*-tests with two samples" function within SPSS. For this type of analysis, you need to have a dichotomous independent variable and an interval/scale for your dependent variable. This lab corresponds with material presented in Chapter 11.

LEARNING OBJECTIVES

The following lab is directed at helping you understand how you can use a *t*-test to compare the means of two independent samples. Specifically, this lab assignment challenges you to clarify your understanding of:

1. Stating the null and research hypotheses
2. Establishing a sampling distribution and critical region

Part 1: Calculating a *t*-Test with Two Samples in SPSS

Let's pretend we are interested in knowing whether the average years of schooling of our sample differs significantly between males and females. Since both are samples and scores on the outcome of interest are independent of each other (presumably, the education level of women has nothing to do with the education level of men), it is most appropriate to conduct a *t*-test on independent samples.

Step 1: State the null and research hypotheses.

In this case, we are going to make the claim that because Alberta is an industry-driven economy where a significant proportion of the male population pursues employment in the oil sands, men and women will differ in the number of years of schooling they obtain.

Therefore, we are claiming that:

H_0: In the population, the mean years of schooling for men and women does not differ ($\mu_{MEN} = \mu_{WOMEN}$).

H_1: In the population, the mean years of schooling for men is not equal to that of women ($\mu_{MEN} \neq \mu_{WOMEN}$).

Step 2: Select the sampling distribution and establish the critical region.

On a *z*-score distribution, the critical region, which corresponds to a p-value < 0.05, will be represented by a *t*-score of ± 1.96 (Figure 10.1).

Should our test statistic be found in the 5 per cent of the distribution that is bounded by the two critical regions of our *t*-score distribution, we can be 95 per cent confident that we can reject our null hypothesis. Since we are only hypothesizing that there will be a difference in H_1, without saying anything about the direction of the difference (such as $\mu_{MEN} < \mu_{WOMEN}$), we're conducting a two-tailed test. If we were conducting a one-tailed test (and we were hypothesizing a direction of the relationship), remember from Chapter 10 that we'd have a critical *t*-score of ± 1.65.

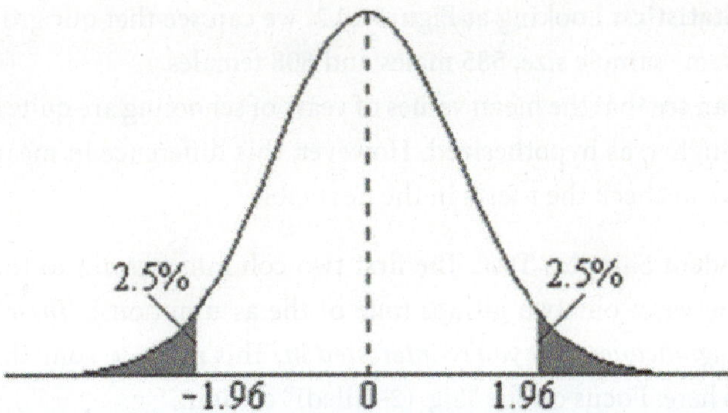

FIGURE 10.1 | The Critical Region at the 95 Per Cent Confidence Interval
Source: © Copyright IBM Corporation 1994, 2016.

Step 3: Using the Data Editor screen, check to see how the two groups of your independent variable are coded in the codebook. In this case, males are coded as "1" and females are coded as "2."

Now, you are ready to run your independent samples *t*-test:

Step 4: In your Syntax file, click on:

Analyze → Compare means → Independent samples t-test

Our test variable will be "Total years of schooling," and our grouping variable will be "Sex of respondent."

Click **Define Groups** and indicate to SPSS which value you would like associated with each group. For right now, leave males as **1** and females as **2.** Then click **Continue** and **Paste** and run your Syntax commands.

You should get the following output:

Group Statistics

	Gender of respondent	N	Mean	Std. Deviation	Std. Error Mean
Total years of schooling	Male	585	15.65	3.273	.135
	Female	608	15.23	3.374	.137

Independent Samples Test

		Levene's Test for Equality of Variances		t-test for Equality of Means						
									95% Confidence Interval of the Difference	
		F	Sig.	t	df	Sig. (2-tailed)	Mean Difference	Std. Error Difference	Lower	Upper
Total years of schooling	Equal variances assumed	.058	.809	2.203	1191	.028	.424	.193	.046	.802
	Equal variances not assumed			2.205	1190.919	.028	.424	.192	.047	.802

FIGURE 10.2 | Output for Independent Samples *t*-Test

Table 1: Group Statistics: Looking at Figure 10.2, we can see that our groups both had approximately the same sample size, 585 males and 608 females.

Already we can see that the mean values of years of schooling are quite similar, although men are slightly higher, as hypothesized. However, this difference in means may be due to chance, so we have to check the *t*-tests in the next table.

Table 2: Independent Samples Test: The first two columns pertain to the Levene's test of equal variances between our two groups (one of the assumptions). *These two columns are not indicators of significance that you're interested in.* This refers to something else, which is beyond our focus here. Focus on the "Sig. (2-tailed)" column.

Earlier, we said that we needed a *t*-score of $> \pm 1.96$. Here we see that we have a *t*-value of 2.203. Since this score falls above our critical score, we can conclude that the mean years of schooling for our two sample groups (males and females) are significantly different. Therefore, we can reject the null hypothesis of no differences between groups. Similarly, if we were conducting a one-tailed test, we would also reject the null hypothesis.

Now it's your turn!

Putting the Information into Practice

Let's pretend we are interested in knowing whether the average number of children under the age of 18 differs significantly between those who were born in Canada and those who were born outside Canada.

We will use variables "Were you born in Canada?" (canb) and "Number of children living in household" (k3b).

1. Identify your dependent and independent variable.
2. State your null and research hypotheses.
3. Run an independent samples *t*-test.
4. What is the average number of children under the age of 18 for those who were born in Canada?
5. What is the average number of children under the age of 18 for those who were not born in Canada?
6. Using a *t*-test, do you accept or reject the null hypothesis? Why?

Lab #11: Bivariate Statistics for Nominal Data

The focus of this lab is to introduce you to the *association or relationship between nominal variables*. Specifically, this lab will help clarify your understanding of independent and dependent variables and how to interpret the chi-square test of statistical significance. This lab corresponds with the material presented in Chapter 12.

LEARNING OBJECTIVES

The following lab is directed at helping you understand how to interpret the relationship between two nominal variables (bivariate relationships). Specifically, this lab assignment challenges you to clarify your understanding of:

1. Dependent and independent variables
2. How to create a cross-tabulation or a contingency table
3. How to interpret your findings and determine statistical significance

Creating Contingency Tables within SPSS

Now, let's learn how to create contingency tables in SPSS. For this example, we will ask this question: "Who Is More Likely to Have Been Diagnosed with High Blood Pressure as an Adult, Males or Females?"

Step 1: Identify your dependent and independent variables.

An independent variable can be thought of as the modifying outcome, the dependent variable can be thought of as the outcome of interest. In this situation, we are interested in seeing if gender (sex) will modify patterns of diagnosis of high blood pressure; therefore, it is our independent variable. The outcome we are interested in is high blood pressure diagnosis (e14_1); therefore, it is our dependent variable.

Step 2: Using the syntax file, click on:

Analyze → Descriptive statistics → Crosstabs

Step 3: Select the dependent and independent variable.

This will bring up the crosstabs dialogue screen (Figure 11.1). Once you are in that screen, bring your dependent variable, "Diagnosed with high blood pressure as an adult" (e14_1), into the "Row(s)" box, and bring your independent variable, sex, into the "Column(s)" box. When you have finished doing that, click on **Statistics icon**.

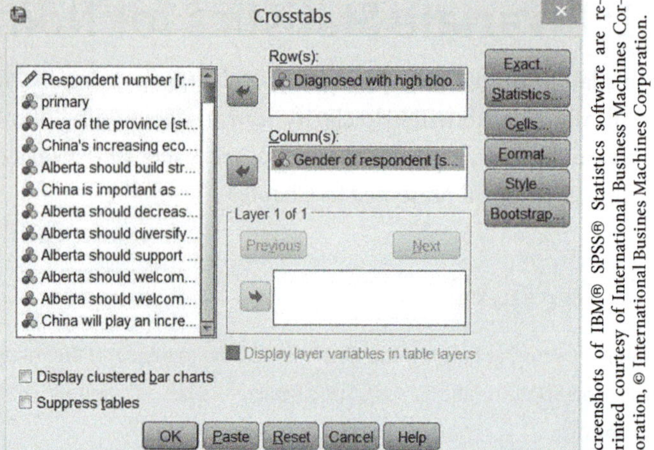

FIGURE 11.1 | Crosstabs Dialogue Screen

Step 4: Select your statistics.

Within this screen, select **chi-square** (Figure 11.2). We will be using a chi-square test to measure the statistical significance of the relationship between genders. Also notice that there's an option to select different measures of association for nominal variables. Select **Phi and Cramer's V**, one of the nominal measures of association covered in Chapter 12 of your text. Then click **Continue**.

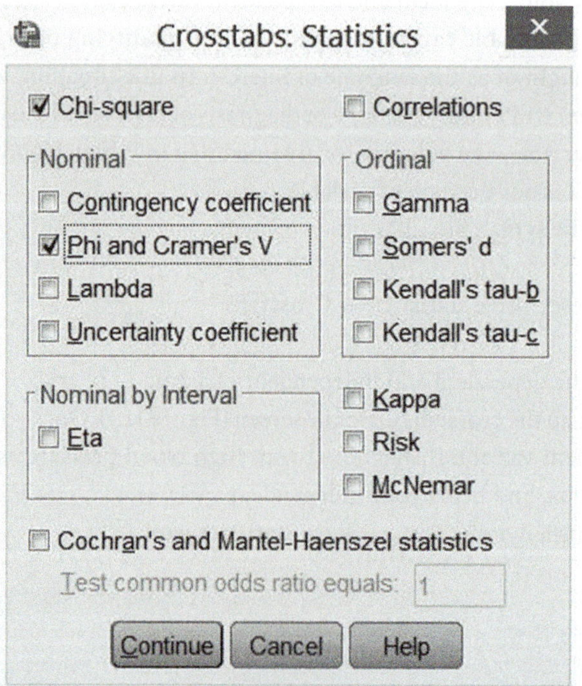

FIGURE 11.2 | Crosstabs: Statistics Dialogue Screen

Step 5: Percentage your columns.

You will return to the crosstabs screen. Once there, select the **Cells** button (Figure 11.3). Within this screen, you will select **Column** under the "Percentages" list. You are selecting columns because you want to percentage by sex so you can compare the differences across males and females. Then hit **Continue**, which will return you to the crosstabs screen. Then click **Continue**, **Paste**, and run your Syntax commands.

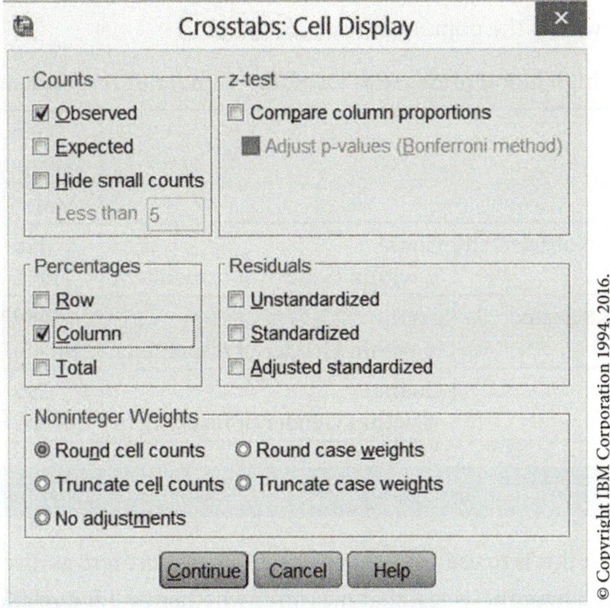

FIGURE 11.3 | Crosstabs: Cell Display Dialogue Screen

You should get the following output:

Case Processing Summary

	Cases					
	Valid		Missing		Total	
	N	Percent	N	Percent	N	Percent
Diagnosed with high blood pressure as an adult * Gender of respondent	1207	100.0%	0	0.0%	1207	100.0%

FIGURE 11.4 | Case Processing Summary

The "Case Processing Summary" tells you the number of respondents, 1207, and the percentage that are valid, 100 per cent. Valid cases are those that are meaningful to your analysis.

Following the case summary report is the contingency table, cross-tabulating gender by whether or not the respondent has been diagnosed with high blood pressure as an adult.

You will notice that you obtain both raw frequencies and percentages. Let's start to interpret these results. There do appear to be differences between men and women in whether they have been diagnosed with high blood pressure as an adult. In particular, we find that females are less likely to report having been diagnosed with high blood pressure than males: 24.8 per cent compared to 28.4 per cent, respectively. However, how do we know if these differences are statistically significant? That is, how do we know if the differences seen in the sample also exist within the population?

Diagnosed with high blood pressure as an adult * Gender of respondent Crosstabulation

			Gender of respondent		
			Male	Female	Total
Diagnosed with high blood pressure as an adult	Not selected	Count	426	460	886
		% within Gender of respondent	71.6%	75.2%	73.4%
	Selected	Count	169	152	321
		% within Gender of respondent	28.4%	24.8%	26.6%
Total		Count	595	612	1207
		% within Gender of respondent	100.0%	100.0%	100.0%

FIGURE 11.5 | Contingency Table

One way to test this is to use a chi-square test. A *chi-square test*, as discussed in Chapter 8 of the textbook, is a hypothesis test that measures whether or not a relationship exists. This measure is suitable for all levels of measurement and all distributions. It tests the *null hypothesis* and measures the discrepancy between observed and expected events. The events are assumed to be independent and have the same distribution, and the outcomes of each event must be mutually exclusive.

Chi-Square Tests

	Value	df	Asymptotic Significance (2-sided)	Exact Sig. (2-sided)	Exact Sig. (1-sided)
Pearson Chi-Square	1.966[a]	1	.161		
Continuity Correction[b]	1.788	1	.181		
Likelihood Ratio	1.966	1	.161		
Fisher's Exact Test				.171	.091
Linear-by-Linear Association	1.964	1	.161		
N of Valid Cases	1207				

a. 0 cells (0.0%) have expected count less than 5. The minimum expected count is 158.24.

b. Computed only for a 2×2 table

FIGURE 11.6 | Chi-Square Tests

To evaluate whether we will reject the null hypothesis that no differences exist between males and females, we need to determine the degrees of freedom, which can be found in the SPSS output. For our results, the degrees of freedom equal 1. Since chi-square has a known distribution, the critical chi-square value for 2 degrees of freedom equals 3.841 at the 0.05 level of statistical significance, and our chi-square value is 1.966. Therefore, since our chi-square value does not exceed the critical chi-square value, we fail to reject the null hypothesis and conclude that there is no statistically significant difference between males and females in terms of whether they have been diagnosed with high blood pressure as adults.

Sometimes, you are interested in identifying the *strength* of a relationship (not just the existence), and this is what measures of association are for. Remember that above we selected phi and Cramer's *V* when we were setting up our analysis. Let's turn to this table now:

Symmetric Measures			
		Value	Approx. sig.
Nominal by nominal	Phi	0.040	0.161
	Cramer's *V*	0.040	0.161
N of valid cases		1207	

There are a number of relevant pieces of information here. First, notice that phi and Cramer's *V* values are identical. This will always be the case in a two by two table, which makes the calculation of phi somewhat unnecessary. (This is why it's included alongside Cramer's *V* in SPSS.) Second, since these are chi-square-based measures of association, you will notice that SPSS reports the significance of the measures. Chi-square is used to calculate significance here, so you needn't check the box for chi-square if you don't want to. The number of valid cases is also reported, and this number should coincide with the number of cases in the database (it does). Sometimes, if people do not answer a question, this number will go down because of missing values.

Remembering the classification criteria covered in Chapter 12; you can see that the relationship is weak and not significant.

Now it's your turn!

Putting the Information into Practice

Continuing with the theme of this chapter, we are going to ask this question: Who Is More Likely to Be Diagnosed with Anxiety Disorder as an Adult, Males or Females?

1. Answer the following questions:
   a. What is your null hypothesis?
   b. What is your research hypothesis?
   c. What is your dependent variable?
   d. What is your independent variable?
   e. What are the levels of measurement of your variables?

2. Produce a contingency table. Be sure to include percentages, the chi-square statistic, and phi and Cramer's *V*.
3. Based on your output, answer the following questions:
   a. What is the total valid sample size for this table?
   b. How many missing cases do you have?
   c. What percentage of females indicate they have been diagnosed with anxiety disorder as an adult? What percentage of males indicate the same?
   d. How much difference (variance) is there between males and females with respect to the percentage who have been diagnosed with anxiety disorder as an adult?
   e. What is the value of your chi-square statistic?
   f. How many degrees of freedom do you have?
   g. What is the critical chi-square value for this table?
   h. Are your findings statistically significant? Why?
   i. What is the nature of the relationship? Why?

Lab #12: Bivariate Statistics for Ordinal Data

The focus of this lab is to introduce you to the *association or relationship between ordinal variables*. Specifically, this lab will help clarify your understanding of independent and dependent variables, how to interpret the chi-square test of statistical significance, as well as other measures of association (i.e., gamma and Somers' *d*). This lab corresponds with material presented in Chapter 13.

LEARNING OBJECTIVES

The following lab is directed at helping you understand how to interpret the relationship between two variables (bivariate relationships). Specifically, this lab assignment challenges you to clarify your understanding of:

1. Dependent and independent variables
2. How to create a cross-tabulation or a contingency table
3. How to interpret your findings and determine statistical significance

Part 1: Establishing Your Research Question and Identifying Your Variables

In this example, we will ask this question: "Does Level of Education Affect an Individual's Opinion on Whether Alberta Should Build Stronger Ties with China?" (b2).

Step 1: Identify your dependent and independent variables.

An independent variable can be thought of as the modifier; the dependent variable can be thought of as the outcome of interest. In this situation, we are interested in seeing if highest level of education will modify opinions on Alberta's ties with China; therefore, it is our independent variable. The outcome we are interested in is opinions on Alberta building stronger ties with China; therefore it is our dependent variable.

Step 2: Recode your independent variable.

In this example, we are going to recode our independent variable from a 15 category variable into a 4 category variable measuring highest level of education. In Lab #4, you were introduced to the recode function. Let's review by recoding our highest level of education variable (k6).

Step 1: Click on:

Transform → Recode into different variables

It is important to select this option because you are interested in creating a new variable from your original variable rather than altering the existing variable (Figure 12.1).

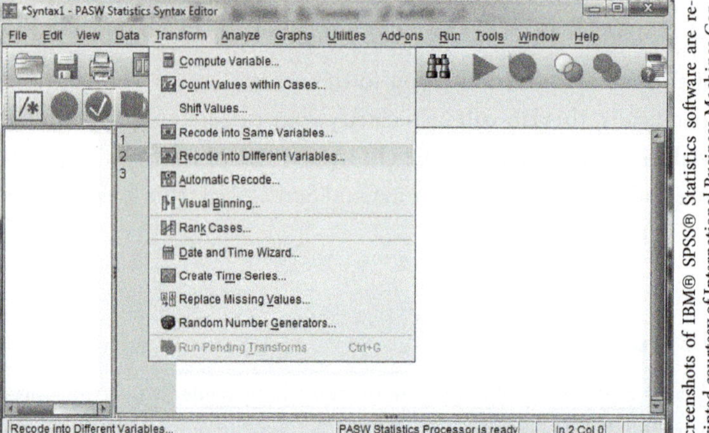

FIGURE 12.1 | Transform into Different Variables

Step 2: The "Recode into Different Variables" dialogue box will appear (Figure 12.2). First, you need to specify the variable you are interested in working with, in this example, Recode "Highest level of education" (k6). Bring that variable into the **Numeric Variable → Output Variable** box. Now, decide on a name and variable label for your new variable and type them into the Output Variable box. Then click on **Change**.

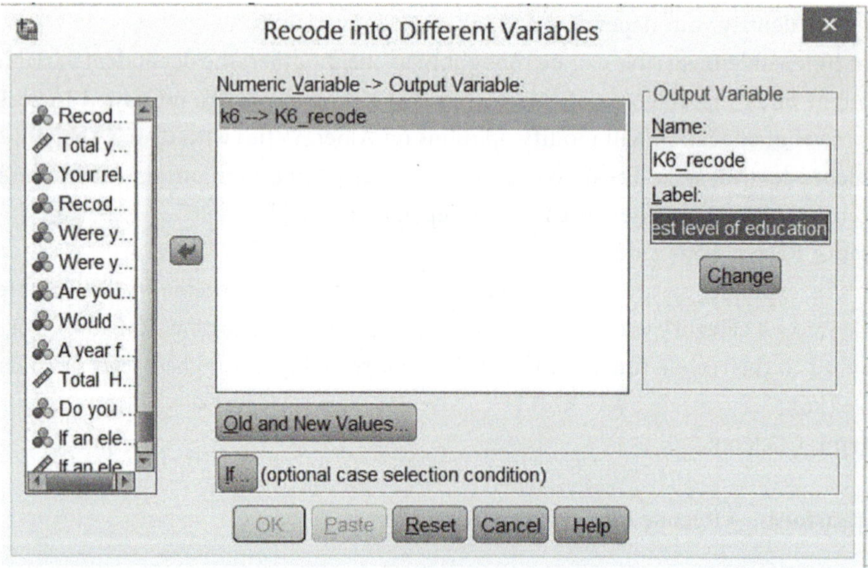

FIGURE 12.2 | Recode into Different Variables Dialogue Box

Step 3: You now want to define the categories within your variable. So, you need to click on **Old and new values** (Figure 12.3). Since you are interested in creating a variable where you want to collapse your variable into four categories:

1. Less than high school
2. High school
3. College/certificate/diploma
4. University degree or higher

You will need to tell SPSS that this is what you want.

Recall from the first lab that you can use your Data Editor screen to determine what values were assigned to particular categories within a variable. In Variable View, go to the row with your variable of interest and click on the cell corresponding with the Values column.

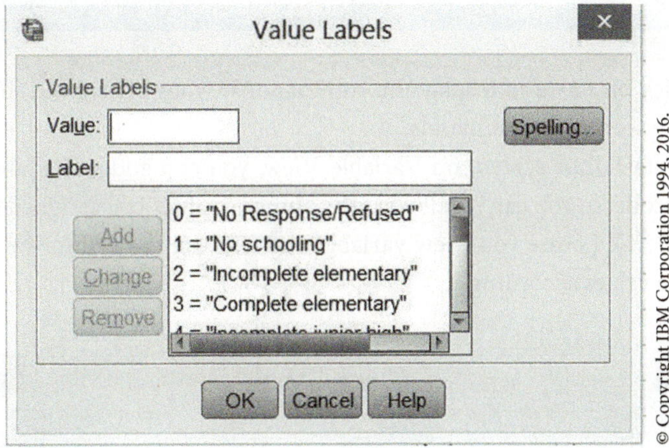

FIGURE 12.3 | Value Labels Dialogue Box

Step 4: By clicking on the Values column for variable k6, we will need to recode our new variable in the following way:

Old value → New value
1, 2, 3, 4, 5, 6 → 1 (Less than high school)
7, 8, 10 → 2 (High school)
9, 11 → 3 (College, certificate, diploma)
12, 13, 14, 15 → 4 (University degree or higher)

Step 5: Within the dialogue box you opened when you clicked on **Old and new values**, you will transform the values in the original k6 variable into the new k6_recode. In

Figure 12.4, you can see that 1 through 6 have been coded into 1; 7, 8, and 10 have been coded into 2; 9 and 11 have been coded into 3; and 12 through 15 have been coded into 4. Once you have finished defining your new variable, click on **Continue**.

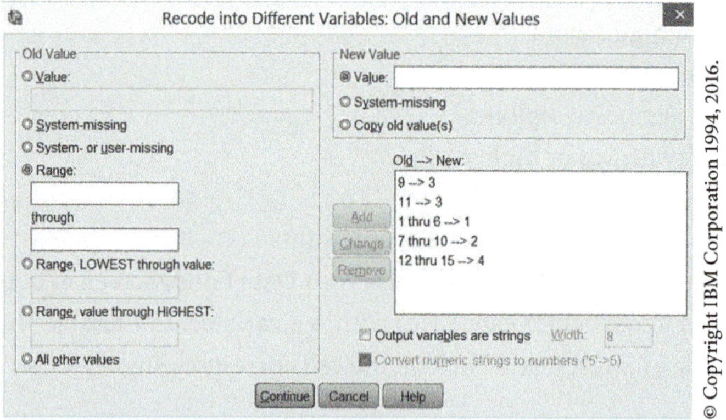

FIGURE 12.4 | Recoding into Different Variables: Old and New Values

Step 6: Click on **Paste** to display the commands in your syntax file. Then click on the green arrow to execute the commands.

In your Data Editor screen, on Variable View, you can add value labels to your new variable, k6_recode, so you can recall what the numeric values you assigned to each category mean (Figure 12.5). Locate your new variable and click on the cell in the Values column. Then, name the values accordingly.

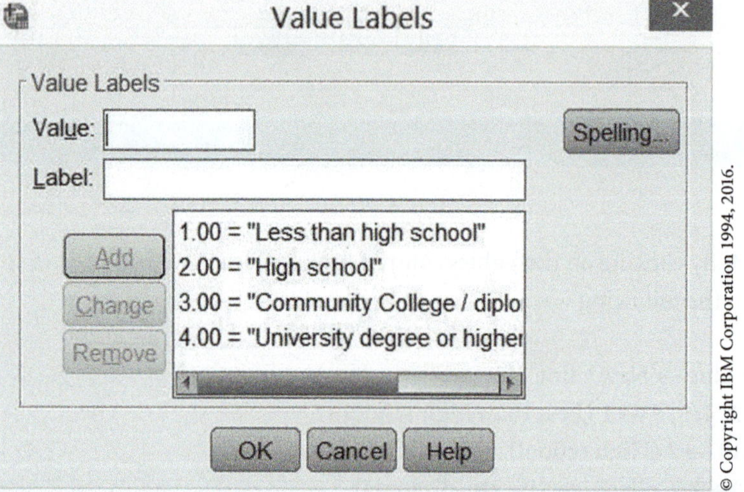

FIGURE 12.5 | Assigning Value Labels

Now, as a final check, run a frequency distribution on the new variable, k6_recode, and the original variable, k6, to confirm your new values add up to the values in the original variable.

Part 2: Creating Contingency Tables within SPSS

Now let's learn how to create contingency tables in SPSS.

Step 1: Using the syntax file, click on:

Analyze → Descriptive statistics → Crosstabs

Step 2: Select the dependent and independent variable.

This will bring up the crosstabs dialogue screen (Figure 12.6). Once you are in that screen, bring your dependent variable, Alberta Should Build Stronger Ties with China (b2) into the Row(s) box, and bring your independent variable, Recode: Highest Level of Education (k6_recode), into the Column(s) box. When you have finished doing that, click on **Statistics icon**.

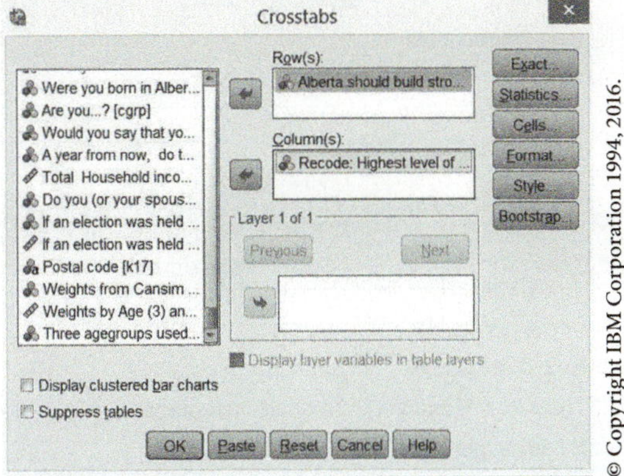

FIGURE 12.6 | Crosstabs Dialogue Screen

Step 3: Select your statistics.

Within this screen (Figure 12.7), select **Chi-square**, **Gamma**, and **Somers'** ***d***. Then click **Continue**.

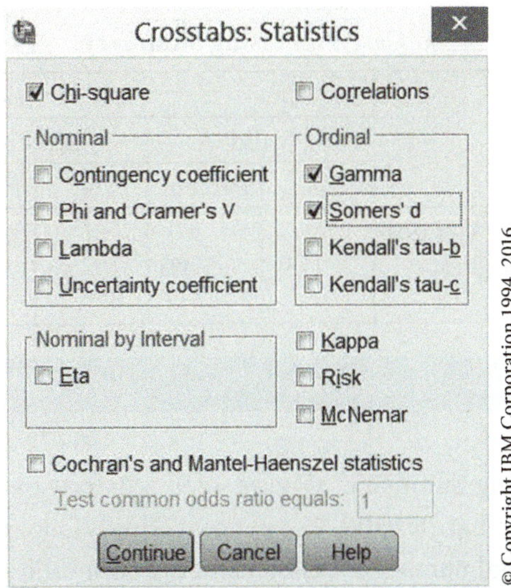

FIGURE 12.7 | Crosstabs: Statistics Dialogue Screen

Step 4: Percentage your columns.

You will return to the crosstabs screen (Figure 12.8). Once there, select the Cells button.

Within this screen, you will select Columns under the Percentages list. You are selecting Columns because you want to percentage along males and females so you can compare the differences across education levels. Then hit **Continue**, which will return you to the crosstabs screen. Then click **Continue, Paste** and run your syntax commands.

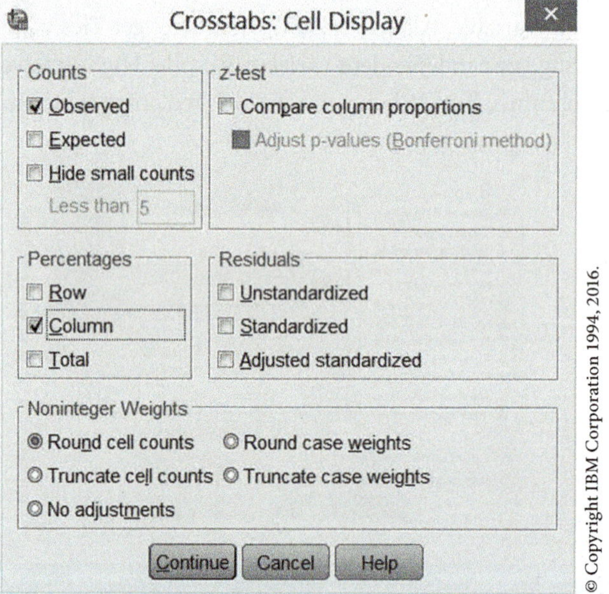

FIGURE 12.8 | Crosstabs: Cell Display Dialogue Screen

You should get the following output:

Case Processing Summary

	Cases					
	Valid		Missing		Total	
	N	Percent	N	Percent	N	Percent
Alberta should build stronger ties with China * Recode: Highest level of education	1151	95.4%	56	4.6%	1207	100.0%

FIGURE 12.9 | Case Processing Summary

The "Case Processing Summary" (Figure 12.9) tells you the number of respondents, 1151, and the percentage that are valid, 95.4 per cent. This is followed by the number of missing cases, 56, and the total number of cases, including both valid and missing cases, 1207.

Following the Case Summary Report is the contingency table (Figure 12.10), cross-tabulating highest level of education by opinion on whether Alberta should build stronger ties with China. Let's start to interpret these results. There do appear to be differences between levels of education and people's opinions. In particular, we find that those with less than a high school education are more likely to strongly disagree that Alberta should build stronger ties with China than those with a university degree or higher: 11.3 per cent compared with 3.3 per cent, respectively. Take a minute to look at the other cells within this table. What else do you notice?

Now we need to determine if the differences seen in the sample also exist within the population. In other words, are these differences statistically significant?

Alberta should build stronger ties with China * Recode: Highest level of education Crosstabulation

			Recode: Highest level of education				
			Less than high school	High school	Community college / diploma or certificate	University degree or higher	Total
Alberta should build stronger ties with China	Strongly disagree	Count	8	15	15	15	53
		% within Recode: Highest level of education	11.3%	4.7%	4.8%	3.3%	4.6%
	Disagree	Count	17	59	66	67	209
		% within Recode: Highest level of education	23.9%	18.5%	21.1%	15.0%	18.2%
	Neither disagree nor agree	Count	9	71	64	88	232
		% within Recode: Highest level of education	12.7%	22.3%	20.4%	19.6%	20.2%
	Agree	Count	36	161	142	235	574
		% within Recode: Highest level of education	50.7%	50.5%	45.4%	52.5%	49.9%
	Strongly agree	Count	1	13	26	43	83
		% within Recode: Highest level of education	1.4%	4.1%	8.3%	9.6%	7.2%
Total		Count	71	319	313	448	1151
		% within Recode: Highest level of education	100.0%	100.0%	100.0%	100.0%	100.0%

FIGURE 12.10 | Contingency Table

One way to test this is to use a chi-square test. To evaluate whether we will reject the null hypothesis that there is no relationship between education and opinion on Alberta's ties with China, we need to determine the degrees of freedom, which can be found in the SPSS output (Figure 12.11). For our results, the degrees of freedom equal 12. Since chi-square has a known distribution, the critical chi-square value for 12 degrees of freedom equals 21.026 at the 0.05 level of statistical significance, and our chi-square value is 30.123. Therefore, since our chi-square value exceeds the critical chi-square value, we can reject the null hypothesis and conclude that there is a statistically significant relationship between highest level of education and opinion on whether Alberta should build stronger ties with China.

The following two tables (Figures 12.12 and 12.13) contain our Somers' *d* statistic and the gamma statistic. We can see we have a Somers' *d* value of 0.089. The positive value means that we have a concordant relationship between our independent and dependent variable, where, for example, a value of Less than High School on the independent variable (Lower Levels of Education) results in a lower expected score on our dependent variable (Strongly Disagree). A score of 0.089 means that the association between our dependent and independent value is weak.

Chi-Square Tests

	Value	df	Asymptotic Significance (2-sided)
Pearson Chi-Square	30.123[a]	12	.003
Likelihood Ratio	30.691	12	.002
Linear-by-Linear Association	13.648	1	.000
N of Valid Cases	1151		

a. 1 cells (5.0%) have expected count less than 5. The minimum expected count is 3.27.

FIGURE 12.11 | Chi-Square Tests

Directional Measures

			Value	Asymptotic Standardized Error[a]	Approximate T[b]	Approximate Significance
Ordinal by Ordinal	Somers'd	Symmetric	.089	.025	3.576	.000
		Alberta should build stronger ties with China Dependent	.088	.025	3.576	.000
		Recode: Highest level of education Dependent	.091	.025	3.576	.000

a. Not assuming the null hypothesis.

b. Using the asymptotic standard error assuming the null hypothesis.

FIGURE 12.12 | Somers' *d* Statistic

We find a similar story with our gamma value (Figure 12.13). The positive value means we have a concordant relationship between our independent and dependent variable in which lower values on the independent variable (Lower Levels of Education) result in lower scores on our dependent variable (Strongly Disagree). The value of .130 means we are 13 per cent better at predicting the score on the dependent variable when we know the value of our independent variable. Furthermore, a score of 0.130 means the association between our dependent and independent values is moderate. Recall from Chapter 13 that the gamma statistic is more liberal in its calculation. If you want to err on the side of caution, it is recommended you use the Somers' *d*.

Symmetric Measures

		Value	Asymptotic Standardized Error[a]	Approximate T[b]	Approximate Significance
Ordinal by Ordinal	Gamma	.130	.036	3.576	.000
N of Valid Cases		1151			

a. Not assuming the null hypothesis

b. Using the asymptotic standard error assuming the null hypothesis.

FIGURE 12.13 | Gamma Value

Now's it's your turn!

Putting the Information into Practice

Continuing with the theme of this chapter, we are going to ask this question: Does Level of Education Affect an Individual's Opinion on Whether China Will Play an Increasingly Significant Role in the Future of Albertans (b10)?

1. Answer the following questions:
   a. What is your null hypothesis?
   b. What is your research hypothesis?
   c. What is your dependent variable?
   d. What is your independent variable?
   e. What are the levels of measurement of your variables?

2. Produce a contingency table. Be sure to include percentages and the chi-square statistic, gamma, and Somers' *d*.
3. Based on your output, answer the following questions:
   a. What is the total valid sample size for this table?
   b. How many missing cases do you have?
   c. What percentage of individuals with less than a high school education indicate they strongly disagree? What percentage of individuals with a university degree or higher indicate the same?
   d. How much difference (variance) is there between those with less than a high school education and those with a university degree or higher with respect to the percentage that strongly agree that China will play an increasingly significant role in the future of Albertans?
   e. What is the value of your chi-square statistic?
   f. How many degrees of freedom do you have?
   g. What is the critical chi-square value for this table?
   h. Are your findings statistically significant? Why?
   i. What is the value of your gamma and your Somers' *d*? What do these statistics tell you?

Lab #13: Bivariate Statistics for Interval/Ratio Data

The focus of this lab is to introduce you to the *association or relationship between interval/ratio level variables*. Specifically, this lab will help clarify your understanding of Pearson's *r* and explained variance. This lab corresponds with the material presented in Chapter 14.

LEARNING OBJECTIVES

The following lab is directed at helping you understand how to interpret the relationship between two variables (bivariate relationships). Specifically, this lab assignment challenges you to clarify your understanding of:

1. Dependent and independent variables
2. How to calculate and interpret Pearson's *r*
3. How to interpret explained variance

Calculating Pearson's *r* in SPSS

When calculating Pearson's *r*, we make the assumption that the two variables we are correlating (evaluation of the extent to which the variables are related) have a linear relationship. That is, the relationship between the two variables is the same, regardless of what the value of either of these variables is. So, the first step in calculating Pearson's *r* is to evaluate whether the relationship between the two variables is linear.

In this example, we are interested in whether the age of individuals is correlated with the number of years of schooling they have completed. In this case, our dependent variable is "Total years of schooling," and our independent variable is an individual's age.

Step 1: Evaluate for linearity.

To evaluate for linearity, we will use a scatterplot. To create a scatterplot in SPSS, click on:

Graphs → Legacy dialogs → Scatter/dot . . .

Select "Simple Scatter," then click on **Define** (Figure 13.1). Place your dependent variable in the *y*-axis and your independent variable in your *x*-axis. Then click **Paste** and run your Syntax commands.

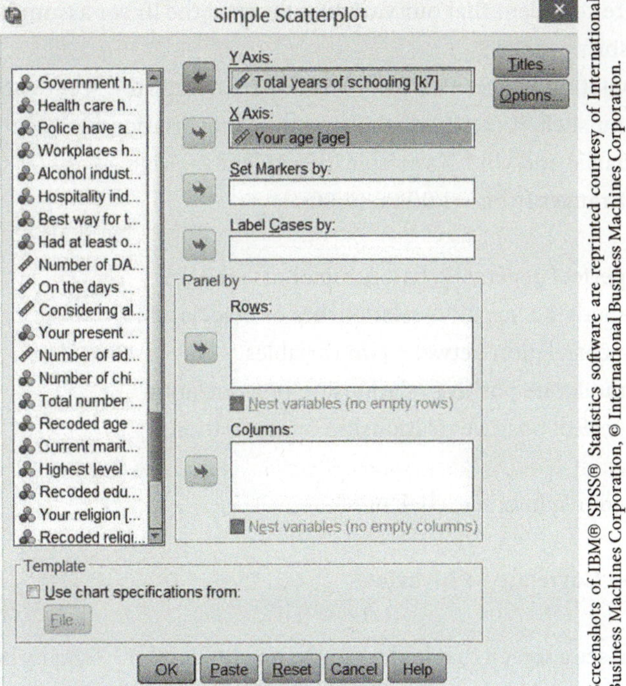

FIGURE 13.1 | Simple Scatterplot Dialogue Screen

This is the scatterplot that should have been produced:

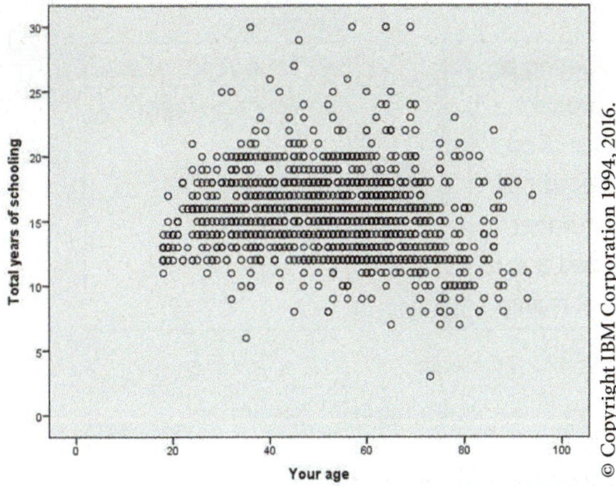

FIGURE 13.2 | Scatterplot Graph

Here are a few observations we could make about this scatterplot:

1. While not as strong as we might have liked to have obtained, the dots appear to be arranged in a linear fashion—thus, for our purposes, satisfying the linear assumption of the Pearson correlation.
2. There are a few outliers both above and below the line. . . .

Now that we are confident that our variables do meet the linear assumption, we are ready to calculate Pearson's r in SPSS.

The computation of our Pearson's r is done "behind the scenes" by SPSS and is quite complex. In a nutshell, it considers the amount of covariation between your X variable (independent variable) and your Y variable (dependent variable).

Pearson's r ranges from −1.00 to +1.00:

−1.0 = a perfect negative relationship or association
−0.5 = a moderate negative relationship or association
0.0 = no correlation between two variables
+0.5 = a moderate positive relationship or association
+1.0 = a perfect positive relationship or association

Step 2: In your syntax file, click on:

Analyze → Correlate → Bivariate

You should see a screen that looks like the one in Figure 13.3. Bring both your dependent and independent variables into the variable list. Be sure to select **Pearson** under the Correlation Coefficients.

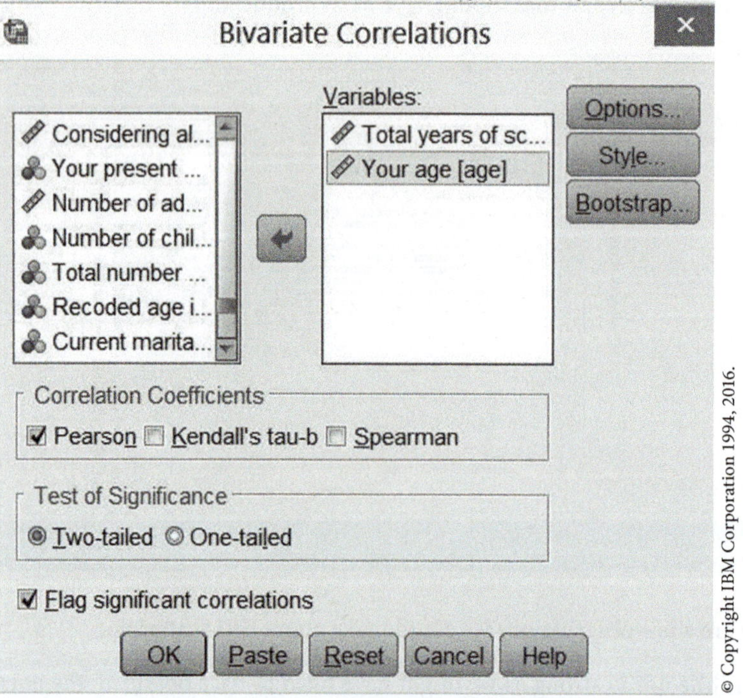

FIGURE 13.3 | Bivariate Correlations Dialogue Screen

Now click on **Options** . . . and select **Means and standard deviations** (Figure 13.4). Then click **Continue, Paste** and run your Syntax commands.

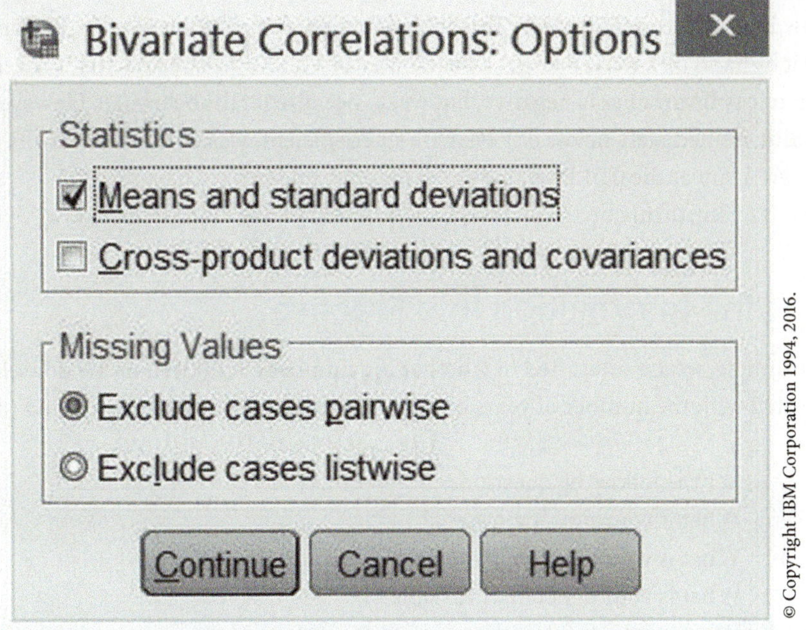

FIGURE 13.4 | Bivariate Correlations: Options Dialogue Screen

You should see two tables that look like these in Figure 13.5:

Descriptive Statistics

	Mean	Std. Deviation	N
Total years of schooling	15.43	3.330	1193
Your age	52.44	16.350	1176

Correlations

		Total years of schooling	Your age
Total years of schooling	Pearson Correlation	1	-.100**
	Sig. (2-tailed)		.001
	N	1193	1167
Your age	Pearson Correlation	-.100**	1
	Sig. (2-tailed)	.001	
	N	1167	1176

**. Correlation is significant at the 0.01 level (2-tailed).

FIGURE 13.5 | Descriptive Statistics and Correlation Matrix Tables

Looking at the first table, we observe that 1193 people answered the schooling question whereas only 1176 reported their age. We can see that, on average, respondents have 15.43 years of schooling and the average respondent age is 52.44 years.

Now look at the second table. This table is called a *correlation matrix*. Referring to the upper-right-hand cell, we see we get a Pearson's r of −0.100. This means that the relationship between our two variables is negative, but weak, because it falls below 0.5. However, looking at the value immediately below our Pearson's r coefficient, we see that our results are statistically significant at the 0.01 level.

Now it's your turn!

Putting the Information into Practice

In this example, we are interested in whether the number of children in the household (k3b) is correlated with the number of years of schooling respondents have completed (k7).

1. Answer the following questions:
   a. What is your null hypothesis?
   b. What is your research hypothesis?
   c. What is your dependent variable?
   d. What is your independent variable?
   e. What are the levels of measurement of your variables?

2. Produce a scatter plot.
   a. Does the relationship between your two variables meet the assumption of linearity?

3. Calculate Pearson's r. Be sure to include the mean and standard deviation of your variables.
   a. What sample size did SPSS use to calculate the Pearson correlation coefficient between k7 and k3b?
   b. What is the mean value of k7? What is the mean value of k3b?
   c. What is the Pearson correlation coefficient? Is it significant? Is it small? Large?
   d. In terms of our two variables, what does the Pearson correlation coefficient mean? (Hint: Is the Pearson correlation coefficient positive or negative?)

Lab #14: Analysis of Variance

The focus of this lab is to introduce you to a procedure known as ANOVA, or *analysis of variance*. Specifically, this lab will help clarify your understanding of when this procedure should be used and how to calculate within-group sum of squares, between-group sum of squares, and the total sum of squares, which are the three major components of ANOVA. Finally, this lab will teach you how to interpret the *F*-distribution with ANOVA. This lab corresponds with the material presented in Chapter 15.

LEARNING OBJECTIVES

The following lab is directed at helping you understand how to interpret the relationship between two variables by using a procedure called ANOVA. Specifically, this lab assignment challenges you to clarify your understanding of:

1. Dependent and independent variables
2. Null and research hypotheses
3. How to interpret your findings and determine statistical significance

Analysis of variance, or ANOVA, is like a *t*-test but allows us to compare more than two groups. Conceptually, ANOVA compares three things:

1. Differences between means
2. Differences in values within samples
3. Differences in values across samples

Essentially, ANOVA is used to compare the variation caused by the independent variable and the variation that occurs at random around the mean within groups to the variation across groups.

Calculating ANOVA with SPSS

In this example, we will ask this question: Are There Significant Differences in Average Years of Schooling (k7) by the Area of the Province Respondents Live in (strata)?

Step 1: Request a one-way ANOVA.

In a new syntax file, click on:

Analyze → Compare means → One-way ANOVA

Step 2: From the left-hand screen, bring over your dependent variable, Total Years of Schooling, into the dependent list (Figure 14.1). Next, bring your independent variable into the Factor List, Area of the Province.

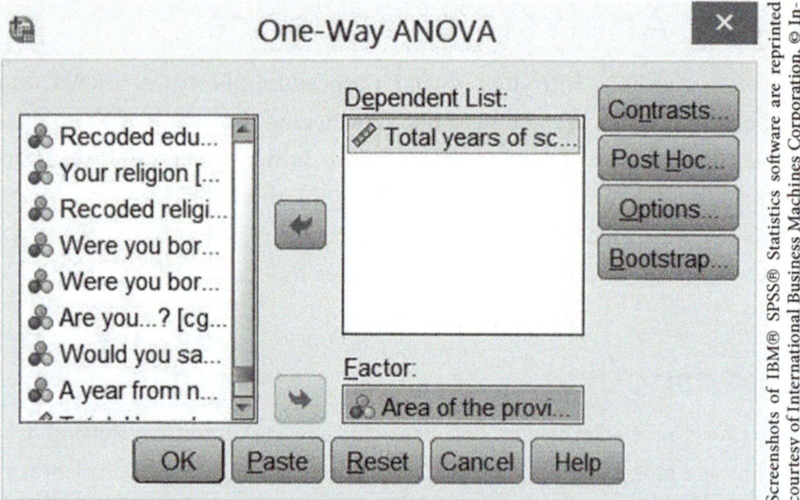

FIGURE 14.1 | One-Way ANOVA Dialogue Screen

Select **Options.** Within Options, click on **Descriptive** (Figure 14.2).

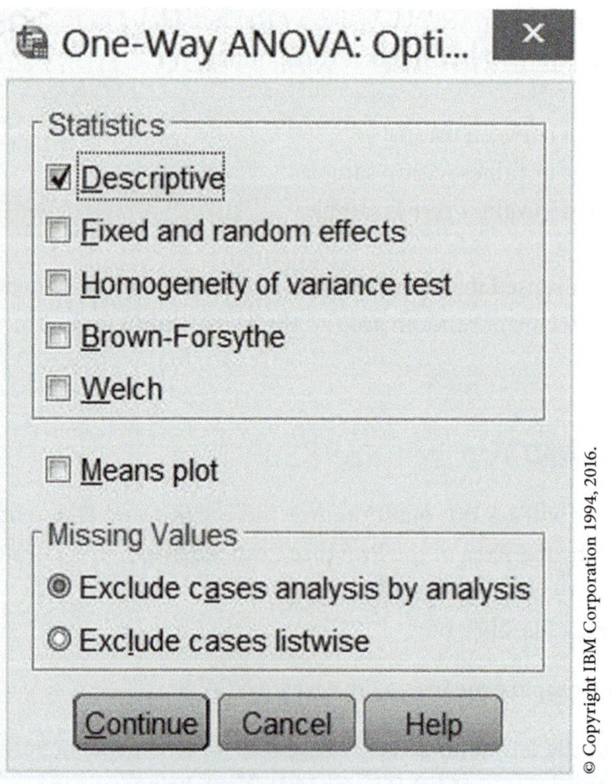

FIGURE 14.2 | One-Way ANOVA Options Dialogue Screen

Now, select **Continue** and **Paste** and run your Syntax commands. You will get the following results in your output file.

Descriptives

Total years of schooling

	N	Mean	Std. Deviation	Std. Error	95% Confidence Interval for Mean Lower Bound	95% Confidence Interval for Mean Upper Bound	Minimum	Maximum
Metro Edmonton	400	15.61	3.436	.172	15.27	15.94	3	30
Metro Calgary	397	15.90	3.144	.158	15.59	16.21	8	30
Other Alberta	396	14.80	3.313	.166	14.47	15.12	7	30
Total	1193	15.43	3.330	.096	15.24	15.62	3	30

FIGURE 14.3 | Descriptives Output

There are a few things to note here. First, let's look at the means. You will notice that respondents from Edmonton (strata = 1) have, on average, 15.61 years of schooling. Those from Calgary (strata = 2) have, on average, 15.90 years of schooling. Respondents from other parts of Alberta (strata = 3) have, on average, 14.8 years of schooling. Although differences are observed, are they meaningful? In other words, can we expect to see them in our population?

ANOVA

Total years of schooling

	Sum of Squares	df	Mean Square	F	Sig.
Between Groups	258.165	2	129.082	11.854	.000
Within Groups	12958.788	1190	10.890		
Total	13216.952	1192			

FIGURE 14.4 | ANOVA Table

To determine this, we need to look at the ANOVA table (Figure 14.4). For our between-group sum of squares we can see that we have two degrees of freedom and a value of 258.165. For our within-group sum of squares, we have 1190 degrees of freedom and a value of 12,958.788. SPSS automatically gives us the level of statistical significance, but we could go to Appendix D of your textbook, just as we have been doing, to get the same result. If we go to Appendix D we find that the critical *F*-distribution score for two degrees of freedom for the between-groups score and for more than 120 degrees of freedom for the within-group value at alpha = 0.01 is 4.61. We can see in our table that our *F*-score (*F*-observed) is 11.854. Because this number exceeds the critical *F*-score of 4.61, we can conclude with 99 per cent confidence that there are significant differences between at least two groups.

Recoding to the Midpoint

Sometimes we have variables that are measured at the ordinal level that we want to turn into interval/ratio variables. An easy way to do this is called *recoding to the midpoint.* We will illustrate this procedure using the variable k12a (income).

First, run a frequency distribution on the original variable.

Total Household income for the past year before taxes and deductions

		Frequency	Percent	Valid Percent	Cumulative Percent
Valid	Under 6000	5	.4	.5	.5
	6000 - 7999	2	.2	.2	.7
	8000 - 9999	2	.2	.2	.9
	10,000 - 11,999	2	.2	.2	1.1
	12,000 - 13,999	6	.5	.6	1.7
	14,000 - 15,999	5	.4	.5	2.2
	16,000 - 17,999	4	.3	.4	2.6
	18,000 - 19,999	9	.7	.9	3.6
	20,000 - 21,999	17	1.4	1.7	5.3
	22,000 - 23,999	14	1.2	1.4	6.7
	24,000 - 25,999	17	1.4	1.7	8.4
	26,000 - 27,999	6	.5	.6	9.0
	28,000 - 29,999	9	.7	.9	10.0
	30,000 - 31,999	24	2.0	2.4	12.4
	32,000 - 33,999	6	.5	.6	13.0
	34,000 - 35,999	21	1.7	2.1	15.1
	36,000 - 37,999	5	.4	.5	15.7
	38,000 - 39,999	5	.4	.5	16.2
	40,000 - 44,999	25	2.1	2.5	18.7
	45,000 - 49,999	22	1.8	2.2	20.9
	50,000 - 54,999	37	3.1	3.8	24.7
	55,000 - 59,999	14	1.2	1.4	26.1
	60,000 - 64,999	42	3.5	4.3	30.4
	65,000 - 69,999	18	1.5	1.8	32.2
	70,000 - 74,999	48	4.0	4.9	37.1
	75,000 - 79,999	26	2.2	2.6	39.7
	80,000 - 84,999	45	3.7	4.6	44.3
	85,000 - 89,999	18	1.5	1.8	46.1
	90,000 - 94,999	24	2.0	2.4	48.6
	95,000 - 99,999	18	1.5	1.8	50.4
	100,000 - 124,999	154	12.8	15.7	66.1
	125,000 - 149,999	74	6.1	7.5	73.6
	150,000+	260	21.5	26.4	100.0
	Total	984	81.5	100.0	
Missing	Don't know	72	6.0		
	No response	151	12.5		
	Total	223	18.5		
Total		1207	100.0		

FIGURE 14.5 | Frequency Distribution on k12a

You can see that there are 33 categories of income and two missing values (Don't Know and No Response).

To recode at the midpoint, we add the endpoints of each category and divide by 2. So the second category, for example, which goes from $6000 to $7999 would be recoded at $6999.50. For the first category (which has no lower limit) and the final category (which has no upper limit), we usually just use the value listed.

First, click on **Transform**, then **Recode into different variable,** select **k12a**, and name the new variable k12a_recode.

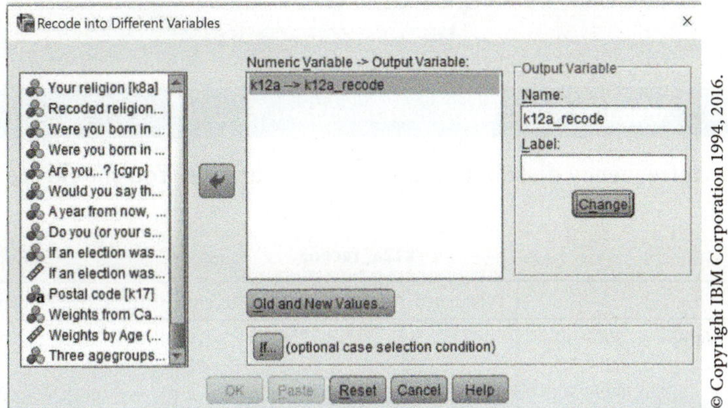

FIGURE 14.6 | Recoding Variable k12a

Next, click on **Old and new values.** You will want to enter the following old and new values:

1 = 6000.0	12 = 26,999.5	23 = 62,499.5
2 = 6999.5	13 = 28,999.5	24 = 67,499.5
3 = 8999.5	14 = 30,999.5	25 = 72,499.5
4 = 10,999.5	15 = 32,999.5	26 = 77,499.5
5 = 12,999.5	16 = 34,999.5	27 = 82,499.5
6 = 14,999.5	17 = 36,999.5	28 = 87,499.5
7 = 16.999.5	18 = 38,999.5	29 = 92,499.5
8 = 18,999.5	19 = 42,499.5	30 = 97,499.5
9 = 20,999.5	20 = 47,499.5	31 = 112,499.5
10 = 22,999.5	21 = 52,499.5	32 = 137,499.5
11 = 24,999.5	22 = 57,499.5	33 = 150,000.0

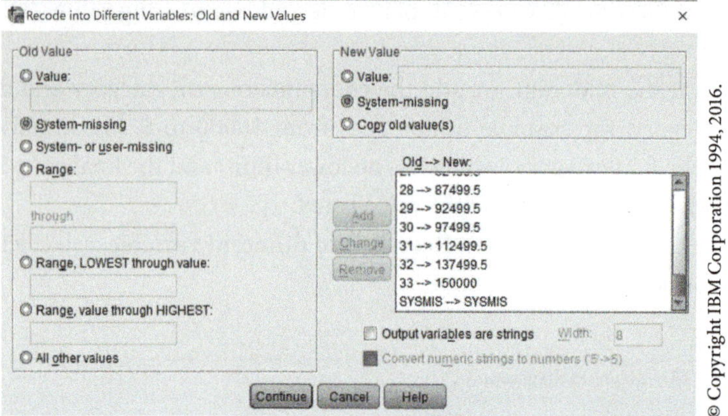

FIGURE 14.7 | Recoding Variable k12a: Old and New Values

Next, run a frequency distribution on your new variable. You should see what appears in Figure 14.8.

k12a_recode

		Frequency	Percent	Valid Percent	Cumulative Percent
Valid	6000.00	5	.4	.5	.5
	6999.50	2	.2	.2	.7
	8999.50	2	.2	.2	.9
	10,999.50	2	.2	.2	1.1
	12,999.50	6	.5	.6	1.7
	14,999.50	5	.4	.5	2.2
	16,999.50	4	.3	.4	2.6
	18,999.50	9	.7	.9	3.6
	20,999.50	17	1.4	1.7	5.3
	22,999.50	14	1.2	1.4	6.7
	24,999.50	17	1.4	1.7	8.4
	26,999.50	6	.5	.6	9.0
	28,999.50	9	.7	.9	10.0
	34,999.50	21	1.7	2.1	15.1
	36,999.50	5	.4	.5	15.7
	38,999.50	5	.4	.5	16.2
	42,499.50	25	2.1	2.5	18.7
	47,499.50	22	1.8	2.2	20.9
	52,499.50	37	3.1	3.8	24.7
	57,499.50	14	1.2	1.4	26.1
	62,499.50	42	3.5	4.3	30.4
	67,499.50	18	1.5	1.8	32.2
	72,499.50	48	4.0	4.9	37.1
	77,499.50	26	2.2	2.6	39.7
	82,499.50	45	3.7	4.6	44.3
	87,499.50	18	1.5	1.8	46.1
	92,499.50	24	2.0	2.4	48.6
	97,499.50	18	1.5	1.8	50.4
	112,499.50	154	12.8	15.7	66.1
	137,499.50	74	6.1	7.5	73.6
	150,000.00	260	21.5	26.4	100.0
	Total	984	81.5	100.0	
Missing	System	223	18.5		
Total		1207	100.0		

FIGURE 14.8 | Frequency Distribution on Variable k12a Recoded

As you can see, you now have an interval/ratio variable, so you can (for example) run summary descriptive statistics on your new variable (see Figure 14.9) and use it as a dependent variable for an ANOVA or a linear regression model.

Descriptive Statistics

	N	Minimum	Maximum	Mean	Std. Deviation
k12a_recode	984	6000.00	150,000.00	95,540.7932	45,652.29501
Valid N (listwise)	984				

FIGURE 14.9 | Summary of Descriptive Statistics, Variable k12a, Recoded

Now it's your turn!

Putting the Information into Practice

In this assignment, we will ask this question: Are There Significant Differences in Income by Area of the Province? Use k12a to measure income (you will have to recode this variable to make it an interval ratio variable—see above) and strata.

1. Answer the following questions:
   a. What is your null hypothesis?
   b. What is your research hypothesis?
   c. What is your dependent variable?
   d. What is your independent variable?
   e. What are the levels of measurement of your variables?

2. Calculate ANOVA (remember to include your descriptive statistics).
3. In your output, using the table called Descriptives, calculate the difference in mean income for
   a. Those who live in Edmonton and those who live in Calgary: ____________
   b. Those who live in Edmonton and those who live in another part of Alberta (not Calgary): ____________
   c. Those who live in Calgary and those who live in another part of Alberta (not Edmonton): ____________

4. What do these data suggest about the relationship between area of residence and income?
5. State the value of the total sum of squares, the within-groups sum of squares, and the between-groups sum of squares.
6. What is the value of your *F*-statistic?
   a. Does this mean your results are statistically significant?
   b. Do you accept or reject your null hypothesis? Why?

Lab #15: OLS Regression: Modelling Continuous Outcomes

The focus of this lab is to introduce you to a form of multivariate analysis called *regression analysis*. Specifically, this lab will help clarify your understanding of when this procedure should be used, how to calculate and interpret ordinary least squares (OLS) regression, and how to compute dummy variables. This lab corresponds with the material presented in Chapter 16.

LEARNING OBJECTIVES

The following lab is directed at helping you understand how to interpret the relationship between multiple variables using OLS regression. Specifically, this lab assignment challenges you to clarify your understanding of:

1. Dummy variables
2. Standardized partial slopes
3. How to interpret your findings and determine statistical significance

Calculating OLS Regression with SPSS

In this example, we are interested in which variables might affect the number of years of schooling an individual has completed. For various reasons, we think that the number of years of schooling an individual has completed may vary by gender (sex) and by what part of Alberta they live in (strata).

Step 1: Code your variables.

Because one of the conditions of OLS regression states that all variables must be interval, ratio, or dummy variables, we will need to recode our variables to meet these conditions.

First, gender is a nominal level variable where the values are coded 1 "Male" and 2 "Female." We will need to recode this variable into a dummy variable where the response categories are 1 "Male" and 0 "Female."

Recall from earlier labs, to recode a variable, click on:

Transform → Recode into different variables

When the dialogue box appears, select the original gender variable (sex) and place it in the box labelled **Numeric Variable → Output Variable**. Then, name your new variable under the box labelled Output Variable. Since we will be making males equal to a value of 1, we will call our new variable male, as we have done in Figure 15.1. Once you have done that, select the button **Change**.

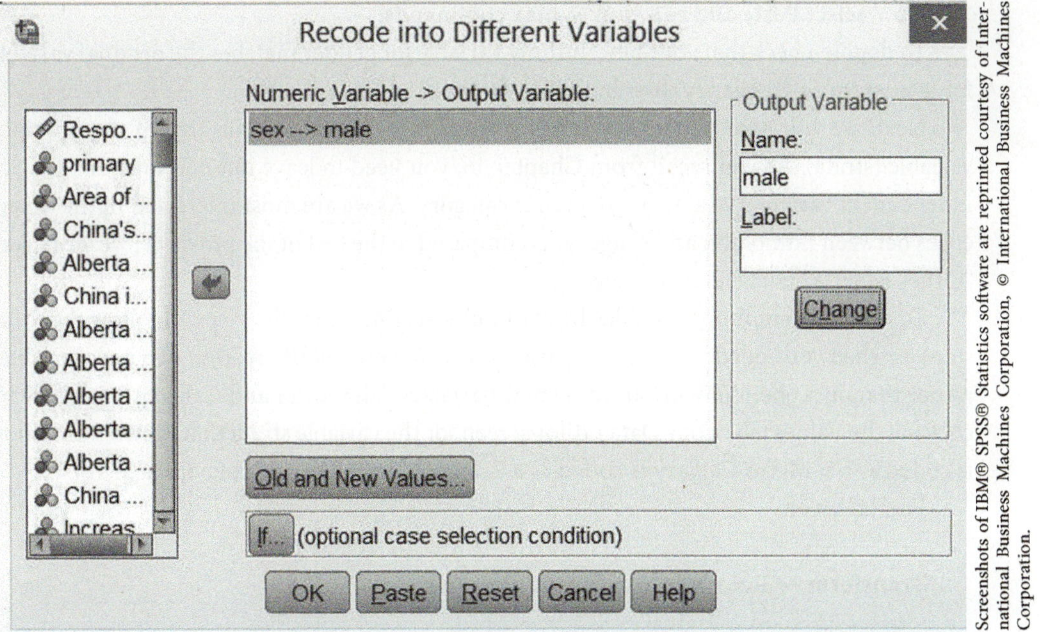

FIGURE 15.1 | Recode into Different Variables Dialogue Screen

Next, click the button **Old and new values.** In this screen you will want to make the old value of 1 into new value 1. In other words, the male value of 1 will remain as 1. Next, make the old value 2 into new value 0. Now, the female value of 2 will become 0. Select **Continue** to return to the **Recode into different variables** dialogue screen (Figure 15.2).

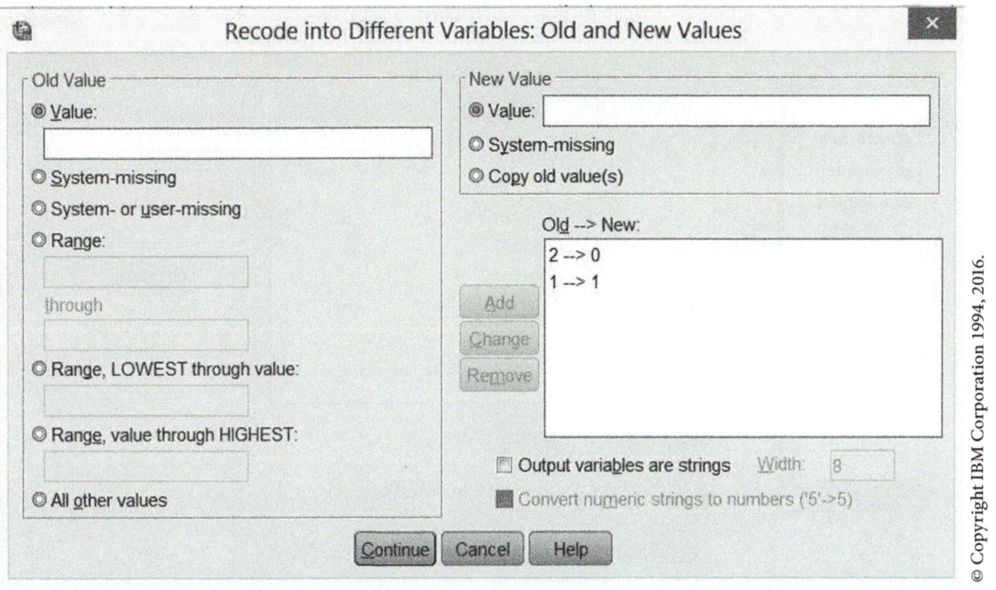

FIGURE 15.2 | Recode into Different Variables: Old and New Values Dialogue Screen

Now, select **Paste** and run your Syntax commands.

To double-check that your new dummy variable for gender matches the original variable for gender, run a frequency distribution on both variables.

Next, we will need to create a series of dummy variables for our area of the province variable (strata). As you recall from Chapter 16, you need to leave out one category of your independent variable (strata) as a reference category. As we are most interested in the differences between Edmonton and Calgary, as compared to the rest of the province, we will make "Other Alberta" our reference category.

To create two dummy variables from our one original variable, you will follow the same steps as when you coded your gender variable, except you will be creating two new variables rather than just one. We will call these new variables Edmonton and Calgary. We know by clicking the Values tab in our Data Editor screen for the variable strata that Metro Edmonton is coded as a 1, Metro Calgary is coded as a 2, and Other Alberta is coded as a 3.

First, click on:

Transform → Recode into different variables

When the dialogue box appears (Figure 15.3), select the original variable (strata) and place it in the box labelled Numeric Variable → Output Variable. Then, name your new variable under the box labelled Output Variable. Since our first dummy variable corresponds with Edmonton, we will be making Edmonton equal to a value of 1 and everyone else equal to 0; we will call our new variable Edmonton, as we have done in Figure 15.3. Once you have done that, select the button **Change**.

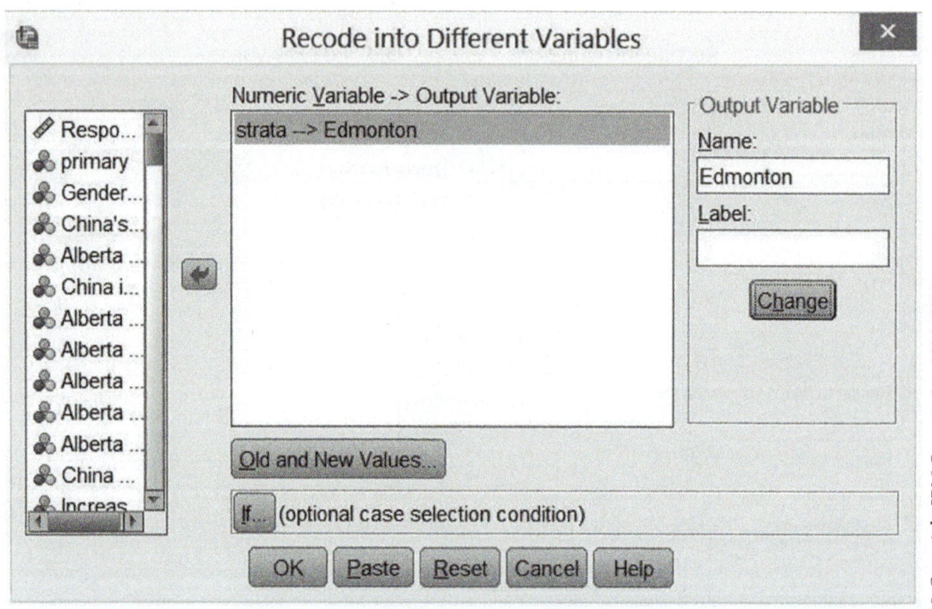

FIGURE 15.3 | Recode into Different Variables Dialogue Screen

Next, click the button **Old and new values.** In this screen, you will want to make the old value of 1 into new value 1. In other words, the Edmonton value of 1 will remain as 1. Next, make the old value 2 into new value 0, and old value 3 into new value 0. Now, the Calgary value of 2 will become 0 and the Other Alberta value of 3 will become 0. Select **Continue** to return to the **Recode into different variables** dialogue screen (Figure 15.4).

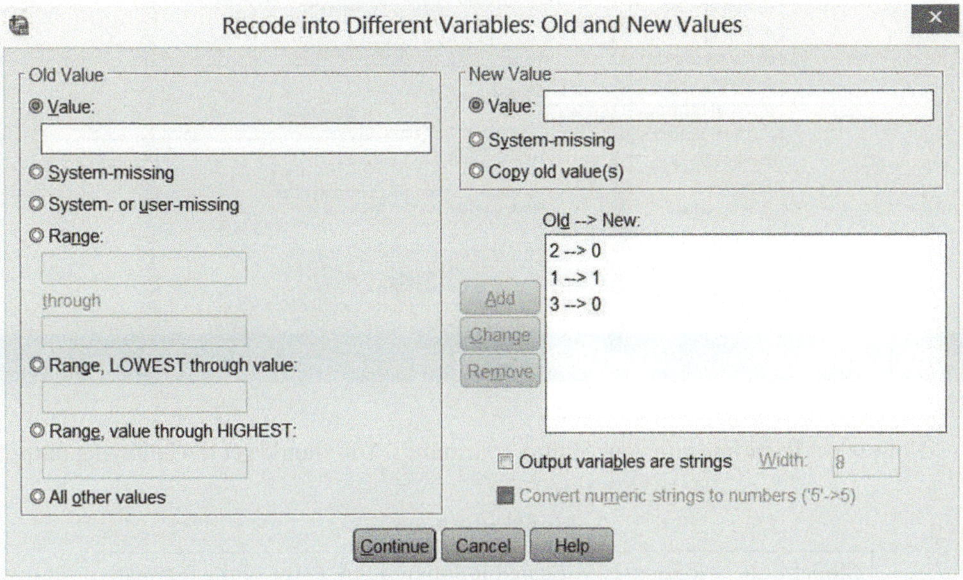

FIGURE 15.4 | Recode into Different Variables: Old and New Values Dialogue Screen

Now, select **Paste** and run your Syntax commands. Repeat this process for Calgary, making a new variable named Calgary, where the old value of 2 becomes the new value of 1, and both other old values have a new value of 0.

Step 2: Conduct an OLS regression in SPSS.

First, click on:

Analyze → Regression → Linear

Bring your dependent variable, Total Years of Schooling (k7), into the Dependent box (Figure 15.5). Then, bring your independent variables (male, Edmonton, Calgary) into the Independent(s) box.

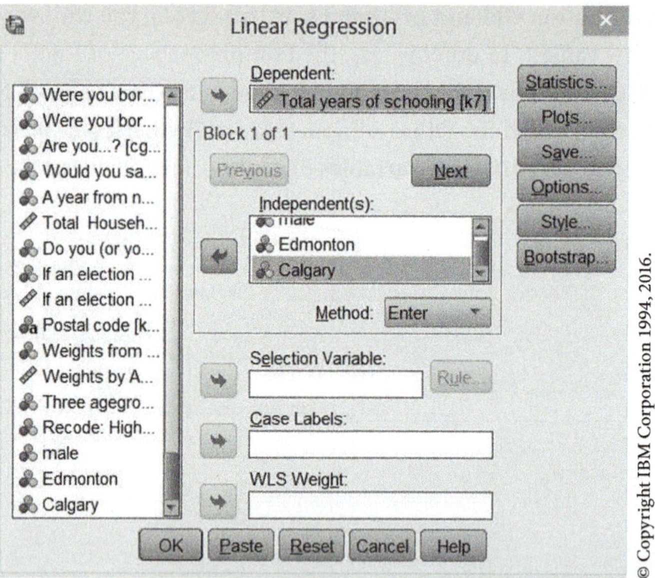

FIGURE 15.5 | Linear Regression Dialogue Screen

Now, select **Paste** and run your Syntax commands. You should get the following output.

Model Summary

Model	R	R Square	Adjusted R Square	Std. Error of the Estimate
1	152[a]	.023	.021	3.295

a. Predictors: (Constant), Calgary, male, Edmonton

FIGURE 15.6 | Model Summary

First, look at the Model Summary table. Notice the "Adjusted R Square" value. *R*-squared measures how well the model fits your data; it tells you how much of the variation in the dependent variable can be explained by all the independent variables. In our example, we have explained about 2.1 per cent of the variation in educational attainment.

ANOVA[a]

Model		Sum of Squares	df	Mean Square	F	Sig.
1	Regression	306.020	3	102.007	9.394	.000[b]
	Residual	12910.933	1189	10.859		
	Total	13216.952	1192			

a. Dependent Variable: Total years of schooling
b. Predictors: (Constant), Calgary, male, Edmonton

FIGURE 15.7 | ANOVA Table

The ANOVA table displays information on the variation, whether random or caused by the independent variables, around the mean within groups and across groups. We find that the differences in variation between and across groups is statistically significant (F = 9.394, Sig. = 0.000, which is less than 0.05).

Finally, we have the piece of output that quantifies our regression equation, the coefficients table. Now, if you recall, the regression equation with the intercept, slope and error term is as follows:

$$y = a + bx + e$$

A **regression equation** expresses the relationship between two or more variables. The variable a is the constant term, the intercept value when all independent values are set to 0, and it equals 14.606. This means that our respondents in the reference category (Females, Other Alberta) have, on average, 14.606 years of schooling, a finding that is statistically significant at the 0.000 level. The bx term in the equation above represents the coefficients (b) and the independent variables (x), and their values are listed below in the column labelled B. We can see from the output that men have 0.401 years more schooling than females, and this finding is statistically significant ($p = 0.036$, which is less than 0.05). Next, individuals living in Edmonton have 0.799 more years of schooling than those living in Other Alberta, and this finding is statistically significant at the 0.001 level. Finally, individuals living in Calgary have 1.090 more years of schooling than those living in Other Alberta, and this finding is statistically significant at the 0.000 level.

Coefficients[a]

Model		Unstandardized Coefficients		Standardized Coefficients		
		B	Std. Error	Beta	t	Sig.
1	(Constant)	14.606	.189		77.469	.000
	male	.401	.191	.060	2.099	.036
	Edmonton	.799	.234	.113	3.421	.001
	Calgary	1.090	.234	.154	4.655	.000

a. Dependent Variable: Total years of schooling

FIGURE 15.8 | Coefficients Table

Now it's your turn!

Putting the Information into Practice

In this assignment, we are interested in whether age, gender, and education affect income. You should use your recoded income variable (which you created in lab 14 above—k12a_recode).

1. Answer the following questions:
   a. What is your null hypothesis?
   b. What is your research hypothesis?
   c. What is your dependent variable?
   d. What are your independent variables?

2. Conduct an OLS regression analysis using SPSS.
3. What is the value of your *R*-squared? What does this mean?
4. Write the equation for your regression equation.
5. What is the effect of age on income, controlling for gender and education? Is this finding statistically significant?
6. What is the effect of education on income, controlling for gender and age? Is this finding statistically significant?
7. What is the effect of gender on income, controlling for age and education? Is this finding statistically significant?

An Introduction to Statistics
for Canadian Social Scientists

STATA Lab Manual

Contents

Preface

How Does This Manual Work?

This manual is intended to give you the opportunity to apply the concepts and principles you learn in each chapter of the text. Through practical examples and step-by-step instructions, this manual seeks to help you improve your understanding of statistical practices and introduce you to STATA, a key statistical tool used in the social sciences. Often we learn through practice, and by actively engaging with the examples and assignments covered in this lab, you will improve your understanding of the course material and the basics of STATA.

For nearly all of the chapters in your textbook, there is a corresponding lab exercise. The labs are approached in a workbook style where you will need to participate in each step to successfully complete the final assignment. You may find the labs challenging at times, even frustrating. Take a deep breath! Rome wasn't built in a day and neither is expertise in statistics. To master a skill, one must practise, be challenged, and then try again. It is the intention of the lab manual to challenge you in this way. The knowledge is cumulative, so if you find a particular section tricky, return to earlier sections and brush up on the basics. You may find you need to do this several times before you are able to make sense of the material and the technical aspects of STATA. Stay with it! By the end of this lab manual, you will have the ability to use STATA efficiently and apply the concepts and principles taught in this course in a meaningful way.

The Model

Each lab contains three sections:

1. The first section will briefly reintroduce the topic covered by the corresponding chapter and set the learning objectives for the lab.
2. The second section will provide an example to help focus your thinking and help you to understand the key concepts to be covered. The examples are intended to illustrate the real-world applications of the concepts.
3. The third section will include a lab assignment to test your understanding of the concepts and skills covered in the lab and corresponding chapter.

What's Covered?

This manual is meant to complement the textbook used in this course, not replace it. Through the use of concrete examples, each lab is intended to solidify your understanding of the general concepts and principles taught in each chapter. To help teach you how to analyze data, this lab manual will provide step-by-step instructions on how to use the STATA analysis software.

The labs cover the following key topics:

- Lab 1: Introduction to STATA
- Lab 2: Identifying Types of Variables: Levels of Measurement
- Lab 3: Univariate Statistics
- Lab 4: Introduction to Probability
- Lab 5: The Normal Curve
- Lab 6: Measures of Central Tendency and Dispersion
- Lab 7: Standard Deviations, Standard Scores, and the Normal Distribution
- Lab 8: Sampling
- Lab 9: Hypothesis Testing: Testing the Significance of the Difference between Two Means
- Lab 10: Hypothesis Testing: One- and Two-Tailed Tests
- Lab 11: Bivariate Statistics for Nominal Data
- Lab 12: Bivariate Statistics for Ordinal Data
- Lab 13: Bivariate Statistics for Interval/Ratio Data
- Lab 14: Analysis of Variance
- Lab 15: OLS Regression: Modelling Continuous Outcomes

The Data Set

The 2013 Alberta Survey (AS) is the twenty-fourth annual provincial survey administered by the Population Research Laboratory (PRL) at the University of Alberta. This annual omnibus survey of households in the province of Alberta enables academic researchers, government departments, and non-profit organizations to explore a wide range of research topics in a structured research framework and environment. Sponsors' research questions are asked together with demographic questions in a telephone interview of Alberta households. In 2013, topics covered by the survey included China's role in Alberta's economy, peoples' interactions with financial institutions/debt management, life experiences and health, opinions about temporary foreign workers, opinions about public health and injuries, and alcohol use. Demographic characteristics that were measured include age, gender, marital status, highest level of education, household income, religion, place of birth, employment status, home ownership, and political party support.

The target population was all persons 18 years of age or older who, at the time of the survey, were living in Alberta and could be contacted by direct dialing. From this population, three samples were drawn to cover Alberta: Edmonton Metropolitan Area, Calgary Metropolitan Area, and the rest of the province. The final data set contains 1207 cases. We have created a version of this data set for use with this textbook, which can be found on the textbook website at www.oupcanada.com/Haan3e. The codebook for the data set can be found in the same place.

This data set is also available online as a public-use data file. You can find it on the Population Research Laboratory's website: www.prl.ualberta.ca/en/AlbertaSurvey.aspx. Click on **Alberta Survey,** then click on the link for **Public Release Alberta Survey Dataset 2013**. You will also be able to find some previous Alberta Surveys on that website. Note that the public-use data file includes variables and incomplete data that was removed for the purposes of clarity; when completing the labs, it will be easier to use the data file found on the textbook website.

Lab #1: Introduction to STATA

The focus of this lab is to introduce you to STATA. To use STATA to analyze data, you will need to become familiar with the technical components of this software package. This lab will help familiarize you with the STATA software, including how to access data files, the various base components (i.e., the Syntax, Data Editor, and Output), how to define new variables, and how to enter data.

You should be able to find STATA on many of the computer terminals at your university, but if you can't, your instructor for the course should be able to help you find STATA.

LEARNING OBJECTIVES

The following lab is directed at helping you understand how to orient yourself in STATA. Specifically, this lab assignment challenges you to clarify your understanding of:

1. The basic components of STATA
2. How to define variables
3. How to enter data

Defining STATA

What Is STATA and How Does It Work?

STATA is a statistical software package commonly used by social scientists. It works by taking a series of commands, supplied by you, and applying them to a set of data, also supplied by you.

STATA will produce output displaying the results of the commands. The commands you supply STATA will determine your output results. Within STATA, commands are either given in the form of menu selections and by filling in dialogue boxes, or by writing your own programming syntax. We will examine both methods in this lab manual.

The sequence of commands will occur in the following order:

- Enter your data into STATA, or open already existing data (which is what we'll do here).
- Tell STATA to apply commands to the data (menus and dialogue boxes).
- STATA then produces the output.

How Do I Begin a STATA Session?

To begin, you need to access a computer that contains the STATA software. Many universities have the STATA software inside computer labs on their public use computers. Once you have found a computer with a copy of the STATA software, you will need to click on the **Start menu** (on the bottom left-hand side of your computer screen), find the STATA program, and open it.

Once you have opened the STATA program, you should see a screen that looks something like the one in Figure 1.1.

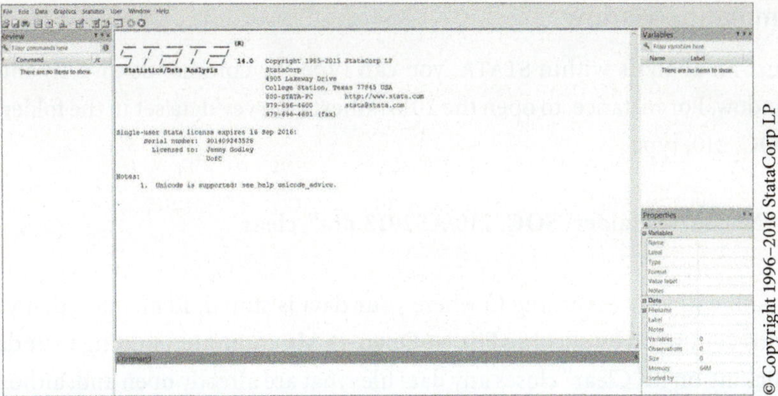

FIGURE 1.1 | STATA Opening Screen

To open your data, select **File → Open → My computer**. Find your data in "My Computer" and double-click on the file. For the examples in this manual, you will be using AS2013.dta. Once you open the data set, all the variables in the AS2013.dta data set are listed along the right side of the screen in the box labelled "Variables."

The Various Components of STATA

The STATA desktop contains several components, which are described below:

- The menu bar
- The command window
- The STATA results window
- The variables window
- The review window

STATA has additional windows that you will be using, also described below:

- The data editor
- The data browser
- The log
- The do-file editor

The Menu Bar

Along the top of the screen you will see the menu bar, which gives you access to all the available commands. The menu bar has a number of headings, dividing the commands into categories with similar functions (File, Edit, Data, Graphics, Statistics, User, Window, and Help). Browse through these pull-down menus to become familiar with their contents.

The Command Window

To conduct an analysis within STATA, you can type the commands directly into the command window. For instance, to open the 2013 Alberta Survey data set in the folder N:\courses folder\ SOC. 210, type

Use "N:\courses folder\ SOC. 210/AS2013.dta", clear

The path will vary according to where your data is stored. Remember that you can also execute this command by clicking **File → Open → My computer**, finding your data file, and double clicking on it. "Clear" closes any data files that are already open and, although it is not necessary here since there are no open files, it is a good habit to have when opening new data. When you type commands into the command window, they are only saved for the current STATA session. If you want to save your commands so that you can run them again, you need to open a do file (see below).

The Results Window

The Results window displays the output from the statistical analysis you perform.

The Variables Window

The Variables window lists the variables contained in your data set. Double click on any variable to bring it over to the commands window.

The Review Window

The Review window lists all of the commands that have been executed in the current session. You can repeat these commands by double clicking on them, clicking anywhere in the command window, and hitting Enter.

Saving Your STATA Data Files

To save your STATA data files, click on **File → Save as**. Move to the directory in which you want to save the file and give your document a name. If you have previously saved your file and you have merely modified it, select **Save** from the File menu or click the icon in the menu bar that looks like a floppy disk. STATA data files are saved with the extension .dta.

Data Editor

For the purposes of this lab, we are going to create a new data set in STATA (we will look at the Alberta Survey later).

Click the "Data" button on the top of the screen and then select "Data Editor" from the options provided. The Data Editor window is like a spreadsheet (as shown in Figure 1.2). Within the Data Editor screen, you can both create and edit pre-existing data sets.

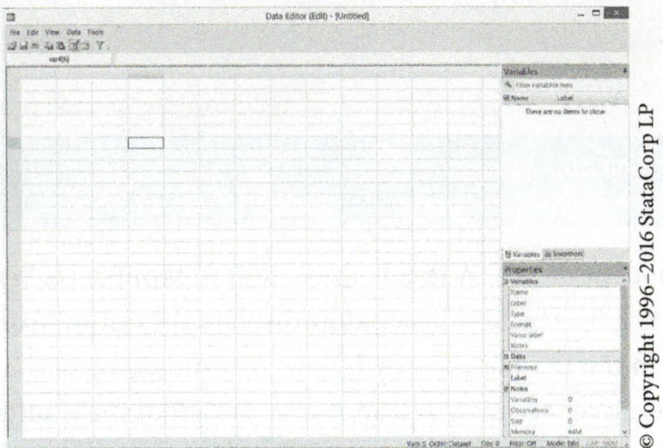

FIGURE 1.2 | Data Editor Window

Do-File Editor

The "Do-file" is a text file containing a list of STATA commands. You can run a Do-file at any time by entering the STATA prompt **Do filename.do**, where "filename" is the name of your Do-file. These files are useful for organizing your records and analyses. As mentioned, there are two approaches to working in STATA: using a point-and-click approach or manually inputting program commands. In this manual, you will mostly be using the point-and-click approach. To open a new syntax file click on the Do-file Editor icon in the main menu. Syntax files can be saved by selecting **File → Save as**. The saved document will have the suffix .do.

FIGURE 1.3 | The Do-File Editor

The Do-file Editor allows you to execute a long list of commands and modifications to your data set. If you plan only to execute a few quick commands, it is also possible to use the command window at the bottom of the main STATA interface screen, but remember those commands will not be saved (Figure 1.4).

FIGURE 1.4 | The Command Window
Source: © Copyright 1996–2016 StataCorp LP

The next component of STATA that is important to familiarize yourself with is the Results window (Figure 1.5). The Results window displays the output from the statistical analysis you undertake in the Do-file Editor window. A running log of the commands you have executed during the current session will appear to the left of the results window.

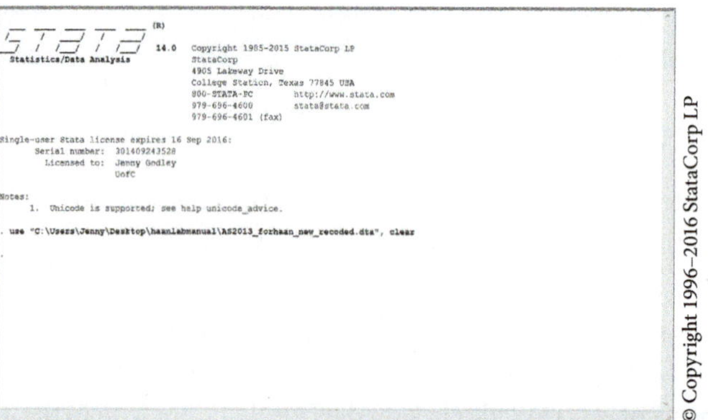

FIGURE 1.5 | Results Window

Entering STATA Commands

There are three ways of entering commands into STATA to execute your procedures. The first is known as the point-and-click procedure. This procedure can be done via the buttons along the top of the main interface (Figure 1.6).

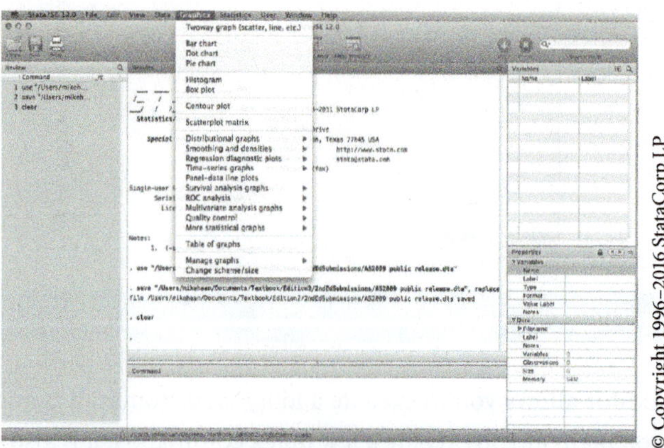

FIGURE 1.6 | The Point-and-Click Technique

The second way is to enter Syntax in the command window (the example in Figure 1.7 asks for a tabulation of the variable sex).

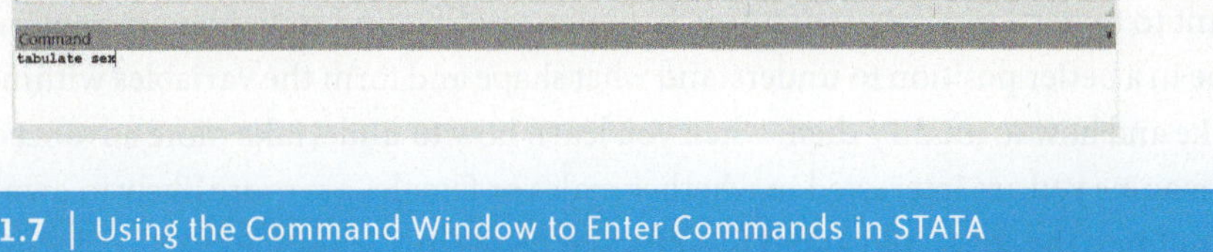

FIGURE 1.7 | Using the Command Window to Enter Commands in STATA
Source: © Copyright 1996–2016 StataCorp LP

The third way is to enter command programming into the syntax file (Figure 1.8).

FIGURE 1.8 | Using the Do-File Editor to Enter Commands in STATA

The Log File (*.smcl)

The Log file is an output file—it records whatever appears in your STATA Results window. You can ask STATA to keep a log of your session by typing **Log using filename.smcl**, and STATA will name the Log file filename.smcl. When you are done, type **Log close**, and the Log file will then be ready for you to view, edit, or print. You can ask STATA to open and close your Log file within your do-file.

Creating a Data File

Although most of the lab exercises in this manual use data from the Alberta Survey, it is important to understand data structure. By understanding how data files are constructed, you will be in a better position to understand what shape and form the variables within your data set take and how to modify them when you learn how to undertake more advanced analysis.

Imagine you are interested in whether males or females are more likely to own a cat. You went out into the street and asked five random people two questions:

1. Are you a male or a female? (Male = 1 and Female = 2)
2. How many cats do you own? (0 to 5)

Person 1: Male (1) owns 2 cats
Person 2: Female (2) owns 1 cat
Person 3: Female (2) owns 0 cats
Person 4: Male (1) owns 5 cats
Person 5: Female (2) owns 1 cat

Now that you have your data, you can enter the information into an STATA data file.

Step 1: Open a new Data Editor window (remember that this is on the top of the main interface window).

Step 2: Enter your data into the cells, using rows to represent each observation (there are five) and columns to contain a piece of information about each individual (sex of respondents and # of cats they own). Once you are done, you should have a small data matrix that looks like Figure 1.9.

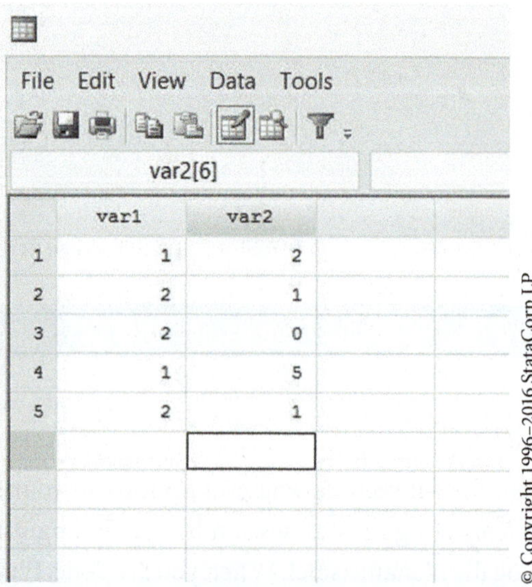

FIGURE 1.9 | The Data Matrix

Step 3: Name your variables by using the Properties box in the lower right-hand corner of the screen (Figure 1.10). You must first decide on the names of your variables; these names must not have spaces and must start with a letter. As good practice, keep the number of characters to a minimum where possible. Often variable names will correspond with the survey question number from which they were derived, such as Q1 or Var01; however, this is not necessary here. For our purposes, we will call our variables sex and cats.

Step 4: Define your variable labels ("Label" under the Variable heading), also within the Properties screen. Although we have given our variables names that clearly indicate what the variable is measuring, your variable label can help clarify what the variable is measuring, or, if you have named the variable after survey questions (i.e., Q1), it will allow other users to know what information Q1 is measuring. For sex, we will put "Sex of respondent" and for cats, we will put "Number of cats owned."

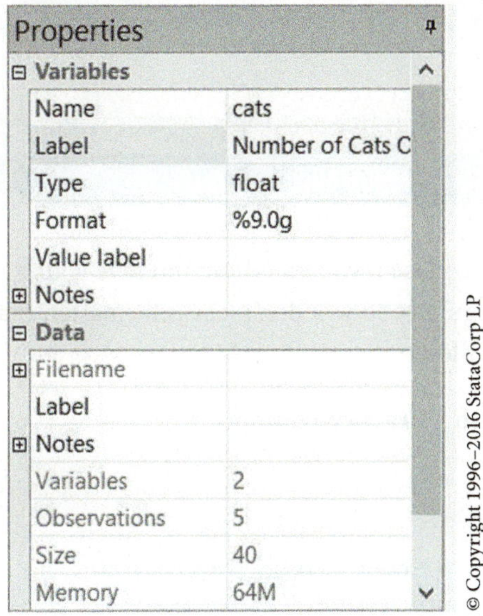

FIGURE 1.10 | Defining Variable Labels within the Properties Screen

Step 5: Define your value labels ("Value label" under the variable heading) within the Properties screen. Often value labels are attached with ordinal and nominal level variables (concepts you will learn about as you proceed through this course). Without labels, it is difficult to know what the values mean. So, for example, we have attached a value of "1" to our male respondents and a value of "2" to our female respondents. You must specify this under the Value label column. To do this, click on **Value label** when you have the variable of interest highlighted. Click on the grey button with three dots in the cell. A dialogue box will open. Click on **Edit label** and another dialogue box will open (Figure 1.11). Following our example, assign labels Male and Female to numeric values "1" and "2," respectively, by entering 1 into the Value box and its label, **Male,** into the Label box. Click **Add** to make the changes for each value of your variable, then click **OK** when you are finished.

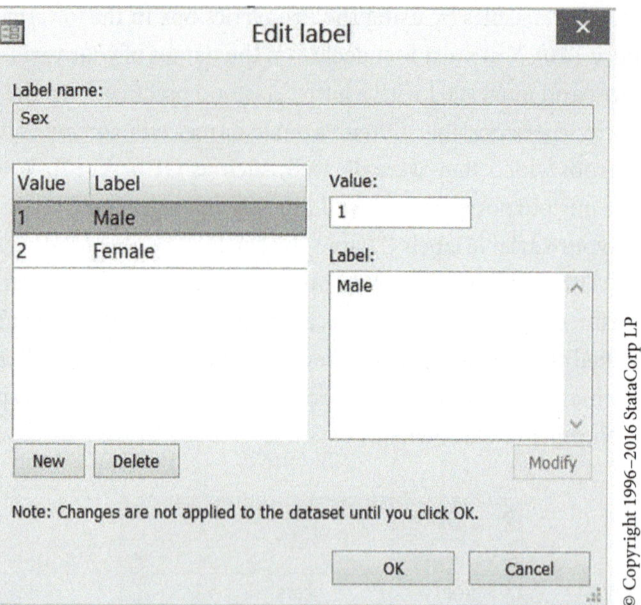

FIGURE 1.11 | Edit Label Dialogue Box

Finally, don't forget to close the Data Editor and save your new data set. It is good practice to save your work regularly so you don't lose information that you have worked to produce. Remember to save files:

File → Save as → Lab 1 Practice Example

Now, it's your turn!

Putting the Information into Practice

For this lab, you will be required to complete a short survey and input the data into the Data Editor screen. For the remainder of the lab assignments, data will be provided for you, but this exercise seeks to make you more familiar with the nature of data organization and storage within the STATA software package.

The following questions were taken from the Alberta Survey 2013.

Part 1: Take the time to answer each question as it relates to your life. Then, ask two friends to answer these questions. Pretend a fourth respondent refused to give you any answers to your survey. You should have a total of four respondents for your short survey.

1. What is your gender?
   1. Male
   2. Female

2. Do you presently have a paid job or are you self-employed?
   1. Yes, paid job
   2. Yes, self-employed
   3. Yes, paid job and self-employed
   4. No, neither

3. What is your *current* marital status?
   1. Never Married (Single)
   2. Married
   3. Common-Law Relationship/Live-In Partner
   4. Divorced
   5. Separated
   6. Widowed

4. Do you presently live in . . .
   1. A City
   2. A Town
   3. A Village
   4. A Rural Area

5. How safe do you feel from crime walking alone in your area after dark? Do you feel . . .

   1. Very safe
   2. Reasonably safe
   3. Somewhat unsafe
   4. Very unsafe

Part 2: Create a data set by using the skills you just learned. Remember to (1) name your variables; (2) define your variable labels; (3) define your value labels; and (4) declare any missing values (hint, this would be the person who refused to answer any of your survey questions). When you have finished creating your data set, which will include four respondents and five variables, save your file. Congratulations, you have just created your first data set. Now, let's learn what to do next!

Lab #2: Identifying Types of Variables: Levels of Measurement

The focus of this lab is to introduce you to the four different levels of variable measurement, how to identify different types of variables within STATA, and how different levels of measurement are coded and organized within STATA. This material corresponds with the material presented in Chapter 2.

LEARNING OBJECTIVES

The following lab is directed at helping you understand levels of measurement and how to identify different types of variables within a data set. Specifically, this lab assignment helps you clarify your understanding of:

1. Variable measurement
2. How variables are organized within STATA

Understanding Levels of Measurement

Variables are measured at four different levels: nominal, ordinal, interval, and ratio. Each of these levels has unique characteristics that define them. For *nominal data*, numeric values are typically used for identification purposes. It is not possible to rank the response categories and there is not a quantifiable difference between categories. For *ordinal data*, the numeric values can signify an inherent ordering because you can rank the response categories but you cannot measure the distance between those categories. For *interval data*, the data can be organized into an order that can be added or subtracted but not multiplied or divided, because there is no true zero value. *Ratio data* are similar to interval data except they have a true zero value.

There are two ways to collect information about how a variable is measured in STATA. One is to examine the variable values in the Data Editor screen and the other is to produce a frequency distribution. Using data from the Alberta Survey 2013, this lab will demonstrate the first way to identify variable values, then give you a chance to practise identifying variables within STATA. In the next chapter of this lab manual, you will learn how to produce frequency distributions.

To open the Alberta Survey data, simply select **File → Open → Data,** then select the Alberta Survey 2013 (AS2013) (your instructor will need to tell you where he/she has placed the file).

When you have opened the file, take a moment to scroll through the list of variables and observations in Data Editor. You will notice there are many variables (134 to be exact) contained within this data set. If you scroll down the observations, you will see that there are 1207 respondents. Return to the main interface screen. You might notice that the names of the variables don't make a lot of sense without a thorough understanding of the background of this survey (if you want more information about the Alberta Survey 2013, the codebook is included online at www.oupcanada.com/Haan3e). However, in most cases, the variable labels provide key insight into what each of those variables is measuring. For the most part, you will be able to make an informed guess as to what the level of measurement would be for most variables. However, this may not always be accurate.

For example, variable k12a has the label "Total Household Income for the Past Year before Taxes and Deductions." Given that this is an income variable and a person can feasibly have no income, you might be inclined to guess this is a ratio level variable. However, when you tabulate this variable (tab k12a), a different story emerges.

As we can see (Figure 2.1), the response categories for this variable are grouped into discrete categories. Therefore, this variable is actually an *ordinal* level variable because you can rank the categories but the values associated with the discrete categories (i.e., 1 refers to under $6000) cannot be meaningfully measured.

```
. tab k12a
```

Total Household income for the past year before taxes and deductions	Freq.	Percent	Cum.
Under 6,000	5	0.41	0.41
6,000 - 7,999	2	0.17	0.58
8,000 - 9,999	2	0.17	0.75
10,000 - 11,999	2	0.17	0.91
12,000 - 13,999	6	0.50	1.41
14,000 - 15,999	5	0.41	1.82
16,000 - 17,999	4	0.33	2.15
18,000 - 19,999	9	0.75	2.90
20,000 - 21,999	17	1.41	4.31
22,000 - 23,999	14	1.16	5.47
24,000 - 25,999	17	1.41	6.88
26,000 - 27,999	6	0.50	7.37
28,000 - 29,999	9	0.75	8.12
30,000 - 31,999	24	1.99	10.11
32,000 - 33,999	6	0.50	10.60
34,000 - 35,999	21	1.74	12.34
36,000 - 37,999	5	0.41	12.76
38,000 - 39,999	5	0.41	13.17
40,000 - 44,999	25	2.07	15.24
45,000 - 49,999	22	1.82	17.07
50,000 - 54,999	37	3.07	20.13
55,000 - 59,999	14	1.16	21.29
60,000 - 64,999	42	3.48	24.77
65,000 - 69,999	18	1.49	26.26
70,000 - 74,999	48	3.98	30.24
75,000 - 79,999	26	2.15	32.39
80,000 - 84,999	45	3.73	36.12
85,000 - 89,999	18	1.49	37.61
90,000 - 94,999	24	1.99	39.60
95,000 - 99,999	18	1.49	41.09

FIGURE 2.1 | Coding Details for Variable k12a

Let's look at one more example. Find the variable named strata with the label "Area of the Province." Without going to the codebook, we do not know how this question was to be answered by respondents. To check to see how the variable is coded, run "Tabulate strata" in the command window.

As we can see from the results (Figure 2.2), this is a nominal level variable with three categories.


```
. tabulate strata

  Area of the |
     province |      Freq.     Percent        Cum.
--------------+-----------------------------------
Metro Edmonton|        404       33.47       33.47
 Metro Calgary|        402       33.31       66.78
 Other Alberta|        401       33.22      100.00
--------------+-----------------------------------
         Total|      1,207      100.00
```

FIGURE 2.2 | Tabulate Strata for Variable Area of the Province

Now it's your turn!

Putting the Information into Practice

For this lab, you will be asked to fill in the missing information in Table 2.1. In the space beside each variable name, provide the Variable label, the Value labels (i.e., 1 = Yes, 2 = No), and try to determine the level of measurement.

TABLE 2.1

Variable	Variable label	Value label	Level of measurement
sex			
d1			
d9			
e11			
ft1			
ft8			
g6e			
k3a			
age			
agex			

Lab #3: Univariate Statistics

The focus of this lab is to begin to introduce you to analysis with one variable. Generating frequencies is a basic procedure used to obtain a summary of a variable by looking at the number of cases associated with each value of the variable. This material corresponds with the material presented in Chapter 3.

LEARNING OBJECTIVES

The following lab is directed at helping you understand ways of studying the characteristics of data. Specifically, this lab assignment challenges you to clarify your understanding of:

1. How to generate and interpret frequency distributions
2. Data presentation
3. The connection between data presentation and levels of measurement

Part 1: Producing Frequency Distributions

We will first learn to produce a frequency table. Throughout this lab manual, we will use the command prompts. To obtain a frequency distribution, use the following steps.

Step 1: Open the Alberta Survey by either using the open button on the toolbar or typing.

Use <<enter the path for your data here>>/AS2013.dta, clear

Step 2: Type the following in the command box:

tabulate sex or tab sex

You should get output that looks like Figure 3.1.

. **tab sex**

Gender of respondent	Freq.	Percent	Cum.
Male	595	49.30	49.30
Female	612	50.70	100.00
Total	1,207	100.00	

FIGURE 3.1 | A Basic Frequency in STATA

What should we note about this output?

1. We can see that we have 1207 total cases.
2. In the table, we can see the distribution of Male versus Female by Frequencies, Per cent, and Cumulative per cent. You can see that there are 595 males and 612 females. Is this the same as what you have on the screen before you?

Part 2: Types of Charts

A *pie chart* is a way of summarizing a set of categorical data. Data are categorical when the values or observations belonging to them can be sorted according to groups but not by values. For example, sex is a categorical variable with two categories, male and female, and people cannot belong to both categories. We can then refer to sex as being mutually exclusive. A pie chart is a circle that is divided into segments, with each segment representing a particular category. The area of each segment is proportional to the number of cases in that category.

The following is an example of a pie chart for a variable called strata that measures in which area of Alberta the respondent lives. Note that in this example we appear to have an equal distribution of respondents from Edmonton, Calgary, and Other Alberta.

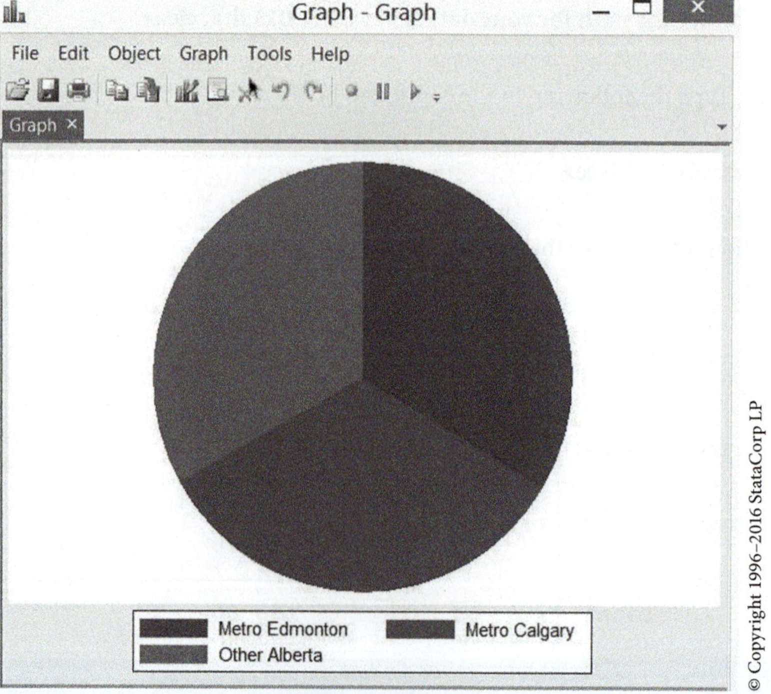

FIGURE 3.2 | A Pie Chart

How can you create a pie chart using STATA?

Type: **Graph pie, over (strata)**

Bar Graphs

A *bar graph* is a way of summarizing a set of categorical data. It displays the data using a number of rectangles of the same width, each of which represents a particular category. The length of each rectangle is proportional to the number of cases in the category it represents. Figure 3.3 is an example of a bar graph for the question "To the Best of Your Knowledge, What Is the Leading Cause of Death for Albertans under the Age of 45?" (g1).

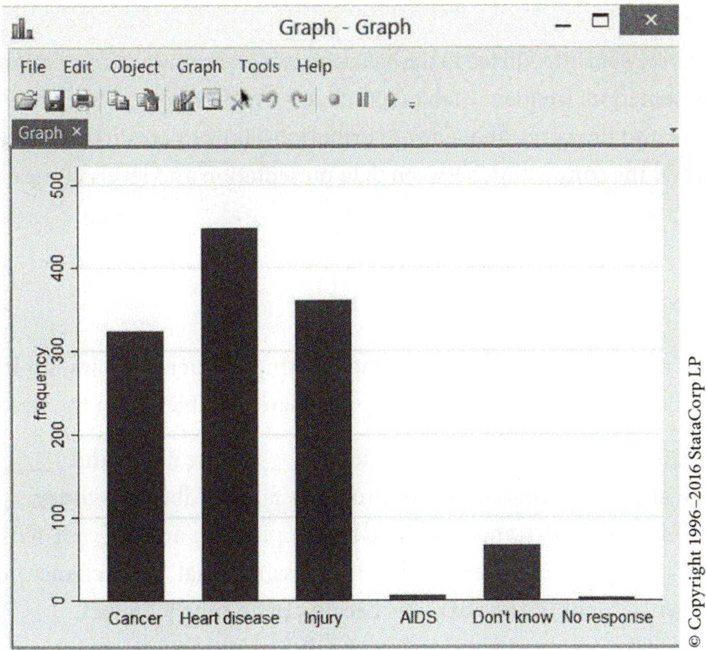

FIGURE 3.3 | A Bar Chart

Bar graphs are created using the following command:

Type: **generate freq = 1**

graph bar (count) freq, over (g1)

Which Mode of Presentation Is Best?

Deciding which format to use to present your data depends on the level of measurement of your selected variable and the clarity of the presentation. For example, if you have a categorical variable, but a large number of categories, a pie chart may make the presentation crowded and confusing. Selecting a bar graph may be more appropriate.

Here are some general suggestions you may want to consider:

- Use tables to display data details that would be lost in graphs or charts.
- Opt for a bar graph to compare data.
- Consider a pie chart to show how percentages relate to each other within a whole.
- Focus on the main point and consider your audience.
- Non-technical audiences often appreciate visual representations of data, so try to use pie and bar charts when you think your audience would prefer them.

Summary

In this section, you were introduced to the basics of data presentation. You explored three methods of data presentation: frequency tables, pie charts, and bar graphs. Specifically, you learned how to generate and interpret frequency distributions, how to create three types of modes of presentation, and the connection between data presentation and levels of measurement.

Now it's your turn!

Putting the Information into Practice

1. Using what you've learned about interpreting frequency tables, fill in the blanks of the following paragraph. Use the appropriate variables from the lesson above.

   In AS2013's sample of 1207 respondents, _____% are males and _____% are females. When asked the different health problems facing Albertans today, _____% felt that injuries were an extremely serious health problem and _____% felt they were not serious. Respondents were varied in guessing what their chances of visiting the emergency room in the next year because of an injury. In fact, _____% felt it would be 1 in 500, compared to _____% who felt it would be 1 in 10. Interestingly, a higher percentage of people thought they would be more likely to go to the emergency room because of a motor vehicle collision. Specifically, _____% felt the chances of going to the emergency room because of a motor vehicle collision were 1 in 500.
2. Produce the specified graphs for the following variables:
   a. e11: Bar chart
   b. d1: Pie chart
   c. g3: Bar chart
   d. k5a: Frequency table

3. Choose one of the variables from Question 2 and provide an interpretation. Specifically, what is the percentage distribution of the categories? What is the total number of respondents who answered the question? How many, if any, missing values are there?

Lab #4: Introduction to Probability

The focus of this lab is to review what you have already learned and to introduce you to the concept of recoding variables. Recoding variables is an important component in conducting analysis because the categories of a variable as they were asked in the questionnaire might not work for your specific needs. For example, if you are interested in comparing individuals who have a high school education or less with those who have a post-secondary education, you may not require a level of detail that looks at all the specific types of post-secondary education available. This lab will show you how to manipulate variables.

We will also look at how to calculate probabilities from STATA output. This material corresponds with the material presented in Chapter 4.

LEARNING OBJECTIVES

The following lab is directed at helping you understand how to manipulate variables. Specifically, this lab assignment challenges you to clarify your understanding of:

1. Why you might want to recode variables
2. How to use STATA to recode variables
3. How to use STATA output to calculate probability

Recoding Variables

Let's assume you are interested in the opinions Albertans have regarding whether temporary foreign workers are needed to fill jobs in the Alberta labour market. Within the current data set, we have a variable that asks, "Indicate how much you agree or disagree with the following statement: Temporary foreign workers are needed to fill jobs in the Alberta labour market" (ft4). After obtaining a frequency distribution (Figure 4.1) of the variable, you realize that you do not require this level of detail. You are interested in whether people disagree, neither agree nor disagree, or agree. Therefore, you realize you will need to collapse the first two categories together (Strongly Disagree and Somewhat Disagree) and the last two categories together (Somewhat Agree and Strongly Agree). You also want to tell STATA not to include the Don't Know and No Response categories in further calculations.

Step 1: To change the coding of this variable to suit your needs, it is easiest in STATA to create a new variable and transform the new variable. It is important to select this option because you are interested in creating a new variable from your original variable rather than altering the existing variable. Although it is possible to recode into the same variable, the original information on the AS2013 will be overwritten, *so always choose to create a new variable!*

Step 2: Create a new variable by typing:

generate ft4_recode = ft4

```
. tab ft4

  Temporary foreign workers |
are needed to fill jobs in  |
  the Alberta labour market |      Freq.     Percent        Cum.
----------------------------+-----------------------------------
          Strongly disagree |         95        7.87        7.87
                   Disagree |        172       14.25       22.12
Neither disagree nor agree  |        288       23.86       45.98
                      Agree |        471       39.02       85.00
             Strongly agree |        150       12.43       97.43
                 Don't know |         29        2.40       99.83
                No response |          2        0.17      100.00
----------------------------+-----------------------------------
                      Total |      1,207      100.00
```


FIGURE 4.1 | Frequency Distribution for Variable ft4

Step 3: Decide on how you want the new variable to be coded. You can use the codebook here to guide you, or you can run the following command to see which numbers correspond to the value labels above (the nol stands for No Labels).

tabulate ft4, nol

```
. tabulate ft4, nol

  Temporary |
    foreign |
workers are |
  needed to |
  fill jobs |
     in the |
    Alberta |
     labour |
     market |      Freq.     Percent        Cum.
------------+-----------------------------------
          1 |         95        7.87        7.87
          2 |        172       14.25       22.12
          3 |        288       23.86       45.98
          4 |        471       39.02       85.00
          5 |        150       12.43       97.43
         98 |         29        2.40       99.83
         99 |          2        0.17      100.00
------------+-----------------------------------
      Total |      1,207      100.00
```


FIGURE 4.2 | Tabulate ft4

Suppose that we are interested in creating a variable where Strongly Disagree and Disagree are in one category and Agree and Strongly Agree are in another. You will need to tell STATA that this is what you want. You will also want to recode No Response and Don't Know to be missing. For our purposes here, the table below demonstrates how we will recode.

TABLE 4.1

Old value	New value
1, 2	1
3	2
4, 5	3
98 99	Missing (use .)

STATA (and all other software packages) typically recognize "." to represent a missing value.

Step 4: Now, to recode the variable, remember to use your new variable:

recode ft4_recode (1 2 = 1) (3 = 2) (4 5 = 3) (98 99 = .)

Remember, we did the recode in two separate steps above, first creating a new variable, then recoding that variable.

Now, as a final check, run a frequency distribution on the new variable, ft4_recode, and the original variable, ft4, to confirm that your new values add up to the values in the original variable (Figure 4.3). For example, the values of Strongly Disagree + Disagree (95 + 172) should sum the total of the first category of the new variable (267).

```
. tab ft4

Temporary foreign workers |
are needed to fill jobs in |
 the Alberta labour market |      Freq.     Percent        Cum.
---------------------------+-----------------------------------
         Strongly disagree |         95        7.87        7.87
                  Disagree |        172       14.25       22.12
Neither disagree nor agree |        288       23.86       45.98
                     Agree |        471       39.02       85.00
            Strongly agree |        150       12.43       97.43
                Don't know |         29        2.40       99.83
               No response |          2        0.17      100.00
---------------------------+-----------------------------------
                     Total |      1,207      100.00

. tab ft4_recode

 ft4_recode |      Freq.     Percent        Cum.
------------+-----------------------------------
          1 |        267       22.70       22.70
          2 |        288       24.49       47.19
          3 |        621       52.81      100.00
------------+-----------------------------------
      Total |      1,176      100.00

.
```


FIGURE 4.3 | Frequency Distribution of Variables ft4 and ft4_recode

Naturally, you'd want to go back and use the information in Lab #2 to attach value Labels to your new variable.

Chapter 4 discusses probabilities and simply requires that you know the number of observations in each category to be able to calculate probabilities. So, if you wanted to know the probability that someone chosen at random in your sample would disagree with the statement about whether people with intellectual disabilities should be parents, you would only need to know the frequency of people in the Disagree category (267) over the total number of people in the sample (1207). This yields a probability of roughly 22 per cent $\left(\frac{267}{1207}\right)$. Calculating the probability of the other outcomes would proceed in a similar way, except that you'd have a different number in the numerator.

Putting the Information into Practice

1. Using the techniques that you have just been taught, recode variable e13, "I usually expect things to go my way." Create a three-category variable where the categories are 1 Disagree, 2 Neither Agree Nor Disagree, and 3 Agree.
2. Interpret your recoded variable. Specifically, what is the percentage distribution of the categories? What is the total number of respondents who answered the question?
3. Based on the frequency distribution of your recoded variable, what is the probability that individuals will agree that they usually expect things to go their way?
4. Based on the frequency distribution of the original variable, what is the probability that individuals will disagree that they usually expect things to go their way?
5. Based on the frequency distribution of your recoded variable, what is the probability that individuals will neither agree nor disagree that they usually expect things to go their way?

Lab #5: The Normal Curve

The focus of this lab is to introduce you to the concept of the normal curve. The distribution of data can take on different forms. Understanding how data are distributed is important for more complex analysis, which you will learn as this course progresses. This lab corresponds with the material presented in Chapter 5.

LEARNING OBJECTIVES

The following lab is directed at helping you understand how to manipulate variables. Specifically, this lab assignment challenges you to clarify your understanding of:

1. How to produce histograms in STATA, and how to use them to see how data are distributed
2. How to describe and identify distributions

Creating a Histogram in STATA

Step 1: In the command window, type

hist k7

You should get output that looks like this (Figure 5.1):

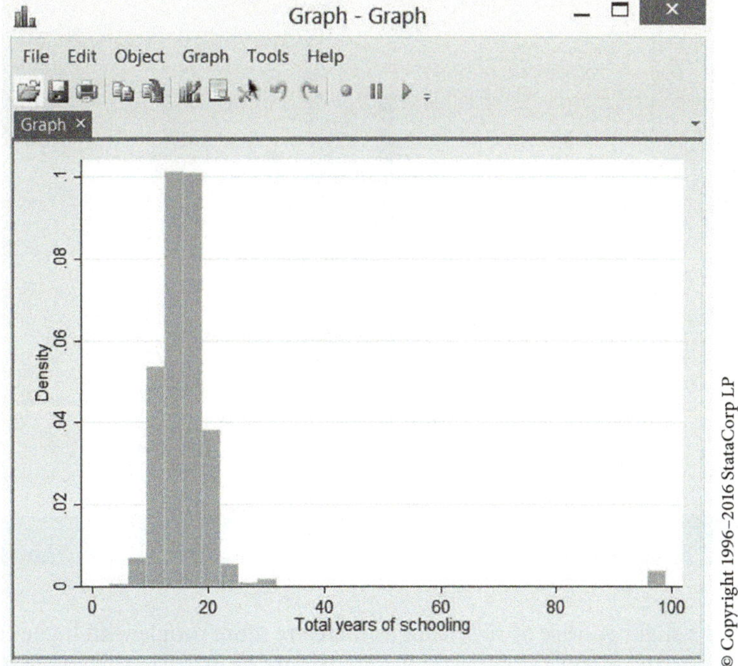

FIGURE 5.1 | Histogram of k7

There are a few problems with this output. First, there appear to be either extreme outliers or missing values on the right side of the distribution. Second, there is no overlaid normal curve.

Both of these problems are easy to fix. To figure out the valid values for k7, consult the codebook. There you will see that there are missing values coded as 99. So, we only have to eliminate them.

To do this, modify the code above to read:

hist k7 if k7<99

That tells STATA that you are interested only in values that are valid.

To overlay the normal curve, you need to add the following:

hist k7 if k7<99, normal

Now, what can we say about this distribution? Well, it is *bimodal* because there are two spikes, one at the 12 years of schooling mark and the other at the 16 years of schooling mark.

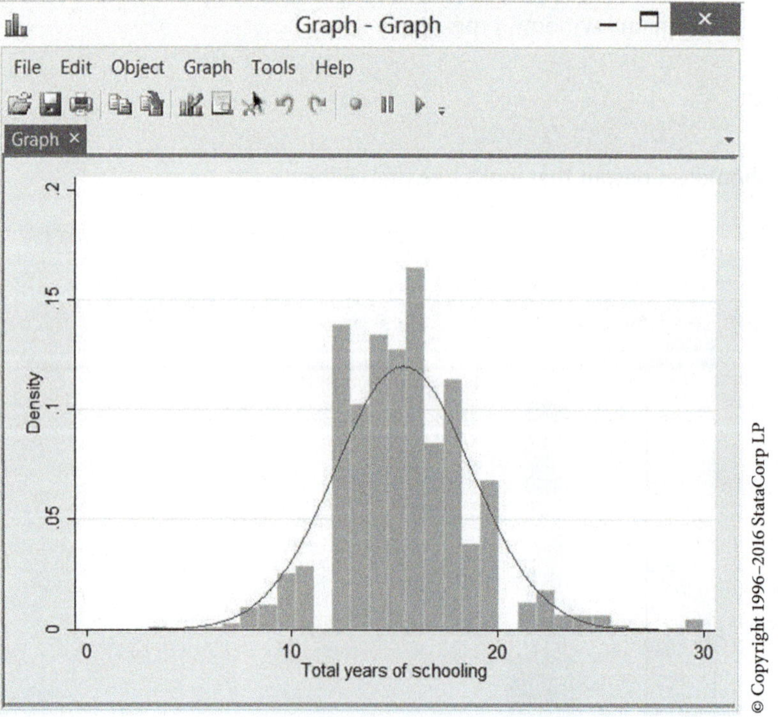

FIGURE 5.2 | Histogram of k7 with Normal Curve

The curve has a slight *positive* or *right skew* as there are more people with higher levels of education than lower levels. The curve is a little more "peaked" than normal. This is hard to know from seeing a distribution, but having two peaks that are higher, and none that are dramatically lower than the rest of the distribution provides a clue that the distribution has positive kurtosis.

To get an even better sense of the distribution of k7, you might choose to ask STATA for a summary of the variable. This can be done by entering the following command:

sum k7 if k7<99, detail

This will give you output that looks like the following (Figure 5.3):
There is quite a bit of information here in the summary output, most of which we'll learn about in the next lab. For our purposes here, only skewness and kurtosis are of interest. The

```
. summ k7 if k7<99, detail

                    Total years of schooling
-------------------------------------------------------------
      Percentiles      Smallest
 1%            8              3
 5%           11              6
10%           12              7       Obs               1,193
25%           13              7       Sum of Wgt.       1,193

50%           15                      Mean           15.43336
                        Largest       Std. Dev.      3.329872
75%           18             30
90%           20             30       Variance       11.08805
95%           21             30       Skewness       .6134882
99%           25             30       Kurtosis       4.610685

.
```


FIGURE 5.3 | Summary of k7 Variable

skewness value of 0.61 confirms our earlier hunch of a slightly positive skew, and the kurtosis value of 4.61 tells us that the distribution is more peaked than normal.

OK, now it's your turn.

Putting the Information into Practice

1. Obtain a histogram for the following variables:
   a. age
   b. k3a
   c. k3b
   d. k3c

2. For each variable, describe the symmetry of the distribution, the skewness of the distribution, its type of kurtosis, and whether it is unimodal, bimodal, or multimodal. For now, you can only do this visually, but we'll learn how to do more than this in the coming chapters.

Lab #6: Measures of Central Tendency and Dispersion

The focus of this lab is to introduce you to various measures of central tendency. Measures of central tendency allow you to further understand the distribution of a variable. In this lab, you will learn how to generate various measures of central tendency within STATA. This lab corresponds with the material presented in Chapter 6.

LEARNING OBJECTIVES

The following lab is directed at helping you understand how to use STATA to generate measures of central tendency. Specifically, this lab assignment challenges you to clarify your understanding of:

1. The differences between the mean, median, and mode
2. The relationship between levels of measurement and the mean, median, and mode
3. How to generate measures of central tendency within STATA

Generating Measures of Central Tendency in STATA

Step 1: Ensure that you have the 2013 Alberta Survey file open.
Step 2: Within the command box, type the following:

sum k7

This will allow you to get basic information on central tendency, and will produce output that resembles Figure 6.1 for the variable k7, which is the total years of schooling variable used in the previous lab.

. **sum k7**

Variable	Obs	Mean	Std. Dev.	Min	Max
k7	1,207	16.40265	9.543924	3	99

FIGURE 6.1 | Summary Statistics for Variable k7

Now, before we continue, we need to ask ourselves if these numbers make sense; in other words, do the measures of central tendency make sense in relation to the level of measurement of our variable? Years of schooling is a ratio level variable because we have a true zero value: individuals can obtain no years of schooling. The lowest value in our summary table is 3. This seems reasonable. However, the maximum value seems unreasonably high, at 99.

In lab #5, we learned about how to remove observations with missing values and how to ask for greater detail in STATA within Sum. Here's a reminder of the syntax:

sum k7 if k7<99, detail

```
. sum k7 if k7<99, detail

                   Total years of schooling
-------------------------------------------------------------
      Percentiles      Smallest
 1%            8              3
 5%           11              6
10%           12              7       Obs               1,193
25%           13              7       Sum of Wgt.       1,193

50%           15                      Mean           15.43336
                        Largest       Std. Dev.      3.329872
75%           18             30
90%           20             30       Variance       11.08805
95%           21             30       Skewness       .6134882
99%           25             30       Kurtosis       4.610685
```


FIGURE 6.2 | Detailed Summary Statistics for Variable k7

The table in Figure 6.2 tells us that, on average, our respondents have 15.43 years of schooling. The median tells us that if we were to line up all our respondents from the lowest level of schooling to the highest, the middle score, or that of the fiftieth percentile, we would have 15 years of schooling. Oddly, STATA doesn't report the mode in this procedure. This number must be retrieved by plotting a histogram and looking for the highest bar or running a frequency and picking the largest category (Figure 6.3).

. tab k7 if k7<99

Total years of schooling	Freq.	Percent	Cum.
3	1	0.08	0.08
6	1	0.08	0.17
7	3	0.25	0.42
8	11	0.92	1.34
9	12	1.01	2.35
10	27	2.26	4.61
11	31	2.60	7.21
12	149	12.49	19.70
13	110	9.22	28.92
14	144	12.07	40.99
15	137	11.48	52.47
16	177	14.84	67.31
17	91	7.63	74.94
18	122	10.23	85.16
19	42	3.52	88.68
20	73	6.12	94.80
21	13	1.09	95.89
22	19	1.59	97.49
23	7	0.59	98.07
24	7	0.59	98.66
25	7	0.59	99.25
26	2	0.17	99.41
27	1	0.08	99.50
29	1	0.08	99.58
30	5	0.42	100.00
Total	1,193	100.00	

FIGURE 6.3 | Frequencies of Variable k7

We see now that the mode is 16 years of schooling.

Returning to Figure 6.2, we have a standard deviation of 3.33 and a variance of 11.09. What does this mean? As you progress through this course, the meaning of these statistics will become clearer. However, for now, knowing how to generate these statistics in STATA is sufficient. We also see that we have a range of 27 by subtracting the highest value (30) from the smallest value (3) for this variable. This means that the distance between the highest level of education to the lowest level of education is 27 years. Since this number is plausible, we should have faith that the data have no errors, or that we didn't do anything wrong (not trivial occurrences in statistics!).

Let's look at one more example to illustrate the difference between the measures of central tendency. Figure 6.4 shows the frequency distribution for the variable "If an Election Were Held Today, How Would You Vote Federally?" (k16a).

```
. tab k16a
```

If an election was held today, how would you vote federally?	Freq.	Percent	Cum.
PC/Tory	502	41.59	41.59
Green Party	28	2.32	43.91
Liberals	191	15.82	59.73
NDP	91	7.54	67.27
Other specified	15	1.24	68.52
Would not vote	58	4.81	73.32
Not eligible	21	1.74	75.06
Don't Know	248	20.55	95.61
No response/Refused	53	4.39	100.00
Total	1,207	100.00	

FIGURE 6.4 | Frequency Distribution of Variable k16a

This variable is a nominal level variable because the categories cannot be ranked. Referring back to the material presented in Chapter 6, we know that the mode is most commonly used for nominal or ordinal level data. This is a good rule to memorize, and here is the reason. When we look at the mean for this variable (removing the Don't Know and Missing values) (Figure 6.5), we have a value of 2.28. If we were to translate that value into words, it would mean that, on average, Albertans would vote for the Green Party with a slight tendency toward the Liberals. If you have ever voted in a federal election, you know that you have to pick one, and only one, party or your ballot will be deemed invalid. Furthermore, the frequency in Figure 6.4 reveals a strong preference for the PC/Tory party. Therefore, the mathematical mean does not make sense for nominal level data. Instead, we should pick the mode, which tells us that the most frequently selected category is 1, the PC/Tory party. Can you see why this is the case?

```
. summ k16a if k16a<98

    Variable |        Obs        Mean    Std. Dev.       Min        Max
-------------+---------------------------------------------------------
        k16a |        906    2.279249     1.68659          1          7
```

FIGURE 6.5 | Summary Statistics for Variable K16a

Now it's your turn!

Putting the Information into Practice

1. For each variable, identify the level of measurement of the variable.
   a. strata
   b. age
   c. sex
   d. b5
   e. b11
   f. e10
   g. ft9
   h. h5
   i. mrelig
   j. k12a

2. For each of the 10 variables, run both a frequency distribution and *appropriate* measures of central tendency and dispersion (i.e., mean, median, mode, standard deviation, variance, and/or range).
3. Based on your analysis of your output, complete the following table by writing in the values of the *appropriate* measures of central tendency and dispersion. Leave the boxes blank where the statistic is inappropriate.

Variable	Mean	Median	Mode	Standard Deviation	Variance	Range
strata						
age						
sex						
b5						
b11						
e10						
ft9						
h5						
mrelig						
k12a						

Lab #7: Standard Deviations, Standard Scores, and the Normal Distribution

The focus of this lab is to introduce you to z-scores and help you further understand how the standard deviation relates to the normal curve. The most commonly used standard score, the z-score, is a measure of the relative location in a distribution. Specifically, z-scores, in standard deviation units, give the distance that a particular score is from the mean. In this lab, you will learn how to generate various z-scores within STATA. This lab corresponds with the material presented in Chapter 7.

LEARNING OBJECTIVES

The following lab is directed at helping you understand how z-scores relate to the normal curve and the standard deviation. Specifically, this lab assignment challenges you to clarify your understanding of:

1. The shape of distributions and the normal curve
2. The relationship between z-scores and the normal curve
3. How to use z-scores to mathematically calculate the percentage of cases that fall between two values

Part 1: Reviewing the Shape and Characteristics of Distributions

Before learning to calculate z-scores, let's first refresh our memories on the shape and characteristics of distributions.

In Figure 7.1, we see a histogram for the variable (k7) "In Total, How Many Years of Schooling Do You Have?"

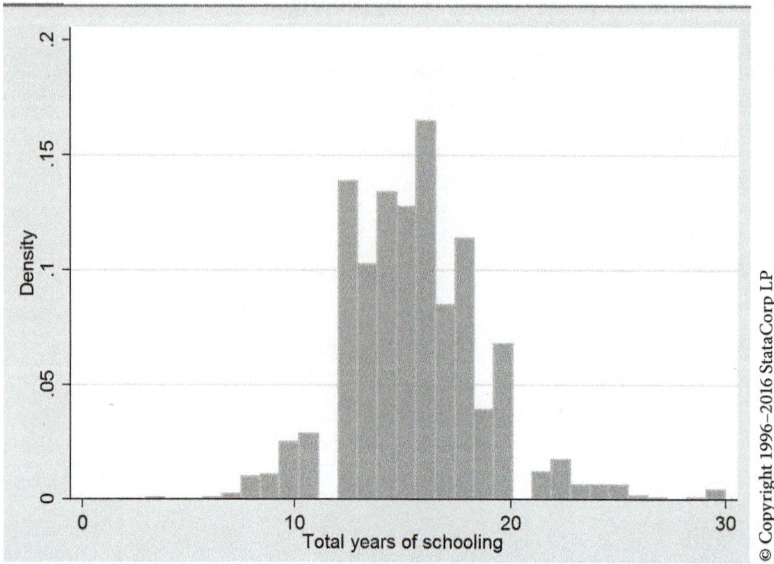

FIGURE 7.1 | Histogram with Normal Curve

Now, what can we say about this distribution? It is *bimodal* because there are two spikes, one at the 12 years of schooling mark and the other at the 16 years of schooling mark. The curve appears to have a slight *positive skew* because there are more people with higher levels of education than lower levels. The two peaks of the distribution are higher than the rest of the distribution, suggesting that the distribution has positive kurtosis.

When we look at the statistics box in Figure 7.2, we see that, on average, respondents have 15.43 years of school, with a standard deviation of 3.33.

```
. summ k7 if k7<99

    Variable |        Obs        Mean    Std. Dev.       Min        Max
-------------+---------------------------------------------------------
          k7 |      1,193    15.43336    3.329872          3         30
```

FIGURE 7.2 | Statistics for "In Total, How Many Years of Schooling Do You Have?"
Source: © Copyright 1996–2016 StataCorp LP

All right, so now we have refreshed our memory regarding the shape and characteristics of distributions. Keeping these elements in mind will help us to further understand *z*-scores and what it means to standardize a distribution.

Part 2: Calculating *z*-Scores

Calculating a *z*-score is very similar to other commands you have learned up to this point, except that we must rely on one of STATA's extensions to complete it:

Step 1: Within your command box, enter:

egen k7z = std(k7) if k7<99

Notice this command gets rid of the 14 missing values.

Step 2: Make sure that your new variable contains *z*-values by running a frequency (Figure 7.3):

tab k7z

```
. tab k7z

Standardize
d values of
       (k7)        Freq.     Percent        Cum.
------------------------------------------------
  -3.733886            1        0.08        0.08
   -2.83295            1        0.08        0.17
  -2.532638            3        0.25        0.42
  -2.232327           11        0.92        1.34
  -1.932015           12        1.01        2.35
  -1.631703           27        2.26        4.61
  -1.331391           31        2.60        7.21
  -1.031079          149       12.49       19.70
  -.7307673          110        9.22       28.92
  -.4304554          144       12.07       40.99
  -.1301435          137       11.48       52.47
   .1701683          177       14.84       67.31
   .4704802           91        7.63       74.94
    .770792          122       10.23       85.16
   1.071104           42        3.52       88.68
   1.371416           73        6.12       94.80
   1.671728           13        1.09       95.89
   1.972039           19        1.59       97.49
   2.272351            7        0.59       98.07
   2.572663            7        0.59       98.66
   2.872975            7        0.59       99.25
   3.173287            2        0.17       99.41
   3.473599            1        0.08       99.50
   4.074223            1        0.08       99.58
   4.374534            5        0.42      100.00
------------------------------------------------
      Total        1,193      100.00
```


FIGURE 7.3 | Frequency of k7z

Step 3: Run summary stats for your new variable k7z (Figure 7.4). Be sure to ask for details so that you get the mean, standard deviation, and variance.

sum k7z, detail

```
. sum k7z, detail

                  Standardized values of (k7)
-------------------------------------------------------------
      Percentiles      Smallest
 1%    -2.232327      -3.733886
 5%    -1.331391       -2.83295
10%    -1.031079      -2.532638       Obs               1,193
25%    -.7307673      -2.532638       Sum of Wgt.       1,193

50%    -.1301435                      Mean          -5.63e-09
                        Largest       Std. Dev.             1
75%      .770792       4.374534
90%     1.371416       4.374534       Variance              1
95%     1.671728       4.374534       Skewness       .6134881
99%     2.872975       4.374534       Kurtosis       4.610685
```


FIGURE 7.4 | Summary Statistics for k7z

In our statistics box, you will first notice that we have a mean of (nearly) 0 and a standard deviation of 1 and a variance of 1.

To convince yourself of the accuracy of the calculations, feel free to manually calculate the z-score for an observation by using the equation presented below (and taken from Chapter 7). Remember that you need to use the standard deviation from the actual variable, not the z-score variable.

$$z = \frac{X - \mu}{\sigma}$$

Now it's your turn!

Putting the Information into Practice

1. Run a frequency distribution for age. Be sure to include a histogram, the mean, standard deviation, and variance.
   a. Are your respondents' ages normally distributed? What evidence supports your answer? (Consider the shape and characteristics of your distribution.)

2. Create a new variable called zage.
3. In your Data Editor window, on Data View, find respondent #61. (Hint: Use the variable called respnum. This variable assigns a number that is unique to each respondent who participated in the survey.) What are respondent #61's values for age and zage?
4. In your Data Editor window, on Data View, find respondent #33. What are respondent #33's values for age and zage?
5. What percentage of respondents are older than respondent #61?
6. What percentage of respondents are younger than respondent #33?
7. What percentage of respondents are older than respondent #61 but younger than respondent #33?

Lab #8: Sampling

The focus of this lab is to introduce you to case selection within STATA and to help you to further understand how larger sampling distributions improve data accuracy. A sample that is accurately and carefully selected without a lot of sampling error allows for a more precise analysis, without including the full population. Because we are often unable to survey every individual, we make decisions about how much of the population to include based on our knowledge of the population parameter. In this lab, we are going to pretend that the total number of respondents who participated in this survey represent the entire population of Alberta (as though the survey were a census). This lab corresponds with the material presented in Chapters 8 and 9.

LEARNING OBJECTIVES

The following lab is directed at helping you understand what effect sample size has on the accuracy of sample values. Specifically, this lab assignment challenges you to clarify your understanding of:

1. The relationship between samples and populations
2. How increasing a sample size will reduce sampling error

How to Select Cases in STATA

Before we begin, run a frequency distribution on the sex variable.

```
. tab sex
```

Gender of respondent	Freq.	Percent	Cum.
Male	595	49.30	49.30
Female	612	50.70	100.00
Total	1,207	100.00	

FIGURE 8.1 | Frequency Distribution on Sex Variable
Source: © Copyright 1996–2016 StataCorp LP

Note that we have 595 males and 612 females. This will be the basis of our population parameters. However, remember, these numbers, in reality, do not represent the true population of Alberta. We are only using them as a population for illustrative purposes.

Step 1: Within your command box, type the following:

sample 5

This will give you a 5 per cent random sample of observations. To prove it, rerun Frequencies.

```
. sample 5
(1,147 observations deleted)

. tab sex
```

Gender of respondent	Freq.	Percent	Cum.
Male	27	45.00	45.00
Female	33	55.00	100.00
Total	60	100.00	

FIGURE 8.2 | Frequency Distribution on the Sex Variable with 5 Per cent Random Sample

Can you see how your sample size is much smaller than what is in the full data set? What's more, given that the sample is randomly drawn each and every time, it is possible that your frequencies differ from those above. Can you see why this might be the case?

Standard Error of a Sample Mean

The next thing we're going to do is use STATA to help us calculate the standard error of a sample mean. Recall from Chapter 9 that the equation is

$$S_{\bar{x}} = \frac{S_X}{\sqrt{n-1}}$$

Using this, it is possible to estimate the distance that your sample is likely to be from a population mean. You can do this even though you don't know what the population mean actually is, using statistical theory and what we know about the normal distribution (which is how we're assuming the data from the Alberta Survey are distributed).

Suppose that you wanted to know the average age of the respondents. Remember that you would do this by using the summary command (often abbreviated as sum), then selecting your variable of interest (age). The resulting output will give you the mean, the standard deviation, and the number of observations necessary to calculate the standard error.

It is important to note that STATA discards all unselected observations and that the only way to get them back is to reopen the data set. Since we selected a sample above, we need to first return to the whole data set. (You will need to close and re-open the data set—make sure you don't save the 5 per cent sample over your complete data set!) Then run a sum on the age variable. Remember to get rid of the missing values by specifying that you don't want the values labelled 99.

```
. summ age if age < 99
```

Variable	Obs	Mean	Std. Dev.	Min	Max
age	1,176	52.44133	16.34969	18	94

FIGURE 8.3 | Average Age of Respondents Summary

Now we want to look at a sample. If you want to temporarily create a sample, simply type **Preserve** in the command box, then sample your observations and perform your analysis on the subset. Once you are finished with your sample and want to return to the larger data set, type **Restore** and STATA will retrieve the original data set. Figure 8.4 shows the summary statistics for the age variable for the 5 per cent sample.

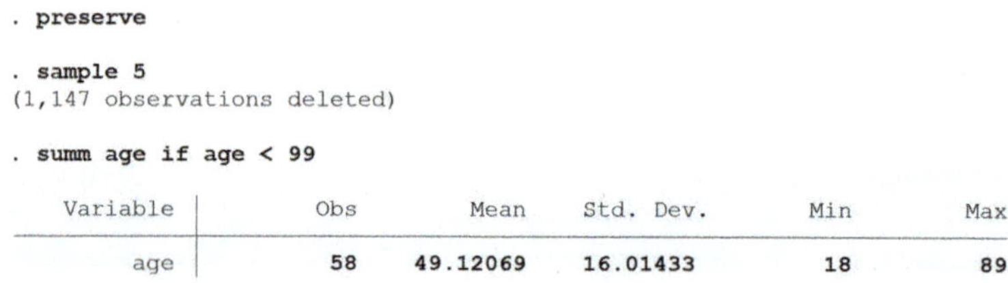

```
. preserve

. sample 5
(1,147 observations deleted)

. summ age if age < 99

    Variable |        Obs        Mean    Std. Dev.       Min        Max
-------------+---------------------------------------------------------
         age |         58    49.12069    16.01433         18         89
```

FIGURE 8.4 | Summary Statistics for the Age Variable for the 5 Per Cent Sample
Source: © Copyright 1996–2016 StataCorp LP

Now it's your turn!

Putting the Information into Practice

Run a frequency distribution for the "Total Years of Schooling" variable (k7) for the population.

Use the select cases procedure to take a random sample of 2 per cent of the cases from the Population (hint: Use Preserve and Restore each time to avoid having to reopen the data set multiple times).

Run summary statistics on k7 (make sure you recode the variable to get rid of missing values first), and record the mean and standard deviation. Calculate the standard error of the sample mean, using the standard deviation estimates generated by STATA.

Repeat this process to complete Table 8.1.

TABLE 8.1

%	Mean	Sampling error
5%		
10%		
25%		
50%		
75%		

1. Which percentage of cases would you choose if you were under budget and time constraints to complete your survey? In other words, which percentage begins to most closely resemble your population and at what point does your distribution start to level off?

Lab #9: Hypothesis Testing: Testing the Significance of the Difference between Two Means

The focus of this lab is to introduce you to the "one sample *t*-tests" function within STATA and help you to further understand how we use confidence intervals to determine generalizing our samples to populations. This lab corresponds with the material presented in Chapter 10.

LEARNING OBJECTIVES

The following lab is directed at helping you understand what effect sample size has on the accuracy of sample values. Specifically, this lab assignment challenges you to clarify your understanding of:

1. How to use a *t*-test to approximate the mean for a population from your sample
2. The relationship between confidence intervals and statistical significance

Calculating a One-Sample *t*-Test in STATA

Let's pretend we are interested in knowing whether the average years of schooling of our sample differs significantly from the population mean of 16 years (high school plus an undergraduate degree) at the 95 per cent confidence level (Figure 9.1). Suppose that we got this number from the Canadian census and that it accurately represents the entire population.

Step 1: Enter the following syntax in your command window:

ttest k7 = 16 if k7 < 99

(Remember that the k7 < 99 part of the command gets rid of the missing values on the education variable).

```
. ttest k7=16 if k7<99

One-sample t test
------------------------------------------------------------------------------
Variable |     Obs        Mean    Std. Err.   Std. Dev.   [95% Conf. Interval]
---------+--------------------------------------------------------------------
      k7 |   1,193    15.43336    .0964067    3.329872    15.24422    15.62251
------------------------------------------------------------------------------
    mean = mean(k7)                                               t =  -5.8776
Ho: mean = 16                                    degrees of freedom =     1192

    Ha: mean < 16               Ha: mean != 16                 Ha: mean > 16
 Pr(T < t) = 0.0000         Pr(|T| > |t|) = 0.0000          Pr(T > t) = 1.0000
```


FIGURE 9.1 | One-Sample *t*-Test: Options Dialogue Box

First, we can note that we have 1193 cases and a mean of 15.43 years of schooling. The standard deviation is 3.33. In the next piece of output, we see we have a *t*-value of −5.8776 with 1192 degrees of freedom. This is the equivalent of our *t*-obtained value. By looking at the sig. (two-tailed) (the middle of the bottom row where you see Pr (|T| > |t|) = 0.000), we can see that we have a significance level of 0.000. This means that we can be more than 99 per cent confident that the average of our sample is significantly different from a sample with a mean of 16 years of schooling. Can you use these numbers to do the *t*-test manually, by looking at the student's *t*-table in Appendix B of your textbook?

Don't despair if you can't, because STATA does the work for you. If you go to the top right-hand side of your output, you can see at the 95 per cent confidence interval that our lower boundary is 15.24 and our upper boundary is 15.62. This means that 95 times out of 100, our sample mean difference will be between these two values. (Notice that this confidence interval does not include 16.)

If you wanted to (and who wouldn't want to?), you could use the information in the STATA output to calculate the confidence intervals yourself and compare them to the numbers that STATA generated.

OK, now it's your turn!

Putting the Information into Practice

1. Let's imagine that the average age of someone living in Alberta is 27 years old. We want to know if the average age of our sample differs significantly from the average age of someone living in Alberta at the 95 per cent confidence level. Conduct a one-sample *t*-test.
2. Can you be 95 per cent confident that the average age of our respondents differs significantly from the average age of someone living in Alberta?
3. What is your confidence interval? What does this mean?

Lab #10: Hypothesis Testing: One- and Two-Tailed Tests

The focus of this lab is to introduce you to the "*t*-tests with two samples" function within STATA. For this type of analysis, you need to have a dichotomous independent variable and an interval/scale for your dependent variable. This lab corresponds with the material presented in Chapter 11.

LEARNING OBJECTIVES

The following lab is directed at helping you understand how you can use a *t*-test to compare the means of two independent samples. Specifically, this lab assignment challenges you to clarify your understanding of:

1. Stating the null and research hypotheses
2. How to establish a sampling distribution and critical region

Calculating a *t*-Test with Two Samples in STATA

Let's pretend we are interested in knowing whether the average years of schooling of our sample differ significantly between males and females. Since both are samples and since scores on the outcome of interest are independent of each other (presumably, the total years of schooling among women has nothing to do with the total years of schooling among men), it is most appropriate to conduct a *t*-test on independent samples.

Step 1: State the null and research hypotheses.

In this case, we are going to test the claim that men's mean years of schooling will differ from that of women.

Therefore, we are claiming that:

H_0: In the population, the mean years of schooling for men and women does not differ $\mu_{MEN} = \mu_{WOMEN}$).

H_1: In the population, the mean years of schooling for men differs from that of women $\mu_{MEN} \neq \mu_{WOMEN}$).

Step 2: Select the sampling distribution and establish the critical region.

On a *z*-score distribution, the critical region that corresponds to a p-value < 0.05 will be represented by a *t*-score of ± 1.96 (Figure 10.1).

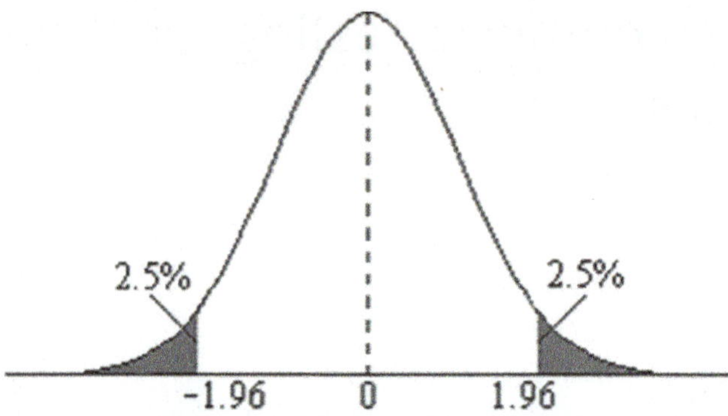

FIGURE 10.1 | The Critical Region at the 95 Per Cent Confidence Interval

Should our test statistic be found in the 5 per cent of the distribution that is bounded by the two critical regions of our *t*-score distribution, we can be 95 per cent confident that we can reject our null hypothesis. Since we are only hypothesizing that there will be a difference in H_1, without saying anything about the direction of the difference (such as $\mu_{men} < \mu_{women}$), we're conducting a two-tailed test. If we were conducting a one-tailed test (and we were hypothesizing a direction of the relationship), remember from Chapter 10 that we'd have a critical *t*-score of ±1.65.

Step 3: In your command box, type the following:

ttest k7 if k7<99, by(sex)

Our test variable will be k7 or "Total Years of Schooling," and our grouping variable is "Sex" or "Sex of respondent."

You should get the output shown in Figure 10.2:

```
. ttest k7 if k7<99, by(sex)

Two-sample t test with equal variances
```

Group	Obs	Mean	Std. Err.	Std. Dev.	[95% Conf.	Interval]
Male	585	15.64957	.1353116	3.27275	15.38382	15.91533
Female	608	15.22533	.1368151	3.37354	14.95664	15.49402
combined	1,193	15.43336	.0964067	3.329872	15.24422	15.62251
diff		.4242437	.1925382		.0464919	.8019955

```
    diff = mean(Male) - mean(Female)                              t =   2.2034
Ho: diff = 0                                     degrees of freedom =     1191

    Ha: diff < 0                 Ha: diff != 0                 Ha: diff > 0
 Pr(T < t) = 0.9861         Pr(|T| > |t|) = 0.0278          Pr(T > t) = 0.0139
```


FIGURE 10.2 | Independent Sample *t*-Test

Looking at Figure 10.2, we can see that both our groups had approximately the same sample size, 585 males and 608 females.

Already we can see that the mean values of years of schooling are quite similar, although men's are slightly higher. However, this difference in means may be due to chance so we have to check the *t*-tests in the next table.

The tables give us the 95 per cent confidence intervals for both groups, then both the groups combined. Next, the *t*-value is listed as 2.2034 with 1191 degrees of freedom. Earlier we said that we needed a *t*-score of $> \pm 1.96$. Here we see that we have a *t*-value of 2.203. Since this score falls above our critical score, we can conclude that the mean years of schooling for our two sample groups (males and females) are significantly different. Therefore, we can reject the null hypothesis of no differences between groups. Now it's your turn!

Putting the Information into Practice

Let's pretend we are interested in knowing whether the average number of children under the age of 18 differs significantly between those who were born in Canada and those who were not.

We will use variables "Were you born in Canada?" (canb) and "Number of children living in household" (k3b).

1. Identify your dependent and independent variables.
2. State your null and research hypotheses.
3. Run an independent samples *t*-test. (Remember to get rid of the missing values on both variables—both variables have missing values coded as 99.)
4. What is the average number of children under the age of 18 living in the household for those who were born in Canada?
5. What is the average number of children under the age of 18 living in the household for those who were not born in Canada?
6. Using a *t*-test, do you accept or reject the null hypothesis? Why?

Lab #11: Bivariate Statistics for Nominal Data

The focus of this lab is to introduce you to the *association or relationship between nominal variables*. Specifically, this lab will help clarify your understanding of independent and dependent variables and how to interpret the chi-square test of statistical significance. This lab corresponds with the material presented in Chapter 12.

LEARNING OBJECTIVES

The following lab is directed at helping you understand how to interpret the relationship between two nominal variables (bivariate relationships). Specifically, this lab assignment challenges you to clarify your understanding of:

1. Dependent and independent variables
2. How to create a cross-tabulation or a contingency table
3. How to interpret your findings and determine statistical significance

Creating Contingency Tables within STATA

Now let's learn how to create contingency tables in STATA. For this example, we will ask this question: "Who Is More Likely To Have Been Diagnosed with High Blood Pressure, Males or Females?"

Step 1: Identify your dependent and independent variables.

An independent variable can be thought of as modifying the outcome. The dependent variable can be thought of as the outcome of interest. In this situation, we are interested in seeing if sex (sex) will modify patterns of high blood pressure (e14_1); therefore, it is our independent variable. The outcome we are interested in is high blood pressure; therefore, it is our dependent variable.

Step 2: Enter the following in your command box:

tab sex e14_1

```
. tab sex e14_1
```

Gender of respondent	Diagnosed with high blood pressure as an adult: Not selec	Selected	Total
Male	426	169	595
Female	460	152	612
Total	886	321	1,207

FIGURE 11.1 | Results of tab sex e14_1

Step 3: Select the appropriate measure of association.

Since we are dealing with nominal data, we want to use an appropriate measure of association, such as *phi* and *Cramer's* V. It is also good practice to request *chi-square* so that we know if the differences are statistically significant. To do this, all we need to do is modify the above syntax slightly:

tab sex e14_1, chi2 V

Note that chi2 after the comma refers to chi-square and V refers to Cramer's *V*. Recall that Cramer's *V* reverts to phi in a two by two table, so it is possible to obtain phi values by requesting only Cramer's *V*.

```
. tab sex e14_1, chi2 V

             |   Diagnosed with high
             |  blood pressure as an
   Gender of |         adult
  respondent |  Not selec   Selected |     Total
-------------+-----------------------+----------
        Male |       426        169 |       595
      Female |       460        152 |       612
-------------+-----------------------+----------
       Total |       886        321 |     1,207

          Pearson chi2(1) =   1.9660   Pr = 0.161
               Cramér's V =  -0.0404
```


FIGURE 11.2 | tab sex e14_1, chi2 V

Let's look first at chi-square. A *chi-square test*, as discussed in Chapter 12 of the textbook, is a hypothesis test that measures whether or not a relationship exists. This measure is suitable for all levels of measurement and all distributions. It tests the *null hypothesis* and measures the discrepancy between observed and expected events. The events are assumed to be independent and have the same distribution, and the outcomes of each event must be mutually exclusive.

To evaluate whether or not we will reject the null hypothesis that no differences exist between males and females, we need to determine the degrees of freedom, which can be found in the STATA output. Our chi-square value is 1.966, which does not exceed the critical chi-square value (not shown, but in Appendix C of your text), so we fail to reject the null hypothesis and conclude that there is no statistically significant difference between males and females in terms of being diagnosed with high blood pressure as an adult.

Now it's your turn!

Putting the Information into Practice

Continuing with the theme of this chapter, we are going to ask this question: Who Is More Likely to Have Been Diagnosed with Anxiety, Males or Females?

1. Answer the following questions:
   a. What is your null hypothesis?
   b. What is your research hypothesis?
   c. What is your dependent variable?
   d. What is your independent variable?
   e. What are the levels of measurement of your variables?

2. Produce a contingency table. Be sure to include the chi-square statistic and Cramer's *V*.

3. Based on your output, answer the following questions:

   a. What is the total valid sample size for this table?
   b. How many missing cases do you have?
   c. What percentage of females indicate they have been diagnosed with anxiety? What percentage of males indicate the same?
   d. How much difference (variance) is there between males and females with respect to the percentage who have been diagnosed with anxiety?
   e. What is the value of your chi-square statistic?
   f. What is the critical chi-square value for this table?
   g. Are your findings statistically significant? Why?
   h. What is the nature of the relationship? Why?

Lab #12: Bivariate Statistics for Ordinal Data

The focus of this lab is to introduce you to the *association or relationship between ordinal variables*. Specifically, this lab will help clarify your understanding of independent and dependent variables, how to interpret the chi-square test of statistical significance, as well as other measures of association (i.e., gamma and Kendall's tau-*b*). This lab corresponds with the material presented in Chapter 13.

LEARNING OBJECTIVES

The following lab is directed at helping you understand how to interpret the relationship between two variables (bivariate relationships). Specifically, this lab assignment challenges you to clarify your understanding of:

1. Dependent and independent variables
2. How to create a cross-tabulation or a contingency table
3. How to interpret your findings and determine statistical significance

Establishing Your Research Question and Identifying Your Variables

In this example, we will ask this question: Does Level of Education Affect an Individual's Opinion about Whether Alberta Should Build Stronger Ties with China? (b2).

Step 1: Identify your dependent and independent variables.

An independent variable can be thought of as modifying the outcome, and the dependent variable can be thought of as the outcome of interest. In this situation, we are interested in seeing if highest level of education will modify opinions on Alberta's ties with China; therefore, it is our independent variable. The outcome we are interested in is opinions on Alberta building stronger ties with China; therefore, it is our dependent variable.

Step 2: Recode your independent variable.

In this example, we are going to recode our independent variable from a 15 category variable into a 4 category variable measuring highest level of education. In Lab #4 you were introduced to the recode function. Let's review by recoding our highest level of education variable (k6) according to the specifications below.

Old value → New value
2, 3, 4, 5, 6 → 1 (Less than high school)
7, 8, 10 → 2 (High school)
9, 11 → 3 (College, certificate, diploma)
12, 13, 14, 15 → 4 (University degree or higher)
99 → (Missing)

This will give you a variable that is collapsed into four categories:

1. Less than high school
2. High school
3. College/certificate/diploma
4. University degree or higher

In addition to this, the people with no response (there are 10) will need to be set to Missing. You will need to tell STATA that this is what you want.

Step 1: In the command box, enter:

recode k6 (2 3 4 5 6=1) (7 8 10=2) (9 11=3) (12/15=4) (99=.), gen(k6recode)

A few things to note here: first, a shortcut was used to denote the value 12 through 15 (12/15). This can be done to speed things up, or you can type in every value that you want to recode (as was done with values 1, 2, 3, 4, 5, and 6). Second, we set all the people who gave no response or refused to answer the question about education to be Missing with (99=.).

Step 2: As a final check on your coding, run a frequency distribution on the new variable, k6recode, and the original variable, k6, to confirm your new values add up to the values in the original variable. This is done by creating a contingency table and entering the following syntax:

tab k6 k6recode

You should get the following output (Figure 12.1):

. tab k6 k6recode

Highest level of education	RECODE of k6 (Highest level of education) 1	2	3	4	Total
Incomplete elementar	1	0	0	0	1
Complete elementary	2	0	0	0	2
Incomplete junior hig	8	0	0	0	8
Complete junior high	15	0	0	0	15
Incomplete high schoo	51	0	0	0	51
Complete high school	0	184	0	0	184
Incomplete college/te	0	63	0	0	63
Complete college/tech	0	0	290	0	290
Incomplete university	0	87	0	0	87
University - diploma/	0	0	38	0	38
University - bachelor	0	0	0	285	285
University - professi	0	0	0	40	40
University - master's	0	0	0	111	111
University - doctorat	0	0	0	22	22
Total	77	334	328	458	1,197

FIGURE 12.1 | Tab k6 k6recode

Notice that the total number of observations equals 1197 instead of 1207. This is because the missing values have been removed by STATA. Otherwise, the coding looks good.

Step 3: Recode variable b2.

A quick look at the codebook suggests that we need to remove some values from variable b2 (Alberta Should Build Stronger Ties with China.). The value 98 means Don't Know and the value 99 means No Response. All other values can remain as they are, since they're already appropriately structured as ordinal data.

recode b2 (98 99 =.), gen(b2recode)

A quick crosstab confirms that our recode was done properly (Figure 12.2):

. tab b2 b2recode

Alberta should build stronger ties with China	RECODE of b2 (Alberta should build stronger ties with China) 1	2	3	4	5	Total
Strongly disagree	53	0	0	0	0	53
Disagree	0	213	0	0	0	213
Neither disagree nor	0	0	232	0	0	232
Agree	0	0	0	578	0	578
Strongly agree	0	0	0	0	84	84
Total	53	213	232	578	84	1,160

FIGURE 12.2 | Tab b2 b2recode

Now we're ready to measure association with gamma and Kendall's tau-*b*.

Step 4: Select your statistics.

Remember from the last lab that we used tab to measure association. The same is true here, except that instead of asking for Cramer's *V* we'll ask for gamma and tau-*b*. We'll also ask for chi-square with the following command:

tab k6recode b2recode, chi2 gamma taub

We should have output that resembles the following (Figure 12.3):

. tab k6recode b2recode, chi2 gamma taub

RECODE of k6 (Highest level of education)	RECODE of b2 (Alberta should build stronger ties with China) 1	2	3	4	5	Total
1	8	17	9	36	1	71
2	15	59	71	161	13	319
3	15	66	64	142	26	313
4	15	67	88	235	43	448
Total	53	209	232	574	83	1,151

```
Pearson chi2(12) =  30.1227   Pr = 0.003
           gamma =   0.1304  ASE = 0.036
  Kendall's tau-b =   0.0892  ASE = 0.025
```


FIGURE 12.3 | Tab k6recode b2recode, chi2 gamma taub

The first thing to look at is the chi-square test. Our chi-square value is 30.1227 with $p = 0.003$. Therefore, since our chi-square value is statistically significant, we can reject the null hypothesis and conclude that there is a statistically significant relationship between highest level of education and opinion on whether Alberta should build stronger ties with China. This should give us confidence as we move on to look at the measures of association.

Under the chi-square value are the gamma and tau-*b* statistics. Looking at gamma first, the positive value means that we have a concordant relationship between our independent and dependent variable where high values of the independent value correspond with high values of the dependent variable. In other words, as educational values increase, opinions on building stronger ties with China increase, and vice versa.

The value of 0.1304 means that we are 13.04 per cent better at predicting the score on the dependent variable when we know the value of our independent variable. Furthermore, a score of 0.1304 means that the association between our dependent and independent value is moderate. Recall from Chapter 13 that the gamma statistic is more liberal in its calculation. If you want to err on the side of caution, it is recommended that you use tau-*b*.

We can see that we have a tau-*b* value of 0.0892. The positive value means that we have a concordant relationship between our independent and dependent variable, where a high score on the independent value (higher levels of education) results in a higher score on our dependent variable (more agreement). The value of 0.0892 suggests that the association between our dependent and independent value is weak.

Now it's your turn!

Putting the Information into Practice

Continuing with the theme of this chapter, we are going to ask this question: Does Level of Education Affect an Individual's Opinion on whether China Will Play an Increasingly Significant Role in the Future Opportunities of Albertans (b10)?

1. Answer the following questions:
   a. What is your null hypothesis?
   b. What is your research hypothesis?
   c. What is your dependent variable?
   d. What is your independent variable?
   e. What are the levels of measurement of your variables?

2. Produce a contingency table. Be sure to include percentages and the chi-square statistic, gamma, and Somers' *d*.
3. Based on your output, answer the following questions:
   a. What is the total valid sample size for this table?
   b. How many missing cases do you have?

c. What percentage of individuals with less than a high school education indicate that they strongly disagree? What percentage of individuals with a university degree or higher indicate the same?
d. How much difference (variance) is there between those with less than a high school education and those with a university degree or higher with respect to the percentage that strongly agree that China will play an increasingly significant role in the future opportunities of Albertans?
e. What is the value of your chi-square statistic?
f. How many degrees of freedom do you have?
g. What is the critical chi-square value for this table?
h. Are your findings statistically significant? Why?
i. What is the value of your gamma and your Somers' *d*? What do these statistics tell you?

Lab #13: Bivariate Statistics for Interval/Ratio Data

The focus of this lab is to introduce you to the *association or relationship between interval/ratio level variables*. Specifically, this lab will help clarify your understanding of Pearson's *r* and explained variance. This lab corresponds with the material presented in Chapter 14.

LEARNING OBJECTIVES

The following lab is directed at helping you understand how to interpret the relationship between two variables (bivariate relationships). Specifically, this lab assignment challenges you to clarify your understanding of:

1. Dependent and independent variables
2. How to calculate and interpret Pearson's *r*
3. How to interpret explained variance

Calculating Pearson's *r* in STATA

When calculating Pearson's *r*, we make the assumption that the two variables we are correlating (evaluating the extent to which the variables are related) have a linear relationship. That is, the relationship between the two variables is the same, regardless of what the value of either of these variables is. So, the first step in calculating Pearson's *r* is to evaluate whether the relationship between the two variables is linear.

In this example, we are interested in whether an individual's age is correlated with the number of years of schooling they have completed. In this case, our dependent variable is "Years of schooling" (k7), and our independent variable is an individual's age (age).

Step 1: Recode variables.

Since both age and education have missing values, we need to recode them:

recode age (99=.), gen(agerecode)
recode k7 (99=.), gen(k7recode)

Don't forget to check your recoding with cross-tabulations.

Step 2: Evaluate for linearity.

To evaluate for linearity, we will use a scatterplot. To create a scatterplot in STATA, click on

scatter k7recode agerecode

Figure 13.1 shows the scatterplot that should have been produced:

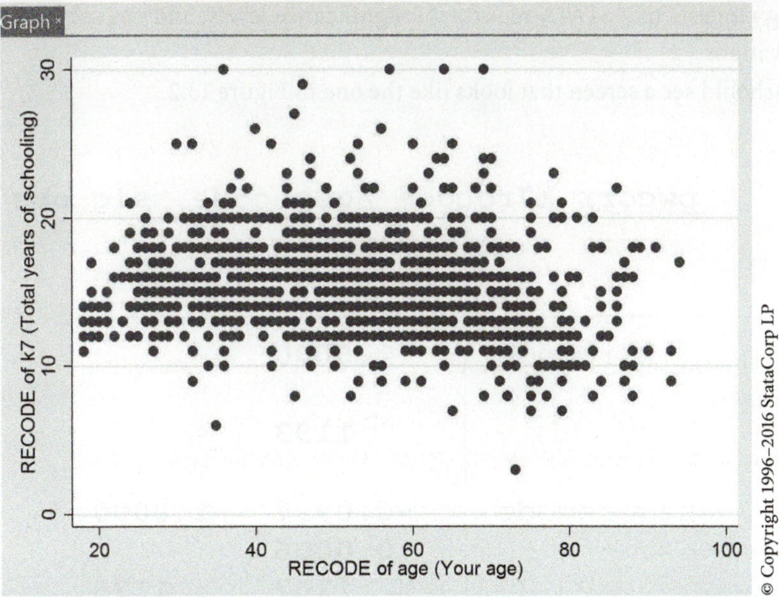

FIGURE 13.1 | Scatterplot

Here are a few observations we could make about this scatterplot:

1. Though not as linear as we may have liked, the dots do not appear to be arranged in a "shotgun" fashion, which means that there is no initial reason to suspect non-linearity. Thus, for our purposes, since there's no reason to suspect non-linearity, we can proceed with calculating Pearson's correlation.
2. There are likely to be quite a few outliers both above and below the line.

Now that we've done the necessary diagnostics, we are ready to calculate Pearson's *r* in STATA.

The computation of our Pearson's *r* is done "behind the scenes" by STATA and is quite complex despite its easy implementation. In a nutshell, it considers the amount of covariation between your *X* variable (independent variable) and your *Y* variable (dependent variable).

Pearson's *r* ranges from −1.00 to +1.00.

−1.0 = a perfect negative relationship or association
−0.5 = a moderate negative relationship or association
0.0 = no correlation between two variables
+0.5 = a moderate positive relationship or association
+1.0 = a perfect positive relationship or association

Step 3: In your command box, type:

pwcorr k7recode agerecode, sig obs

where sig requests that STATA reports the significance levels, and obs asks for the number of observations.

You should see a screen that looks like the one in Figure 13.2.

```
. pwcorr k7recode agerecode, sig obs

             | k7recode agerec~e
-------------+------------------
    k7recode |   1.0000
             |
             |     1193
             |
   agerecode |  -0.0997   1.0000
             |   0.0006
             |     1167     1176
             |
```

FIGURE 13.2 | Pairwise Correlation between Age and Years of Schooling

Source: © Copyright 1996–2016 StataCorp LP

Looking at the first table, we observe that 1167 people have valid values on both variables, a necessary condition for the calculation of correlation.

This table is called a correlation matrix. Referring to the lower left-hand cell, we see that we get a Pearson's r of −0.0997. This means that the relationship between our two variables is negative but weak. However, looking at the value immediately below our Pearson's r coefficient, we see that our results are statistically significant at the 0.0006 level.

Now it's your turn!

Putting the Information into Practice

In this example, we are interested in whether the number of children in the household (k3b) is correlated with the total number of years of schooling respondents have completed (k7).

1. Answer the following questions:
   a. What is your null hypothesis?
   b. What is your research hypothesis?
   c. What is your dependent variable?
   d. What is your independent variable?
   e. What are the levels of measurement of your variables?

2. Produce a scatterplot.
   a. Does the relationship between your two variables meet the assumption of linearity?

3. Calculate Pearson's *r*. Be sure to include the mean and standard deviation of your variables.
   a. What sample size did STATA use to calculate the Pearson correlation coefficient between k7 and k3b?
   b. What is the mean value of k7? What is the mean value of k3b?
   c. What is the Pearson correlation coefficient? Is it significant? Is it small or large?
   d. In terms of our two variables, what does the Pearson correlation coefficient mean? (Hint: Is the Pearson correlation coefficient positive or negative?)

Lab #14: Analysis of Variance

The focus of this lab is to introduce you to a procedure known as ANOVA, or *analysis of variance*. Specifically, this lab will help clarify your understanding of when this procedure should be used *and* how to calculate within-group sum of squares, between-group sum of squares, and the total sum of squares, which are the three major components of ANOVA. Finally, this lab will teach you how to interpret the *F*-distribution with ANOVA. This lab corresponds with the material presented in Chapter 15.

LEARNING OBJECTIVES

The following lab is directed at helping you understand how to interpret the relationship between two variables by using a procedure called ANOVA. Specifically, this lab assignment challenges you to clarify your understanding of:

1. Dependent and independent variables
2. Null and research hypotheses
3. How to interpret your findings and determine statistical significance

Analysis of variance, or ANOVA, is like a *t*-test but allows us to compare more than two groups. Conceptually, ANOVA compares three things:

1. Differences between means
2. Differences in values within samples
3. Differences in values across samples

Essentially, ANOVA is used to compare the variation caused by the independent variable and randomly occurring variations around the mean within groups to the variation across groups.

Calculating ANOVA with STATA

In this example, we will ask this question: Are There Significant Differences in Years of Schooling (k7) by the Area of the Province That Respondents Live In (strata)? The area of the province is coded as Metro Edmonton, Metro Calgary, and Other Alberta. There are no missing values for this variable in the data set.

Step 1: Recode variables as necessary to get rid of missing values.

recode k7 (99=.), gen(k7recode)

Step 2: Run ANOVA by typing the following into your command box:

oneway k7recode strata, tabulate means standard

where "oneway" is the basic command, and "tabulate means standard" requests a table that contains means and standard deviations; this will give you the output shown in Figure 14.1. Please note that if you saved your data set from Lab #13 and are using that version, this step is unnecessary, and you will get the following error message: "k7recode already defined."

```
. oneway k7recode strata, tabulate means standard

             |  Summary of RECODE of k7
             |     (Total years of
 Area of the |       schooling)
    province |      Mean   Std. Dev.
-------------+------------------------
   Metro Edm |    15.605   3.4356198
   Metro Cal | 15.896725   3.1441648
   Other Alb | 14.795455   3.3125889
-------------+------------------------
       Total | 15.433361   3.3298719

                        Analysis of Variance
    Source              SS         df      MS            F     Prob > F
------------------------------------------------------------------------
Between groups      258.16466      2   129.08233      11.85     0.0000
 Within groups      12958.7876   1190   10.8897374
------------------------------------------------------------------------
    Total           13216.9522   1192   11.0880472

Bartlett's test for equal variances:  chi2(2) =   3.1328  Prob>chi2 = 0.209
```

FIGURE 14.1 | ANOVA Output of k7recode and Strata

There are a few things to note here. First, let's look at the means. You will notice that those who live in Edmonton (strata = 1), had, on average, 15.61 years of schooling. Those who lived in Calgary (strata = 2) had, on average, 15.90 years of schooling. On the opposite end of the spectrum, those who lived in a rural part of Alberta had 14.80 years of schooling on average.

There are differences in the means of all the different groups, but are they statistically significant? ANOVA can begin to answer this question by comparing the variance within groups with the variance between groups. To risk overgeneralization, we could say that if the average person is no more similar to someone in their own category than to those in another response category, we would conclude that there is no statistically significant pattern in the data.

To determine this, we need to look at the second table in the output above. For our between-group sum of squares, we can see that we have two degrees of freedom, and for our within-group sum of squares we have 1190 degrees of freedom. In the SPSS output, the last column in this table indicates that the *F*-score is significant at the level $p < 0.000$, suggesting there are significant between-group differences. Alternatively, if we go to Appendix D at the back of the textbook, we find that the critical *F*-distribution score for two degrees of freedom for the between-group score and for more than 120 degrees of freedom for the within-group value is 2.99. We can see in our table that our *F*-score (*F*-observed) is 11.85. Because this number exceeds the critical *F*-score of 2.99, we can conclude with 95 per cent confidence that there are significant differences between at least two groups. This is the same conclusion whether we use the appendix or the value calculated in SPSS.

Recoding to the Midpoint

Sometimes we have variables that are measured at the ordinal level that we want to turn into interval/ratio variables. An easy way to do this is called "recoding to the midpoint." We will illustrate this procedure using the variable k12a (income).

First, run a frequency distribution on the original variable.

income for the past year before taxes and deductions	Freq.	Percent	Cum.
Under 6,000	5	0.41	0.41
6,000 - 7,999	2	0.17	0.58
8,000 - 9,999	2	0.17	0.75
10,000 - 11,999	2	0.17	0.91
12,000 - 13,999	6	0.50	1.41
14,000 - 15,999	5	0.41	1.82
16,000 - 17,999	4	0.33	2.15
18,000 - 19,999	9	0.75	2.90
20,000 - 21,999	17	1.41	4.31
22,000 - 23,999	14	1.16	5.47
24,000 - 25,999	17	1.41	6.88
26,000 - 27,999	6	0.50	7.37
28,000 - 29,999	9	0.75	8.12
30,000 - 31,999	24	1.99	10.11
32,000 - 33,999	6	0.50	10.60
34,000 - 35,999	21	1.74	12.34
36,000 - 37,999	5	0.41	12.76
38,000 - 39,999	5	0.41	13.17
40,000 - 44,999	25	2.07	15.24
45,000 - 49,999	22	1.82	17.07
50,000 - 54,999	37	3.07	20.13
55,000 - 59,999	14	1.16	21.29
60,000 - 64,999	42	3.48	24.77
65,000 - 69,999	18	1.49	26.26
70,000 - 74,999	48	3.98	30.24
75,000 - 79,999	26	2.15	32.39
80,000 - 84,999	45	3.73	36.12
85,000 - 89,999	18	1.49	37.61
90,000 - 94,999	24	1.99	39.60
95,000 - 99,999	18	1.49	41.09
100,000 - 124,999	154	12.76	53.85
125,000 - 149,999	74	6.13	59.98
150,000+	260	21.54	81.52
Don't know	72	5.97	87.49
No response	151	12.51	100.00
Total	1,207	100.00	

FIGURE 14.2 | Frequency Distribution for k12a Recoded

You can see that there are 33 categories of income, and 2 missing values (98 and 99). To recode at the midpoint, we add the endpoints of each category and divide by 2. So the second category, for example, which goes from $6000 to $7999 would be recoded at $6999.50. For the first category (which has no lower limit) and the final category (which has no upper limit), we usually just use the value listed.

The recoding command for this income variable, therefore, would look like this:

recode k12a (1=6000) (2=6999.5) (3=8999.5) (4=10999.5) (5=12999.5) (6=14999.5) (7=16999.5) (8=18999.5) (9=20999.5) (10=22999.5) (11=24999.5) (12=26999.5) (13=28999.5) (14=30999.5) (15=32999.5) (16=34999.5) (17=36999.5) (18=38999.5) (19=42499.5) (20=47499.5) (21=52499.5) (22=57499.5) (23=62499.5) (24=67499.5) (25=72499.5) (26=77499.5) (27=82499.5) (28=87499.5) (29=92499.5) (30=97499.5) (31=112499.5) (32=137499.5) (33=150000) (98=.) (99=.), gen (k12a_recode)

A frequency table of your new variable should look like this (Figure 14.3):

Household income for the past year before taxes and deduc	Freq.	Percent	Cum.
6000	5	0.51	0.51
6999.5	2	0.20	0.71
8999.5	2	0.20	0.91
10999.5	2	0.20	1.12
12999.5	6	0.61	1.73
14999.5	5	0.51	2.24
16999.5	4	0.41	2.64
18999.5	9	0.91	3.56
20999.5	17	1.73	5.28
22999.5	14	1.42	6.71
24999.5	17	1.73	8.43
26999.5	6	0.61	9.04
28999.5	9	0.91	9.96
30999.5	24	2.44	12.40
32999.5	6	0.61	13.01
34999.5	21	2.13	15.14
36999.5	5	0.51	15.65
38999.5	5	0.51	16.16
42499.5	25	2.54	18.70
47499.5	22	2.24	20.93
52499.5	37	3.76	24.70
57499.5	14	1.42	26.12
62499.5	42	4.27	30.39
67499.5	18	1.83	32.22
72499.5	48	4.88	37.09
77499.5	26	2.64	39.74
82499.5	45	4.57	44.31
87499.5	18	1.83	46.14
92499.5	24	2.44	48.58
97499.5	18	1.83	50.41
112499.5	154	15.65	66.06
137499.5	74	7.52	73.58
150000	260	26.42	100.00
Total	984	100.00	

FIGURE 14.3 | Frequency Table for k12a Recoded at the Midpoint

As you can see, you now have an interval/ratio variable, so you can (for example) run summary statistics on your new variable (see Figure 14.4) and use it as a dependent variable for an ANOVA or a linear regression model.

```
. summ k12a_recode

    Variable |        Obs        Mean    Std. Dev.       Min        Max
-------------+---------------------------------------------------------
 k12a_recode |        984    95540.79     45652.3       6000     150000
```

FIGURE 14.4 | Summary Table for k12a Recoded at the Midpoint

Now it's your turn!

Putting the Information into Practice

In this assignment, we will ask this question: Do Albertans Have Different Incomes Depending on Whether They Live in Calgary, Edmonton, or Rural Alberta? Use k12a to measure income (you can recode the categories at the midpoint to create an interval/ratio variable that is suitable for ANOVA—see previous step for recoding income [k12a] at the midpoint), and strata to measure what area of the province the respondent lives in. Be sure to handle all missing values appropriately.

1. Answer the following questions:
   a. What is your null hypothesis?
   b. What is your research hypothesis?
   c. What is your dependent variable?
   d. What is your independent variable?
   e. What are the levels of measurement of your variables?

2. Calculate ANOVA (remember to include your descriptive statistics).
3. In your output, using the table called descriptives, calculate the difference between mean income for
   a. Those who live in Calgary and those who live in Edmonton: ______________
   b. Those who live in Calgary and those who live in a rural area: ______________
   c. Those who live in Edmonton and those who live in a rural area: ______________

4. What do these data suggest about the relationship between where Albertans live and income?
5. State the value of the total sum of squares, the within-group sum of squares, and the between-group sum of squares.
6. What is the value of your F-statistic?
   a. Does this mean your results are statistically significant?
   b. Do you accept or reject your null hypothesis? Why?

Lab #15: OLS Regression: Modelling Continuous Outcomes

The focus of this lab is to introduce you to a form of multivariate analysis called *regression analysis*. Specifically, this lab will help clarify your understanding about when this procedure should be used, how to calculate and interpret ordinary least squares (OLS) regression, and how to compute dummy variables. This lab corresponds with the material presented in Chapter 16.

LEARNING OBJECTIVES

The following lab is directed at helping you understand how to interpret the relationship between multiple variables by using OLS regression. Specifically, this lab assignment challenges you to clarify your understanding of:

1. Dummy variables
2. Standardized partial slopes
3. How to interpret your findings and determine statistical significance

Calculating OLS Regression with STATA

In this example, we are interested in which variables might affect the number of years of schooling an individual has completed. For various reasons, we think that the number of years of schooling an individual has completed may vary by gender (sex) and by what part of Alberta they live in (strata).

Step 1: Code your variables.

Because one of the conditions of OLS regression states that all variables must be interval, ratio, or dummy variables, we will need to recode our variables to meet these conditions.

First, gender is a nominal level variable where the values are coded 1 Male and 2 Female. We will need to recode this variable into a dummy variable where the response categories are 1 Male and 0 Female.

STATA has a very efficient way to create dummy variables:

gen male = (sex==1)

Next, we will need to create a series of dummy variables for our area of the province variable (strata). As you recall from Chapter 16, you need to leave out one category of your independent variable (strata) as a reference category. Because we are most interested in the differences between Edmonton and Calgary, as compared to the rest of the province, we will make Other Alberta our reference category. We will call these new variables Edmonton and Calgary. We know by clicking the Values tab in our Data Editor screen for the variable strata that Edmonton is coded as a 1, Calgary is coded as a 2, and Other Alberta is coded as a 3.

gen edmonton = (strata==1)
gen calgary = (strata==2)

As always, you'll want to check your coding by looking at tabulations of the data.

Step 2: Conduct an OLS regression in STATA.

STATA syntax for estimating regression is very concise and always requires that you enter your dependent variable first, followed by all independent variables:

regress k7recode male edmonton calgary

This will give you the following output:

```
. regress k7recode male edmonton calgary

      Source |       SS           df       MS      Number of obs   =     1,193
-------------+----------------------------------   F(3, 1189)      =      9.39
       Model |  306.019537         3  102.006512   Prob > F        =    0.0000
    Residual |  12910.9327     1,189  10.8586482   R-squared       =    0.0232
-------------+----------------------------------   Adj R-squared   =    0.0207
       Total |  13216.9522     1,192  11.0880472   Root MSE        =    3.2952

------------------------------------------------------------------------------
    k7recode |      Coef.   Std. Err.      t    P>|t|     [95% Conf. Interval]
-------------+----------------------------------------------------------------
        male |   .4007723   .1909073     2.10   0.036     .0262196     .775325
    edmonton |   .7994148   .2336466     3.42   0.001     .3410093     1.25782
     calgary |   1.089634   .2341011     4.65   0.000     .6303363    1.548931
       _cons |    14.6062   .1885417    77.47   0.000     14.23629    14.97611
------------------------------------------------------------------------------
```


FIGURE 15.1 | Regress k7Recode Male Edmonton Calgary

First, look at the *R*-squared value in the top right-hand corner. *R*-squared measures how well the model fits your data; it tells you how much of the variation in the dependent variable can be explained by all the independent variables. In our example, we have explained 2.32 per cent of the variation in total years of schooling. Adjusted *R*-squared values, though we didn't cover them in the text, are a modification of *R*-square that adjusts for the number of terms in a model. *R*-square almost always increases when a new term is added to a model, but adjusted *R*-square increases only if the new term improves the model more than would be expected by chance.

Second, the bottom table of the output above provides the regression coefficients. The regression equation for the intercept, slope, and error term is as follows:

$$y = a + bx + e$$

A regression equation expresses the relationship between two or more variables. The variable a is the constant term (expressed as _cons in STATA), the intercept value when all independent values are set to 0, and it equals 14.6062. This means that our respondents in the reference category (Females, Other Alberta) have, on average, 14.61 years of schooling, a finding that is statistically significant at the 0.000 level. The other coefficients in the equation above represent the effect that each independent variable has on the dependent variable when all other independent variable values are 0. We can see from the output that men have 0.401 years more schooling than females. This finding is statistically significant ($p = 0.036$, which is less than 0.05). Next, individuals living in Edmonton have 0.799 more years of schooling than those living in Other Alberta, and this finding is statistically significant at the 0.001 level. Finally, individuals living in Calgary have 1.09 more years of schooling than those living in Other Alberta, and this finding is statistically significant at the 0.000 level. Confidence intervals are provided for each coefficient.

Now it's your turn!

Putting the Information into Practice

In this assignment, we are interested in whether sex and years of schooling affect income. (Use your recoded interval/ratio income variable—the recode of k12a that you constructed in Lab #14—as your dependent variable.)

1. Answer the following questions:
   a. What is your null hypothesis?
   b. What is your research hypothesis?
   c. What is your dependent variable?
   d. What are your independent variables?

2. Recode variables to get rid of any missing values.
3. Conduct an OLS regression analysis using STATA.
4. What is the value of your *R*-squared? What does this mean?
5. Write the equation for your regression equation.
6. How does sex affect income, controlling for years of schooling? Is this finding statistically significant?
7. How does total years of schooling affect income, controlling for sex? Is this finding statistically significant?

Glossary[1]

absolute value The value of a number, disregarding its positive or negative sign. It is denoted by a pair of "|" symbols: thus the absolute value of –2.5 is $|-2.5| = 2.5$.

addition rule of probabilities To determine the probability of either of two mutually exclusive (i.e., independent) events occurring, add the independent probabilities. For example, the probability of rolling a die to get a 1 or a 6 is determined by adding the two probabilities together ($\frac{1}{6} + \frac{1}{6} = \frac{1}{3}$).

alternative hypothesis *See* **research hypothesis**.

ANOVA (or **analysis of variance**) A measure of the total variability in a set of data is given by the sum of squared differences of the observations from their overall mean. This is the total sum of squares (TSS). It is often possible to subdivide this quantity into components that are identified with different causes of variation (referred to as the within-group sum of squares and the between-group sum of squares). *See* **sum of squares**.

arithmetic average *See* **mean**.

associations Two variables are associated if they are not independent (i.e., if the value of one variable affects the value, or the distribution of the values, of the other). Thus, for a human population, height and weight are associated and so are occupational prestige and income.

asymptotic normality The distribution of a statistic is said to be asymptotically normal if the distribution of the statistic approaches a normal distribution as the sample size increases.

axis scales The values that appear along the *x*- and *y*-axis of any chart.

axis titles The titles used to describe the data on the *x*- and *y*-axis of any chart.

bar chart A diagram where the quantity of an item is expressed through the size of bars.

BEDMAS A mnemonic device for remembering the order of operations that stands for Brackets, Exponents, Division, Multiplication, Addition, Subtraction.

bell curve A term used to describe the bell-like resemblance of a normal distribution.

between-group sum of squares *See* **ANOVA** *and* **sum of squares**.

bimodal Having two modes or modal classes.

binary variables Any variable that has only two response categories.

bivariate analysis Any analytical technique that examines the relationship between two variables.

bivariate relationships *See* **associations**.

bivariate statistics A suite of statistical procedures used to describe the relationship between two variables.

categorical variables Variables whose values are not numerical. Examples include gender (male, female), paint colour (red, white, blue), and type of bird (duck, goose, owl).

causation The belief that some actions or behaviours produce certain outcomes.

central limit theorem Proposed by Laplace, explaining the importance of the normal distribution for a large random sample of observations from a distribution with mean [μ] and variance. The distribution of the sample mean is approximately normal with mean [μ] and variance $(\frac{1}{n})\sigma^2$, and the distribution of the sample total is approximately normal with mean [μ] and variance [σ^2]. The phrase "central limit theorem" appears in a 1919 article by von Mises.

chi-square test A statistical test to determine the similarity of the number of occurrences being investigated to the expected occurrences under the assumption of no relationship between the two variables. The symbol for chi-square is X^2.

collinearity (or **multicollinearity**) Any instance where two independent variables in a regression model are highly correlated with one another. Collinearity makes it difficult to isolate the effect of any one independent variable.

concordant and discordant pairs Concordant pairs (in bivariate association) refer to a situation in which a positive (or negative) score on one variable corresponds with a positive (or negative) score on another variable. Discordant pairs describe the opposite situation, where a positive score on one variable corresponds with a negative score on another. Often used with ordinal data.

confidence interval A confidence interval for an unknown population parameter is an interval calculated from sample values by a procedure such that if a large number of independent samples is taken, a certain percentage of the intervals obtained will contain the unknown population parameter.

The term "confidence interval" was introduced in 1934 by Neyman.

confidence limits The end points of a confidence interval.

confidentiality A set of rules and procedures that are put in place to ensure that any individual who participates in a study cannot be identified by the information he or she provides.

contingency table A table displaying the frequencies for each combination of two or more variables. The variables are either categorical variables or numerical variables for which the possible outcomes have been arranged in groups. The term was first used by Karl Pearson in 1904. Each location in a table is called a cell, and the corresponding frequency is the cell frequency. Also called cross-classification, or cross-tabulation, tables.

continuous variable A variable whose set of possible values is a continuous interval of real numbers *x*, such that $a < x < b$, in which *a* can be $-\infty$ and *b* can be $+\infty$.

control A variable that has an effect that is of no direct interest. The analysis of the variable of interest is made more accurate by controlling for variation in the control variable.

convenience sample A cheap method of obtaining a sample. An example would be interviewing supermarket customers in the ice-cream aisle. This method of collecting data would result in a non-random sample and the results would not be generalizable to all supermarket customers. (You would have an over-representation of ice-cream lovers).

correlation The relationship between two variables, without stipulating a causal or temporal order.

correlation matrix A square symmetric matrix in which the element in row *j* and column *k* is equal to the correlation coefficient between random variables X_j and X_k. The diagonal elements are always equal to 1 because the diagonal denotes a correlation between a variable and itself, and a variable is always perfectly correlated with itself.

covariance The covariance of two random variables is the difference between the expected value of their product and the product of their separate expected values. For random variables *X* and *Y*, $\text{Cov}(X, Y) = E(XY) - E(X) * E(Y)$.

Cramer's *V* In 1946, Cramer suggested that a measure of association could be based on the value of X^2. This is Cramer's *V*; it reduces to phi in a two by two table.

critical value of *z* (or **z(critical)**) or $z_{critical}$) An end point of a critical region. In a hypothesis test, comparison of the value of a test statistic with the appropriate critical value determines the result of the test. For example, 1.96 is the critical value for a two-tailed test in the case of a normal distribution and a 5 per cent significance level: thus, if the test statistic *z* is such that $|z| > 1.96$, then the null hypothesis is rejected and the alternative hypothesis is accepted in preference to the null hypothesis.

degrees of freedom A parameter that appears in some probability distributions used in statistical inference, particularly the *t*-distribution, the chi-squared distribution, and the *F*-distribution. The phrase "degrees of freedom" was introduced by Sir Ronald Fisher in 1922.

dependent variable The outcome of interest in a bivariate or multivariate analysis.

dichotomization The process of transforming a categorical variable into a series of dummy variables.

dichotomous (or **dummy**) **variable** A variable, taking only the values 0 and 1, derived from a polytomous categorical variable. If the categorical variable has *k* categories, then $(k - 1)$ dummy variables are required. For example, with four categories, the three dummy variables ($X1$, $X2$, $X3$) could be assigned the values (1, 0, 0) for category one, (0, 1, 0) for category two, (0, 0, 1) for category three, and (0, 0, 0) for category four. Dummy variables enable the inclusion of categorical information in regression models.

direction Used to describe association; it can be either positive or negative.

discrete (**outcome**) **variables** A nominal dependent variable.

dispersion The amount of variety in a distribution of scores.

distribution The set of values of a set of data, possibly grouped into classes, together with their frequencies or relative frequencies. In the case of random variables, the distribution is the set of possible values together with their probabilities in the discrete case and the probability density function in the case of a continuous variable.

dummy variable *See* **dichotomous variable.**

empirical probability The estimated probability of an event calculated by using real data from a conducted experiment instead of theoretical sample space.

experiment Any design that investigates the effects of a single explanatory variable in highly controlled conditions. Rarely used in social science research.

explained variation The amount of variation in a dependent variable explained by one or more independent variables.

face validity A test is said to have face validity if a reading of the items appears to reflect the areas that the test purports to measure.

F-distribution A theoretical relative frequency distribution of the ratio of two independent sample variances.

frequency The number of times that a particular data value is obtained in a sample. For example, the frequency of 5 in the sample 4, 6, 5, 7, 4, 5, 2, 5 is 3. The sum of the frequencies is the sample size. The term is also used in connection with a set of values. For example, the number of people aged between 20 and 30, or the number of people with blue or green eyes.

Gaussian curve *See* **normal curve.**

grand mean (or **overall sample mean**) When the data comes from different groups (e.g., "males" and "females"), the grand mean is the mean of all the values, regardless of their group.

heteroscedasticity An assumption that the dependent variable does not exhibit similar amounts of variance across the range of values for an independent variable. Variances can either increase or decrease in a patterned way across the range of values of the independent variable.

histogram A specific type of bar chart where the bars are pushed together to illustrate the distribution of a variable. Histograms are useful for visualizing variance and standard deviation.

homoscedasticity An assumption that the dependent variable exhibits similar amounts of variance across the range of values for an independent variable.

hypothesizing relationship A formal statement about the possibility of an association between two or more variables. Hypotheses must be mutually exclusive and exhaustive.

independent variable Any variable that is believed to affect or explain the values of an outcome of interest in a bivariate or multivariate analysis.

inferential statistical tests Involve generalizing from samples to populations, performing hypothesis testing, determining relationships among variables, and making predictions.

interval-level of measurement A scale of measurement that can be used to measure the difference, or distance, between two general states or points, for example, the use of a ruler to measure length, the use of a stopwatch to measure a time interval, or the measurement of a musical interval (octave, fifth, etc.).

inverse function If A is a function of B for variable f, then the inverse function for f is in the opposite direction, from B to A.

Kendall's tau-b A measure of association often used with but not limited to two-by-two tables. It is computed as the excess of concordant over discordant pairs. It is often used in two by two tables.

Kendall's tau-c Like Kendall's tau-*b*, except that adjustments are made for table size.

Kruskal's gamma (or **Kruskal and Goodman's gamma**) A symmetric measure that varies from +1 to −1, based on the difference between concordant pairs (N_{same} and discordant pairs ($N_{\text{different}}$). Gamma is calculated as $\frac{(p - Q)}{(p + Q)}$.

Kruskal and Goodman's lambda For two categorical variables (A and B having, respectively, J and K categories), a measure with a probabilistic interpretation is Kruskal and Goodman's lambda, suggested by Goodman and Kruskal in 1954. Suppose that we are asked to guess the category of B for the next observation. An intelligent guess would be the category that was the most common so far. Lambda measures the improvement made in predicting the dependent variable with information on the independent variable.

kurtosis A measure of whether data distribution is peaked or flat relative to a normal distribution.

latent (or **unobserved**) **concept** An unobserved variable that may account for variation in the data and/or for apparent relations between observed variables.

law of large numbers An empirical probability that will increasingly resemble its theoretical probability as the number of trials increases.

least squares regression line Typically used with OLS regression. It is a line that best approximates the relationship between dependent variables and one or more independent variables.

legend A box that describes the contents of a graph or chart.

levels of measurement A term used to describe the relationship between response categories in any variable. *See* **nominal**, **ordinal**, **interval**, and **ratio** levels of measurement.

leverage A measure of how much the predicted value of the dependent variable changes when an observation is removed.

line graph A graph that shows values as single points that are joined by a line.

line of best fit *See* **least squares regression line.**

logarithm An alternative notation for expressing an exponent; the inverse of exponentiation. It is often used with logistic regression.

longitudinal survey Any survey that observes the same respondent at more than one point in time.

marginals If the cell frequencies of a (multidimensional) contingency table are totalled over one or more of the categorizing variables, the result is a set of marginal totals. For a two-dimensional table, the marginal totals are the row and column totals.

maximum likelihood A commonly used method for obtaining an estimate of an unknown parameter of an assumed population distribution. The likelihood of a data set depends on the parameter(s) of the distribution or probability density function from which the observations have been taken.

mean The mean of a set of N items of data $X_1, X_2, \cdots, X_n$ is $\bar{x} = \frac{\sum X_N}{N}$, which is the sum of all values of a variable, divided by the total number of observations used to calculate the sum. The mean is the most common measure of central tendency, and is usually denoted by placing a bar over the symbol for the variable being measured.

mean square Used in ANOVA, it can be conceived as a standardized sum of squares that can be assessed against known distribution F.

measures of central tendency Any measure of the tendency of quantitative data to cluster around some central value. The central value is commonly estimated by the mean, median, or mode, whereas the closeness with which the values surround the central value is commonly quantified using the standard deviation or variance. The phrase "central tendency" was first used in the late 1920s.

median The middle value in any vector of numbers. When the number of numbers is even, the mean of the two middlemost numbers is used as the median.

missing data Data that have not been collected from the respondent for a variety of reasons (refusal, uncertainty, etc.).

mode The most common value in any vector of numbers.

multimodal Any variable with more than one mode. Usually refers to variables with more than two modes.

multiple correlation coefficient A measure of the linear dependence of more than one numerical random variable on another.

multiplication rule of probabilities Observing two independent outcomes in succession is equal to the product of the probability of the two individual outcomes.

multivariate statistics Any technique that involves more than one variable. Usually refers to analysis with more than two variables.

mutually exclusive A statistical term used to describe a situation where two events or scenarios are entirely independent of one another.

nominal-level of measurement Variable whose values are not numerical. Examples include gender (male, female), paint colour (red, white, blue), and type of bird (duck, goose, owl). A variable with just two categories is said to be dichotomous, whereas one with more than two categories is described as polytomous. The corresponding nouns are *dichotomy* and *polytomy.*

non-integer Any number that is not whole.

non-parametric test Any test that makes no distributional assumptions about the sample or population under investigation.

non-probability (or **non-random**) **sampling strategies** Any sampling technique where individuals in a population do not have an equal probability of selection.

normal (or **Gaussian** or **bell**) **curve** Any curve that resembles the axial cross-section of a bell.

normal score *See* **standard score**.

null hypothesis Any hypothesis that states that there is no difference between two samples on a variable of interest, for example, positing that there is no difference in income between men and women or that there is no relationship between two variables under consideration.

odds ratio The ratio of the odds of something occurring in one situation to the odds of the same event occurring in a second situation. An odds ratio of 1 implies that the odds of an event occurring (and hence the probability of its occurrence) are unaffected by the change in situation: the odds are independent of the situation.

one-tailed assessment (or **one-tailed test**) A statistical hypothesis test in which the values for which we can reject the null hypothesis are located entirely in one tail of the probability distribution.

order of operations A protocol for the order in which equations are solved. *See* **BEDMAS**.

ordinal-level of measurement A categorical variable in which the categories have an obvious order (e.g., strongly disagree, disagree, neutral, agree, strongly agree), but the distances between categories cannot be accurately measured.

ordinary least squares (**OLS**) **regression** The simplest and most frequently used of all statistical regression models. The model states that the random variable Y is related to the variable X by $Y = \alpha + \beta_X + e$, where the parameters α and β correspond to the intercept and the slope of the line, respectively, and e denotes a random error.

outliers Observations that are very different from other observations in a set of data.

paired samples *t*-test (or **repeated measures *t*-test** or ***t*-test for dependent samples**) Any *t*-test with samples that are not completely independent from one another. An example of this would be people who are measured at different points in time.

partial slope coefficients The slope coefficient between two variables (usually an independent variable and a dependent variable) after allowing for the effect of other variables.

Pearson's *r* A measure of the degree to which n pairs of values of random variables X and Y are related. When the correlation between two variables is positive, the values of one variable rise as the values of the other variable rise. The correlation is negative if the values of one variable rise as the values of the other fall.

percentile A $\frac{1}{100}$ slice of a sample or population that's been ranked and divided according to scores on one variable.

phi *See* **Cramer's *V*.**

pie chart A type of chart where the distribution of categories in a variable is illustrated in a circle of different colours.

pilot testing A preliminary test or study of a program or questionnaire used to try out procedures and make any needed changes or adjustments. A pilot test should always be conducted on people who are not part of the final sample.

population The complete set of all people in a country, a town, or any region under study (or just the number of such people). By extension the term is used for the complete set of objects of interest.

post-hoc tests Any tests that occur after analysis has occurred.

predicted values The values predicted by a model fitted to a set of data.

probability The probability of an event is a number lying in the interval $0 <= p <= 1$, with 0 corresponding to an event that never occurs and 1 to an event that is certain to occur.

probability samples Any sampling technique where individuals in a population have an equal (or roughly equal) probability of selection.

proportional reduction of error Any measure that indicates the degree to which an estimate based on information on the independent variable is superior to an estimate made with no information on the independent variable. It may also be interpreted as the per cent of variation in a dependent variable explained by the independent variable.

quota sample Any sample where the number of people in particular groups is used as criteria for selection.

random process Any process in which results may not be certain.

range The area of variation between the upper- and lowermost values of a variable.

rate A special kind of ratio of two measurements with different units but usually with an intuitive denominator (such as kilometres per hour, cents per kilogram).

ratio-level of measurement A scale of measurement where the difference, or distance, is measured between a state or point of interest and a standard state or point. For example, height above sea level, distance from London, frequency of a musical note (in cycles per second), temperature in Kelvins (above absolute zero), clock time (13:05 on 5 November 2001 CE). Confusingly, comparison of ratio scale measurements gives an interval scale, and in music an interval is measured by the ratio of the frequencies.

ratio A stated relationship between two quantities. Rates are a special type of ratio.

reference group, with dummy variables An omitted group for whom values can be derived from other variables in a model.

regression equation An equation that represents a formal statement about a hypothesized relationship between a series of independent variables and one dependent variable.

relationships *See* **associations.**

repeated measures *t*-test *See* **paired samples *t*-test.**

research hypothesis A specific, testable prediction about a relationship between two or more variables that is expected to emerge from an analysis.

sample A portion of a population, often chosen for the purpose of statistical analysis.

sample distribution of means A distribution that describes the variation in the values of the mean over a series of samples. Tends to asymptotically resemble the normal distribution.

sample distribution of proportions A distribution that describes the relative frequency of an occurrence by comparing the number of occurrences to the overall sample size.

sample space All theoretical possible outcomes of an event. Each probability is a fraction of the sample space, with the sum of all probabilities being 1.

sampling error The degree to which a sample "misses" its population on quantities of interest.

sampling frame A list of members of the population of interest.

simple random sample The most basic form of probability sample, where members are chosen with reliance on randomly generated numbers.

skewness If the distribution of a variable is not symmetrical about the median or the mean, it is said to be skewed. The distribution has positive skewness if the tail of high values is longer than the tail of low values, and negative skewness if the reverse is true.

snowball sampling A non-probability sampling method where each person interviewed may be asked to suggest additional people for interviewing.

Somers' *d* A non-proportional reduction in error measure of association for ordinal data.

sparsity The inverse of density, or the property of being scanty or scattered. In statistics, it typically refers to an area in a distribution where there are too few observations to confidently generalize from a sample to a population.

Spearman's *rho* A rank correlation coefficient that may be used as an alternative to Kendall's tau-*b*. Individuals are arranged in order according to two different criteria (or by two different people). The null hypothesis is that the two orderings are independent of one another. It is based on the differences in the ranks given in two orderings.

standard deviation The square root of the variance. Karl Pearson introduced the term in 1893, using the symbol σ in the following year.

standard error of the mean The square root of the variance divided by the square root of the sample size, used to detect the accuracy with which a variable is measured in a particular analysis.

standardized A variable that has the same unit of measurement as other variables and is therefore appropriate for comparison. Converting all currencies to the US dollar, for example, allows you to compare currencies with a common metric. Standard deviations, rates, and ratios are also examples of standardization.

standard score (or **normal score** or **z-score**) The normal score corresponding to the kth largest of n observations is the expected value of the kth largest of n independent observations from a standard normal distribution.

statistics The practice and science of collecting and analyzing numerical data in large quantities.

statistical significance The probability that an event or difference observed in a sample occurred by chance alone.

stratified (or **hierarchical**) **random sample** The process of separating a population into several groups, then randomly sampling from those groups.

Student's t-distribution The form of the distribution was published in 1908 by Gosset, writing under the pen-name "Student," in the context of a random sample of size n from a population having a normal distribution. Often used with smaller samples because it is asymptotically equal to the normal distribution.

sum of squares A measure of the variability in a set of data is given by the sum of squared differences of the observations from their overall mean. This variability can either be explained by a grouping variable (within-group sum of squares), or unexplained by a grouping variable (between-group sum of squares). The sum of these two is the total sum of squares. *See* **ANOVA**.

symmetrical Identical in size, shape, and relative position. Usually used to describe distributions on either side of a mean.

systematic random sample A type of probability sample that starts at a random position on a list and selects every nth unit of a sampling frame until the desired sample size is reached.

theoretical probability Any probability generated from an infinite number of trials. Said differently, it is what the probability of an occurrence should be. For example, the theoretical probability of a coin landing heads is 0.5.

total sum of squares *See* **sum of squares.**

total variation The amount of variation in a dependent variable that is available to be explained by an independent variable.

t-test A test to assess whether there is an equality of means between two variables having normal distributions and equal variances. Also called student's t-test.

t-test for dependent samples *See* **paired samples t-test.**

two-tailed assessment (or **two-tailed test**) A two-sided test is a statistical hypothesis test in which the values for which we can reject the null hypothesis are located in two tails of the probability distribution.

type one error The chance of accepting the research hypothesis when the null hypothesis is actually true. Often called a false positive.

type two error The chance of rejecting the research hypothesis when it is actually true. Often called a false negative.

unexplained variation The amount of variation in a dependent variable that is not explained by one or more independent variables.

unimodal Any variable with only one mode.

unit of analysis The level at which something is being studied (e.g., an individual, a household, a social organization, or a country).

univariate statistics Pertaining to one variable.

variable The characteristic measured or observed when an experiment is carried out or an observation is made. Variables may be non-numerical (*see* **categorical variables**) or numerical.

variance A measure of the variability in the values of a random variable. It is defined as the expectation of the squared difference between the random variable and its most common value (often the mean).

weights A derived value that denotes how many population observations are represented by a sample observation. For example, if an observation in a sample has a weight of five, then that person represents five people in the total population.

within-group sum of squares *See* **sum of squares** *and* **ANOVA.**

x-axis The scale that runs horizontally across a chart.

y-axis The scale that runs vertically across a chart.

zero-order correlations A correlation between two variables without any assumptions of temporal ordering or causality.

z-score *See* **standard score.**

Note

1. Many of these definitions are used with permission from Graham Upton and Ian Cook. 2006. *Oxford Dictionary of Statistics*. Oxford: Oxford University Press.

References

1996 Census Codebook Online. Accessed from http://www12.statcan.gc.ca/english/census01/info/census96.cfm

Allison, Paul D. 2000. *Logistic Regression Using the SAS System: Theory and Application*. Cary, NC: The SAS Institute.

Anscombe, F. J. 1973. "Graphs in Statistical Analysis." *American Statistician*, 27: 17–22.

Berk, Richard A. 1983. "An Introduction to Sample Selection Bias in Sociological Data." *American Sociological Review*, 48, 3: 386–398.

Campbell, Rachel. 2006. "Teenage Girls and Cellular Phones: Discourses of Independence, Safety, and 'Rebellion.'" *Journal of Youth Studies*, 9: 195–212.

Canadian Centre for Justice Statistics. 2003. "Crime Statistics." *The Daily*. Ottawa: Statistics Canada.

Desrosièrs, A. 1998. *The Politics of Large Numbers: A History of Statistical Reasoning*. C. Naish, trans. Cambridge, MA: Harvard University Press.

Galton, Francis. 1889. *Natural Inheritance*. London: MacMillan & Co.

Gigerenzer, G., Z. Swijtink, T. Porter, L. Daston, J. Beatty, and L. Krüger. 1991. *The Empire of Chance: How Probability Changed Science and Everyday Life*. Cambridge: Cambridge University Press.

Hacking, Ian. 1975. *The Emergence of Probability: A Philosophical Study of Early Ideas about Probability, Induction and Statistical Inference*. London: Cambridge University Press.

Hald, A. 1990. A History of Probability and Statistics and Their Applications before 1750. New York: Wiley.

Huff, Darrell. 1954. *How to Lie with Statistics*. New York: Norton.

King, Gary, James Honaker, Anne Joseph, and Kenneth Scheve. 2001. "Analyzing Incomplete Political Science Data: An Alternative Algorithm for Multiple Imputation." *American Political Science Review*, 95, 1: 49–69.

Kranzler, Gerald, and Janet Moursund. 1999. *Statistics for the Terrified*. Upper Saddle River, NJ: Prentice Hall.

Little, R. J. A., and D. B. Rubin. 1987. *Statistical Analysis with Missing Data*. New York: Wiley.

Lorenz, Frederick O. 1987. "Teaching about Influence in Simple Regression." *Teaching Sociology*, 15: 173–177.

Maistrov, L. E. 1974. *Probability Theory: A Historical Sketch*. S. Kotz, trans. New York; London: Academic Press.

Myles, J., Feng Hou, Garnett Picot, and Karen Myers. 2007. "Why Did Employment and Earnings Rise among Lone Mothers in Canada during the 1980s and 1990s?" *Canadian 5 Public Policy*, 23, 2: 1–26.

Pearson, Karl. 1894. "Contributions to the Mathematical Theory of Evolution. I. On the Dissection of Asymmetrical Frequency-Curves." *Philosophical Transactions* (see p. 80) CLXXXV.

Porter, T. 1986. *The Rise of Statistical Thinking, 1820–1900*. Cambridge, MA: Princeton University Press.

Rowntree, Derek. 2000. *Statistics without Tears: An Introduction for Non-Mathematicians*. London: Penguin.

Rubin, D. B. 1987. *Multiple Imputation for Non-Response in Surveys*. New York: John Wiley & Sons.

Ruggles, Steven, and Susan Brower. 2003. "Measurement of Household and Family Composition in the United States, 1850–2000." *Population and Development Review*, 29, 1: 73–101.

Schafer, J. L. 1997. *Analysis of Incomplete Multivariate Data*. New York: Chapman and Hall.

Statistics Canada. 2001. *Aboriginal Peoples Survey (APS), 2001: User's Guide to the Public Use Microdata File*. Catalogue no. 89M0020GPE.

Statistics Canada. 2002. *Ethnic Diversity Survey: User's Guide*. Catalogue no. 89M0019GPE.

Statistics Canada. 2006 (23 May). "Income of Individuals." *The Daily*. Accessed from http://www.statcan.ca/Daily/ English/060523/d060523c.htm on 16 January 2007.

Stigler, S. M. 1986. *The History of Statistics: The Measurement of Uncertainty before 1900*. Cambridge, MA: Harvard University Press.

Walker, Helen M. 1929. *Studies in the History of Statistical Method, with Special Reference to Certain Educational Problems*. Baltimore: The Williams & Wilkins Company.

Worswick, C. 2001. *School Performance of the Children of Immigrants, 1994–1998* (No. 178). Ottawa: Statistics Canada.

Index